Paul Glasserman
Karl Sigman
David D. Yao
(Editors)

Stochastic Networks

Paul Glasserman
Columbia University
Department of Management Science and
Operations Management
New York City, NY 10027

Karl Sigman
Columbia University
Department of Operations Research and
Industrial Engineering
New York City, NY 10027

David D. Yao
Columbia University
Department of Operations Research and
Industrial Engineering
New York City, NY 10027

CIP data available.
Printed on acid-free paper.

Camera ready copy provided by the author.
Printed and bound by Braun-Brumfield, Ann Arbor, MI.
Printed in the United States of America.

9 8 7 6 5 4 3 2 1

ISBN 0-387-94828-7 Springer-Verlag New York Berlin Heidelberg SPIN 10541749

Preface

Two of the most exciting topics of current research in stochastic networks are the complementary subjects of stability and rare events — roughly, the former deals with the *typical* behavior of networks, and the latter with significant *atypical* behavior. Both are classical topics, of interest since the early days of queueing theory, that have experienced renewed interest motivated by new applications to emerging technologies. For example, new stability issues arise in the scheduling of multiple job classes in semiconductor manufacturing, the so-called "re-entrant lines;" and a prominent need for studying rare events is associated with the design of telecommunication systems using the new ATM (asynchronous transfer mode) technology so as to guarantee quality of service.

The objective of this volume is hence to present a sample — by no means comprehensive — of recent research problems, methodologies, and results in these two exciting and burgeoning areas. The volume is organized in two parts, with the first part focusing on stability, and the second part on rare events. But it is impossible to draw sharp boundaries in a healthy field, and inevitably some articles touch on both issues and several develop links with other areas as well.

Part I is concerned with the issue of stability in queueing networks. For a network to be stable, it is known that the usual traffic condition is *necessary*: the overall input rate must be strictly less than the service rate at each node in the network. The more interesting issue, therefore, is to identify additional conditions under which stability is guaranteed. Finding additional conditions to ensure stability, however, is quite complex in multiclass networks. In particular, service disciplines (or scheduling rules and policies) play a critical role; and many "favorite" disciplines have been identified as possibly unstable. Indeed, for almost every commonly used scheduling rule, including the first-come-first-served discipline and simple priority rules, there have been examples in which the rule is unstable. (See the literature review part in Chapters 1, 2, and 3.)

Recently, it has been shown that the stability of a multiclass queueing network operating under a certain scheduling rule can be studied via a corresponding deterministic fluid network. This is the approach taken in the first three chapters here. Typically, conditions on network configuration, routing pattern, and relations among parameters are needed (in addition to the traffic condition), in order to ensure that certain scheduling rules are stable. Furthermore, deep connections emerge from this line of investigation between evaluating the stability region of a network and bounding its performance, the link arising through mathematical programming formulations of both problems.

The focus of Chapter 1 (Dai and Vande Vate) is on *global stability* — enough service capacity under any non-idling discipline — in a two-station network, and on characterizing the global stability region. Two phenomena, *virtual stations* and *push starts*, are identified to be crucial in determining the capacity of the two-station system, which, in turn, is shown to be solely determined by priority disciplines. In this sense, priority disciplines are the "extreme points" of the space of non-idling policies. Priority disciplines are also the focus of Chapter 2 (Chen and Yao), where the driving problem is to identify, in general multiclass networks, priority disciplines under which stability is guaranteed. Under an *acyclic class transfer* condition, it is shown that a natural priority scheme that resembles (but is more general than) the first-buffer-first-served discipline guarantees stability in a very general class of networks, including those with a deterministic routing mechanism.

Chapter 3 (Kumar and Kumar) studies *efficiency* in closed networks, a counterpart to stability in open networks. The central issue is whether a given scheduling policy will yield a throughput that matches the maximum sustainable (which can be determined by letting the population size tend to infinity). A subclass of the least-slack scheduling policy is identified to be efficient. Chapter 4 (Kumar) presents a linear programming approach for bounding the throughput of closed networks and the number in system of open newtorks.

Chapter 5 (Miyazawa) studies a discrete-time Jackson network with *batch* job transfers among stations. The stationary joint queue-length distribution is bounded by a product of geometric distributions, from which stability is characterized. Chapter 6 (Massey) analyzes non-stationary queues using a general asymptotic method called *uniform acceleration*. Necessary and sufficient conditions for underloading are obtained, and shown to be the time-varing analogue to steady-state stability.

Part II focuses on rare events, which, as we mentioned earlier, are intimately associated with new technologies in communications, such as ATM and broad-band ISDN (integrated service data networks), which allow the simultaneous transmission of different traffic types such as data, voice, and video. This has brought increased attention to engineering networks to meet quality of service specifications — the level of performance guaranteed to each type of traffic. Quality of service is usually assured through bounds on the probability of large delays or buffer overflows. Because network standards are extremely stringent, asymptotics of tail distributions are relevant in assessing very low probability events.

The first three chapters in Part II deal with numerical means, broadly interpreted, in estimating rare-event probabilities. Chapter 7 (Glynn and Torres) considers the general issue of how long a stochastic process needs to be observed in order to accurately estimate the probability of a rare event. Their analysis is relevant to both real systems and simulations. They show that the growth of the required observed period is typically exponential, and the critical growth rate is computed. Chapter 8 (Gong and Nananukul) presents the technique of using rational approximants to evaluate rare-event probabilities, using cell loss probabilities in ATM multiplexers as a primary example. Their approximation method allows them to extend easily calculated probabilities for small systems to estimates for large-scale intractable systems. The application of quasi-Monte Carlo is also discussed. Chapter 9 (McDonald) studies the mean time for a large build-up in a queue when arriving jobs join the shorter of two parallel queues. The asymptotics he develops suggest efficient means of simulation through a change of measure: the key step both for the asymptotics and for efficient simulation is finding the right *twist*, meaning the right exponential change of measure.

Another recent issue in new communications technologies are the somewhat related phenomena of long-range dependence, heavy-tailed behavior, and self-similarity of the traffic in high-speed networks. The remaining chapters in Part II all attempt to understand how system performance changes qualitatively when a queue is fed with inputs having any of these properties.

Chapter 10 (Asmussen) surveys rare events associated with random walks and single-server queues in which the underlying distributions have heavy (in particular, *subexponential*) tails. Asymptotic expressions are derived and the behavior of the processes leading to the occurrence of rare events characterized. The results support the intuition that in the heavy-tail case, rare events occur due largely to a big jump, and otherwise the process evolves in its typical way. And this is in contrast to the exponential case, where rare events occur as a consequence of a build-up over a time period during which there is a drift of the underlying parameters. Chapter 11 (Jelenković and Lazar) studies similar rare events for queue length in single-server queues with subexponential input (as well as the exponential case), focusing on ATM applications. In particular, it is demonstrated that real-time traffic such as video exhibits multiple time scale and subexponential characteristics, and that a multiplexer loaded by such processes has a distinct asymptotic behavior. In Chapter 12 (Anantharam) arrival processes with long-range dependence are fed into a discrete-time queueing network with Bernoulli routing among a set of quasi-reversible queues. It is observed that the internal traffic processes all possess long-range dependence.

The last two chapters in Part II focus on long-range dependence arising from self-similarity of the type observed in *fractional Brownian motion* (FBM). Chapter 13 (Konstantopoulos and Lin) considers a single-class queueing network, where the arrival and service processes are self-similar processes, and proves that the normalized queue-length process converges to a reflected FBM. This parallels, albeit being fundamentally different from, the diffusion approximation of standard queueing networks, in which the arrival and service processes do *not* possess long-range dependence, and the limit is a reflected (standard) Brownian motion. Chapter 14 (Chang, Yao, and Zajic) generalizes the integral relationship between standard Brownian motion and FBM to a "filter" that generates a process with long-range dependence by feeding into the filter a process with short-range dependence. A *moderate deviations principle* (MDP) for the filter output is then established based on the MDP of the filter intput, and used as a basis to analyze both transient and steady-state rare-event behavior of queues fed with the long-range dependent input that is the output of the filter.

Acknowledgments

We have been extremely fortunate to have the ablest assitance of a group of reviewers, who read all chapters critically and carefully, and whose constructive comments and suggestions have clearly enhanced the quality of the volume. They include Hong Chen and Hanqin Zhang of the Univerisity of British Columbia, and our colleagues at Columbia: Predrag Jelenković, Yingdong Lu, Perwez Shahabuddin, Yashan Wang, Alan Sheller-Wolf, Tim Zajic, and Li Zhang. Much of the tedious work associated with LaTeX formating was handled by Li Zhang, with good humor and exemplary professionalism.

This volume originated from a workshop with the same title, held at Columbia on November 3-4, 1995, organized by Columbia's Center for Applied Probability (CAP). It is also meant to be the first of a continuing series of Springer-Verlag publications of CAP intellectual activities and research output. For their support and assistance, in both the workshop and the publication of the volume, we thank Chris Heyde, CAP Director, and John Kimmel, our editor at Springer-Verlag.

New York City, April 1996 P.G., K.S., D.D.Y.

Contributors

VENKAT ANANTHARAM
Department of Electrical Engineering and Computer Science, University of California, Berkeley, CA 94720; ananth@vyasa.eecs.berkeley.edu

SØREN ASMUSSEN
Department of Mathematical Statistics, University of Lund, Box 118, S-221 00 Lund, Sweden; asmus@maths.lth.se

CHENG-SHANG CHANG
Department of Electrical Engineering, National Tsing Hua University, Hsinchu 30043, Taiwan, R.O.C.; cschang@ee.nthu.edu.tw

HONG CHEN
Faculty of Commerce, University of British Columbia, Vancouver, B.C. V6T 1Z2, Canada; chen@hong.commerce.ubc.ca

JIM DAI
School of Industrial and Systems Engineering, Georgia Institute of Technology, Atlanta, GA 30332; dai@isye.gatech.edu

PETER GLYNN
Department of Operations Research, Stanford University, Stanford, CA 94305; glynn@leland.stanford.edu

WEI-BO GONG
Department of Electrical and Computer Engineering, University of Massachusetts at Amherst, MA 01003; gong@ecs.umass.edu

PREDRAG R. JELENKOVIĆ
Department of Electrical Engineering, Columbia University, New York, NY 10027; predrag@ctr.columbia.edu

TAKIS KONSTANTOPOULOS
Department of Electrical and Computer Engineering, University of Texas, Austin, TX 78712; takis@alea.ece.utexas.edu

P.R. KUMAR AND SUNIL KUMAR
Department of Electrical and Computer Engineering, University of Illinois, 1308 West Main Street, Urbana, IL 61801; prkumar@decision.csl.uiuc.edu

Aurel A. Lazar
Department of Electrical Engineering, Columbia University, New York, NY 10027; aurel@ctr.columbia.edu

Si-jian Lin
Department of Electrical and Computer Engineering, University of Texas, Austin, TX 78712; sjlin@alea.ece.utexas.edu

William A. Massey
Bell Laboratories, Office 2C-120, 600 Mountain Ave., Murray Hill, NJ 07974-0636; will@research.att.com

David McDonald
Department of Mathematics, University of Ottawa, Ontario, Ont. K1N 6N5, Canada; dmdsg@mathstat.uottawa.ca

Masakiyo Miyazawa
Department of Information Sciences, Science University of Tokyo, Yamazaki 2641, Noda-city, Chiba 278, Japan; miyazawa@is.noda.sut.ac.jp

Soracha Nananukul
Department of Electrical and Computer Engineering, University of Massachusetts at Amherst, MA 01003; nananuku@despot.ecs.umass.edu

Marcelo Torres
Department of Operations Research, Stanford University, Stanford, CA 94305; marcelo@leland.stanford.edu

David D. Yao
Department of Industrial Engineering and Operations Research, Columbia University, New York, NY 10027; yao@ieor.columbia.edu

Tim Zajic
Department of Industrial Engineering and Operations Research, Columbia University, New York, NY 10027; and IBM Research Division, T.J. Watson Research Center, Yorktown Hts, NY 10598; zajic@ieor.columbia.edu

Contents

1

Global Stability of Two-Station Queueing Networks

J. G. Dai and John H. Vande Vate

ABSTRACT This paper summarizes results of Dai and Vande Vate [15, 14] characterizing explicitly, in terms of the mean service times and average arrival rates, the global pathwise stability region of two-station open multi-class queueing networks with very general arrival and service processes. The conditions for pathwise global stability arise from two intuitively appealing phenomena: virtual stations and push starts. These phenomena shed light on the sources of bottlenecks in complicated queueing networks like those that arise in wafer fabrication facilities. We show that a two-station open multi-class queueing network is globally pathwise stable if and only if the corresponding fluid model is globally weakly stable. We further show that a two-station fluid model is globally (strongly) stable if and only if the average service times are in the interior of the global weak stability region. As a consequence, under stronger distributional assumptions on the arrival and service processes, the queueing network is globally stable in a stronger sense when the mean service times are in the interior of the global pathwise stability region. Namely, the underlying state process of the queueing network is positive Harris recurrent.

1.1 Introduction

Queueing networks offer an appealing method for modeling complex manufacturing processes. Unfortunately, they are themselves generally too complex for successful analysis. For example, the primary tool for evaluating the performance of a given dispatching rule is simulation. In fact, we generally resort to simulation even to determine whether a queueing network is stable under a given dispatching rule.

Even very simple queueing networks exhibit surprising and often counterintuitive behavior. In a surprising series of examples, Kumar and Seidman [28], Lu and Kumar [13] and Rybko and Stolyar [33] demonstrated queueing networks that cannot keep up with customer arrival rates under certain non-idling queueing disciplines even though the traffic intensity at each station is less than one. Bramson [4, 5] and Seidman [34] independently provided examples demonstrating the same remarkable behavior with the popular first-in–first-out (FIFO) queueing discipline.

These examples have inspired a number of investigations into the capacity of general queueing networks. For example, Kumar and Meyn [27], Dai [11], Chen [8], Down and Meyn [17], Chen and Zhang [10], Foss and Rybko [23], Bramson [6, 7] all proposed various sufficient conditions, which if satisfied, ensure that the network has adequate capacity.

In some specific instances, researchers have been able to characterize exact conditions. For example, Botvich and Zamyatin [3] determined exactly those average service rates able to keep up with a given average rate of customer arrivals in a specific two-station queueing network following a particular buffer priority discipline. Dumas [18] accomplished the same for a specific three-station queueing network. In both cases, the capacity of the network was less than that of the busiest server.

While these researchers have studied networks following specific queueing disciplines, we are interested in the capacity of queueing networks that may follow any non-idling queueing discipline, that is, any queueing discipline that requires servers to work whenever there is work for them to do. We summarize results in Dai and Vande Vate [15, 14] describing conditions under which the network is *globally stable*, i.e., has sufficient capacity regardless of the queueing discipline employed as long as it is non-idling. A globally stable queueing network will always have sufficient capacity to meet given customer arrival rates no matter what queueing disciplines the servers follow as long as they keep busy whenever there are customers available to serve. On the other hand, a queueing network that is not globally stable will not have adequate capacity to meet demand under some non-idling queueing disciplines.

These results show that among all non-idling queueing disciplines, the static buffer priority disciplines alone define the capacity of two-station queueing networks. In other words, one corollary of our results is the conclusion that static buffer priority queueing disciplines are extreme or "worst" in the class of all non-idling queueing disciplines for two-station queueing networks.

Two phenomena determine the capacity of two-station queueing systems: virtual stations and push starts. These two phenomena provide insight into the sources of bottlenecks in complicated networks like those in wafer fabrication facilities.

Virtual stations affect the global stability of queueing networks because, under some non-idling queueing disciplines, certain groups of buffers can never be served simultaneously even though they are served at different stations. Thus, just as at stations, the traffic intensities at these groups of buffers called *virtual stations* must be at most one.

Push starts affect the global stability of queueing networks because of their influence on virtual stations. In particular, if we give highest priority to the first few buffers customers visit, they pass through to the rest of the network at the same rate they arrive, but the servers have that much less capacity to dedicate to the rest of network. This does not influence the traffic intensities at the stations, but it dramatically affects the traffic intensities at virtual stations.

Together, these two intuitively appealing phenomena characterize when a two-station queueing network is globally pathwise stable. We show that the corresponding two-station *fluid networks* are globally weakly stable under the same conditions. Thus, a two-station, open multi-class queueing network is globally pathwise stable if and only if the corresponding fluid model is globally weakly stable.

When the arrival and service processes satisfy stronger distributional assumptions, it is possible to establish stronger forms of stability like *Harris recurrence.* This paper also proves that under a non-idling queueing discipline if the mean service times are in the *interior* of the global pathwise stability region, the underlying state process is *positive* Harris recurrent and so the queueing network is stable in a stronger sense. Conversely, when the mean service times are outside the global pathwise stability region, there is a buffer priority discipline under which the total number of customers in the network diverges to infinity. Under this stronger notion of stability, we cannot conclude that the conditions are exact, however, because we do not know whether the network will be stable when the the mean service times are on the boundary of the global pathwise stability region. We conjecture that the network is not positive recurrent in this case.

Although these results provide compelling evidence in support of the belief that fluid models accurately describe the capacity of queueing networks, we do not prove the strong relationships between the two models directly. Rather, we observe the phenomena determining the capacity of the two-station fluid model and argue that they also determine the capacity of the two-station queueing network. This indirect argument highlights a more direct connection between the two models: In each case, the same queueing discipline gives rise to the same constraint on the global stability region.

Our proof of the necessity of our conditions for the global stability of the queueing network is direct. We show that if the system violates one of our conditions, the number of customers in the system goes to infinity.

Our proof of the sufficiency of our conditions to ensure the global stability of the queueing network relies on the fact that the queueing network is globally stable if the fluid model is. We determine when a two-station fluid network is stable by determining when there is a piecewise linear Lyapunov function for it. We formulate the problem of determining the coefficients of the Lyapunov function as a linear programming problem, which has unbounded objective values if and only if the coefficients and hence the Lyapunov function exist. Our linear program arises directly from the piecewise linear Lyapunov function introduced in Dai [12], which generalizes that of Botvich and Zamyatin [3] and is simpler than that independently formulated by Down and Meyn [17].

We transform our linear program into a parametric network flow problem in an acyclic network. The fluid network is globally stable if there is a value of the parameter for which the minimum flow in this network is sufficiently small. Thus, sufficient conditions for global stability arise from the constraints imposed on the upper ideals of the partial order defined by the acyclic network in order to ensure there is a sufficiently small flow.

Recently, Bertsimas, Gamarnik and Tsitsiklis [2] showed that a two-station fluid network is globally stable if and only if a certain linear program has bounded objective value. We extend the results of Bertsimas et al. by stating explicitly in terms of the service times, necessary and sufficient conditions for a two-station fluid network to be stable under all non-idling dispatching policies. The explicit description of necessary and sufficient conditions for the stability of two-station fluid networks provides a number of corollaries not immediately available from the linear programming characterization of Bertsimas et al. [2]. Most important among these is a complete understanding of how virtual stations arise in two-station fluid networks. In addition, our conditions demonstrate that the global stable region of a two-station fluid network is *monotone*, i.e., reducing service times maintains global stability. This is not the case for stability with respect to a given dispatching policy. It is possible for a dispatching policy to be stable for a given fluid network, but instable when the service times are reduced. For fluid networks with more than two stations even the global stable region need not be monotone.

In Section 1.2 we introduce our two-station queueing model and define how to measure its capacity. In Section 1.3 we describe the corresponding fluid model. Section 1.4 states our main results, which we interpret and refine in terms of virtual stations and push starts in Section 1.5. This section proves the necessity of our conditions for global pathwise stability. In Section 1.7 we outline the proof that these conditions are also sufficient. Finally, in Section 1.6, we consider a stronger notion of stability in the queueing network, positive Harris recurrence, under additional assumptions on the arrival and service processes.

1.2 The Queueing Model

We consider a queueing network with two single-server stations, denoted A and B. The system serves a set $I = \{1, \ldots, n\}$ of n different types of customers. Type i customers arrive according to the exogenous arrival process $S_0^i = \{S_0^i(t), t \geq 0\}$, where $S_0^i(t)$ is the cumulative number of exogenous arrivals by time t.

Different types of customers may follow different routes, but each type i customer follows the same deterministic route visiting first one station and then the other a number of times before exiting the system. We number the visits by type i customers consecutively from 1 to c_i and let A_i denote those to station A and B_i, those to station B.

Following Kelly [26], we refer to type i customers during visit k (either waiting or being served) as *class* (i,k) customers. We assume the system can accommodate an unlimited number of class (i,k) customers and treat these customers as though they resided in a dedicated buffer with infinite capacity. Class (i,k) customers receive service at station $\sigma(i,k)$ according to the service process $S_k^i = \{S_k^i(t), t \geq 0\}$, where $S_k^i(t)$ is the cumulative number of service completions for class (i,k) customers if the server dedicates t units of *service* to the class.

The arrival processes and service processes can be random. We assume that they are defined on a probability space and satisfy a *strong law of large numbers*. That is, we assume that for almost every sample path, as $t \to \infty$,

$$S_0^i(t)/t \to \lambda_i = 1, \qquad \text{for each } i \in I, \text{ and} \tag{1.1}$$

$$S_k^i(t)/t \to \mu_k^i, \qquad \text{for each class } (i,k). \tag{1.2}$$

We interpret λ_i as the average arrival rate for type i customers and μ_k^i as the average service rate for class (i,k) customers. Since μ_k^i is the average service rate, we can interpret $m_k^i = 1/\mu_k^i$ as the average service time for class (i,k) customers. We assume that $\lambda_i = 1$ without loss of generality, since measuring the buffer levels for type i customers in units of λ_i and scaling the service processes accordingly (so that we scale the average service rate for class (i,k) customers to μ_k^i/λ_i and the average service time to $\lambda_i m_k^i$), normalizes the rates at which customers enter the system. This scaling amounts to nothing more than changing the units by which we measure the different types of customers.

We let $Q_k^i(t)$ denote the number of class (i,k) customers in the buffer (or being served) at time t and $T_k^i(t)$ the cumulative time in $[0,t]$ that server $\sigma(i,k)$ spends on class (i,k) customers. Note that $S_k^i(T_k^i(t))$ is the number of customers to complete class (i,k) service by time t and so:

$$Q_k^i(t) = Q_k^i(0) + S_{k-1}^i(T_{k-1}^i(t)) - S_k^i(T_k^i(t)) \tag{1.3}$$

for each class (i,k), where we model exogenous arrivals of type i customers by setting $T_0^i(t) = t$.

In general, a station serves many classes and so the server must decide which class and even which customer to serve next. A *queueing discipline* dictates which customer to work on each time the system changes state or experiences an *event*. Events occur when a customer arrives or a service is completed. Whitt [37] shows that simply changing how a system handles simultaneous events can dramatically affect its capacity. Thus, to reach any meaningful conclusions about the capacity of queueing networks, we must adopt some convention on simultaneous events. We assume that the queueing discipline responds to events one at a time, which is consistent with the way a single sequential processor would handle them.

We are primarily interested in preempt-resume, static buffer priority queueing disciplines as these alone determine the global stability of two-station queueing networks. A *preempt-resume* queueing discipline can interrupt service to one customer in order to serve another and later resume the interrupted service where it left off. *Static buffer priority queueing disciplines* simply stipulate that each station serve its different classes according to some fixed rank order and customers within a class are served on a first-come–first-served basis. Under a static buffer priority discipline, a server cannot work on a class unless there are no customers available in any higher priority class at the station.

After normalizing the average arrival rates, the traffic intensities at the stations are given by:

$$\rho_A = \sum_{i \in I} \sum_{k \in A_i} m_k^i \quad \text{and} \quad \rho_B = \sum_{i \in I} \sum_{k \in B_i} m_k^i. \tag{1.4}$$

They measure the nominal work imposed on the stations each unit of time. If the traffic intensity at some station exceeds 1, work for the station arrives faster than the server can complete it and so clearly the server and hence the system does not have sufficient capacity.

Even if each server individually has sufficient capacity, the system as a whole may not because the servers must interact: one station can only serve a customer after another station has finished. Thus, we extend the notion of the capacity of a server to define the capacity of the system. Just as we say a server has sufficient capacity if he can complete the work as quickly as it arrives, we say that the system has sufficient capacity if it can finish serving customers as quickly as they arrive. More formally, we let $D_i(t)$ denote the number of type i customers to complete their last service, namely class (i, c_i) service, by time t. Thus, $D_i(t) = S^i_{c_i}(T^i_{c_i}(t))$.

Definition 1.2.1 We say that the system has *sufficient capacity* or is *pathwise stable* if the long run average arrival and departure rates are equal, that is, if for almost every sample path,

$$\frac{D_i(t)}{t} \to \lambda_i \text{ as } t \to \infty \text{ for each type } i.$$

Thus, a queueing network is *globally pathwise stable* if, under each non-idling queueing discipline, $D_i(t)/t \to \lambda_i$ for each type i.

Pathwise stability, first introduced by El-Taha and Stidham [21] (see also El-Taha and Stidham [21] and Altman, Foss, Riehl and Stidham [1]), is very weak by conventional standards. The well-known $M/M/1$ queueing system, for example, is pathwise stable if and if only the average arrival rate λ does not exceed the average service rate μ. When $\lambda = \mu$, however, the system is *null* recurrent, not positive recurrent. In particular, it does not possess an equilibrium and is often considered "unstable". In Section 1.6, we introduce a stronger notion of stability—***positive Harris recurrence.*** An $M/M/1$ queueing system with $\lambda = \mu$ is not stable under this stronger notion.

Having sufficient capacity is necessary, but not sufficient to ensure that for almost every sample path the number of customers in the system is bounded over time: Even in an $M/M/1$ queueing system with $\rho := \lambda/\mu < 1$, the number of customers in the system is not bounded. Having sufficient capacity is enough to ensure the following properties of the long term behavior of the system. Note this result holds even when there are more than two stations.

Lemma 1.2.2 If an open multi-class queueing network has sufficient capacity, then for each class (i, k)

$$S_k^i(T_k^i(t))/t \rightarrow \lambda_i \quad \text{and} \tag{1.5}$$

$$T_k^i(t)/t \rightarrow \lambda_i m_k^i \tag{1.6}$$

as $t \rightarrow \infty$.

Proof. The definition of sufficient capacity ensures (1.5) for the last class for each type of customer. It is easy to show that if a class (i, k) satisfies (1.5) it also satisfies (1.6). Finally, a class feeding a class (i, k) satisfying (1.5) must itself satisfy this condition. For a more detailed proof, see Lemma 1.1 of Dai and Vande Vate [15]. □

For future reference, we define an ***excursion*** to be a block of consecutive visits to the same station. In the Lu-Kumar network of Figure 1 for example, there are two excursions at station A — each consisting of a single visit — and one excursion at station B consisting of visits 2 and 3.

We let E_i denote the set of excursions for type i customers and number these excursions consecutively from 1 to $|E_i|$. We partition E_i into E_A^i, the set of excursions at station A, and E_B^i, those at station B. Since an excursion at one station must be followed by an excursion at the other (unless it is the last excursion), one of these is the set of odd numbered excursions and the other is the set of even numbered excursions depending on where type i customers first enter the network.

We partition the visits of an excursion into the ***last visit*** and all the rest, which we call ***first visits.*** We let $\ell(i, e)$ denote the last visit and $f(i, e)$ the set of first visits in excursion e for type i customers. If an excursion consists of only one visit, that visit is the last visit and the excursion has no first visits. For example, in the Lu-Kumar network of Figure 1, $\ell(1, 2) = 3$ and $f(1, 2) = \{2\}$ while $\ell(1, 1) = 1$ and $f(1, 1) = \emptyset$.

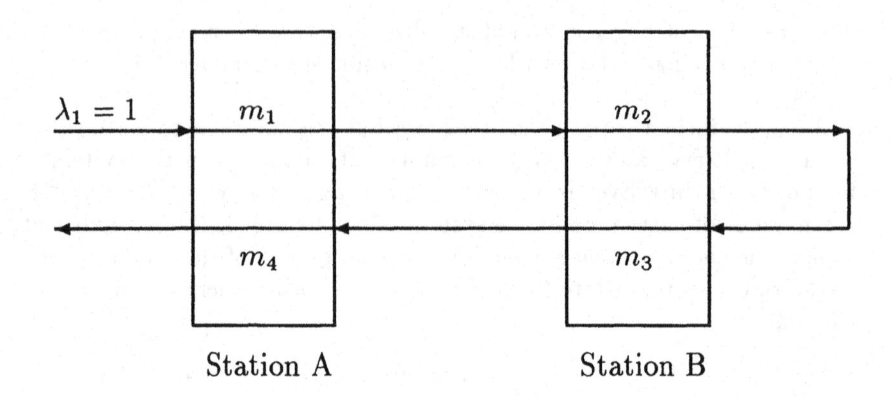

FIGURE 1. The Lu-Kumar Network.

1.3 The Fluid Model and Fluid Limits

Fluid models are continuous, deterministic approximations to discrete, stochastic queueing networks. The fluid model corresponding to our two-station queueing network replaces discrete customers arriving according to a random process with continuous fluids arriving at a constant rate. Type i fluid arrives at the constant rate λ_i and follows the same prescribed route as type i customers before exiting the system. As in the queueing model, we assume that $\lambda_i = 1$ without loss of generality.

We refer to type i fluids during visit k as *class* (i, k) *fluid* and we let $\overline{Q}^i_k(t)$ denote the volume of class (i, k) fluid in the buffer at time t. Server $\sigma(i, k)$ can process class (i, k) fluid at rate μ^i_k. This means that the server depletes class (i, k) fluid from the buffer at rate μ^i_k when he devotes all his efforts to serving that class. Equivalently, each unit of class (i, k) fluid requires service lasting $m^i_k = 1/\mu^i_k$ units of time.

We denote by $\overline{T}^i_k(t)$ the cumulative effort server $\sigma(i, k)$ has dedicated to class (i, k) up to time t. Since, $\overline{T}^i_k(t)$ is the cumulative time allocated to class (i, k) service, we refer to $\overline{T}(\cdot) = (\overline{T}^i_k(\cdot))$ as the *cumulative allocation.* Note that $\overline{T}^i_k(\cdot)$ is a non-decreasing function of time and $\mu^i_k\overline{T}^i_k(t)$ is the cumulative volume of fluid to complete class (i, k) service.

We let $I_A(t)$ denote the cumulative time station A is idle and $I_B(t)$, the cumulative time station B is idle up to time t. Note that $I_A(\cdot)$ and $I_B(\cdot)$ are non-decreasing functions of time as well.

Finally, we let $V_A(t) = \sum_{i\in I}\sum_{k\in A_i}\overline{Q}^i_k(t)$ denote the volume of fluid in the buffers at station A and $V_B(t) = \sum_{i\in I}\sum_{k\in B_i}\overline{Q}^i_k(t)$, the volume of fluid in the buffers at station B. The non-idling conditions can be expressed via the requirement that when $V_A(t) > 0$, i.e., when there is work at station A, the time derivative $\dot{I}_A(t) = 0$ and so station A is not accumulating idle time. Note that since $I_A(\cdot)$ need not be everywhere differentiable, we only impose this condition on the *regular* points or points where $I_A(\cdot)$ is differentiable. Similarly, we express the non-idling condition at station B via the requirement that when $V_B(t) > 0$ and $I_B(\cdot)$ is differentiable at t, then $\dot{I}_B(t) = 0$.

Fluid models can also include other conditions representing the queueing discipline. As we are interested in the network under all non-idling queueing disciplines, however, the following equations define our fluid model:

$$\overline{Q}^i_k(t) = \overline{Q}^i_k(0) + \mu^i_{k-1}\overline{T}^i_{k-1}(t) - \mu^i_k\overline{T}^i_k(t) \text{ for each class } (i,k), \quad (1.7)$$

$$I_A(t) = t - \sum_{i\in I}\sum_{k\in A_i}\overline{T}^i_k(t), \quad (1.8)$$

$$I_B(t) = t - \sum_{i\in I}\sum_{k\in B_i}\overline{T}^i_k(t). \quad (1.9)$$

Note that we model exogenous arrivals of fluid type i by setting $\mu^i_0 = \lambda_i = 1$ and $\overline{T}^i_0(t) = t$. In addition we require that:

$$\begin{array}{lll}
\overline{Q}^i_k(t) \geq 0 & \text{for each class } (i,k) & (1.10)\\
\overline{T}^i_k(0) = 0 & \text{for each class } (i,k), \text{ and } \overline{T}^i_k(\cdot) \text{ is non-decreasing,} & (1.11)\\
I_A(0) = 0 & \text{and } I_A(\cdot) \text{ is non-decreasing,} & (1.12)\\
I_B(0) = 0 & \text{and } I_B(\cdot) \text{ is non-decreasing,} & (1.13)\\
\dot{I}_A(t) = 0 & \text{if } V_A(t) > 0 \text{ and } I_A(\cdot) \text{ is differentiable at } t, \text{ and} & (1.14)\\
\dot{I}_B(t) = 0 & \text{if } V_B(t) > 0 \text{ and } I_B(\cdot) \text{ is differentiable at } t. & (1.15)
\end{array}$$

A *fluid vector* is a vector $(\overline{Q}(\cdot), \overline{T}(\cdot))$, where $\overline{Q}(\cdot) = (\overline{Q}^i_k(\cdot))$ is a vector of buffer levels and $\overline{T}(\cdot) = (\overline{T}^i_k(\cdot))$ is a vector of allocations. A fluid vector $(\overline{Q}(\cdot), \overline{T}(\cdot))$ satisfying (1.7)–(1.15) is a *fluid solution.* The set of fluid solutions describes all possible trajectories of the fluid network under non-idling queueing disciplines.

The fluid network is said to be *weakly stable under non-idling queueing disciplines*, or simply *globally weakly stable*, if starting out empty, it remains empty, i.e., if every fluid solution to the system with

$$V_A(0) + V_B(0) = 0,$$

satisfies

$$V_A(t) + V_B(t) = 0,$$

for all $t \geq 0$.

The fluid network is said to be *(strongly) stable under non-idling queueing disciplines*, or simply *globally stable*, if there is some finite time $\tau > 0$

beyond which every fluid solution $(\overline{Q}(\cdot), \overline{T}(\cdot))$ that begins with one unit of fluid in the system, i.e., with

$$V_A(0) + V_B(0) = 1,$$

will be empty for all $t \geq \tau$, i.e., will satisfy

$$V_A(t) + V_B(t) = 0,$$

for all $t \geq \tau$.

In the remainder of this section, we discuss the relationships between fluid solutions and *fluid limits* or the limits of sample paths in a corresponding queueing network under a time and space scaling. Fluid limits provide a direct link between the discrete, stochastic queueing network and the continuous, deterministic fluid network; see, for example, Chen and Mandelbaum [9] and Dai [11]. To introduce fluid limits, we need to define a convergence notion in the path space. For each positive integer k, let $D^k[0, \infty)$ denote the set of functions from $[0, \infty)$ to $\mathbb{R}^k$ that are right continuous on $[0, \infty)$ and have left limits on $(0, \infty)$. Notice that for almost every sample path ω, the queue length process $\{Q(t, \omega), t \geq 0\}$ is an element in $D^k[0, \infty)$, where k is the total number of customer classes in the network. A sequence of functions $\{f_j\}$ in $D^k[0, \infty)$ is said to converge to $f \in D^k[0, \infty)$ uniformly on compact sets (u.o.c.) if for each $t > 0$,

$$\sup_{0 \leq s \leq t} |f_j(s) - f(s)| \to 0 \quad \text{as } j \to \infty,$$

where for a vector $x \in \mathbb{R}^k$, $|x|$ is the Euclidean norm of x.

For each sample path ω, each class (i, k) and each $t > s \geq 0$,

$$T^i_k(t, \omega) - T^i_k(s, \omega) \leq t - s.$$

Therefore the family

$$\{(T(r\,\cdot, \omega)/r, r \geq 1\}$$

is pre-compact under the u.o.c. topology. That is, for each sequence $\{r_j\}$ with $r_j \to \infty$ as $j \to \infty$, there is a subsequence $\{r_{j'}\} \subset \{r_j\}$ such that $T^i_k(r_{j'}t)/r_{j'}$ converges to a limit $\overline{T}^i_k(t)$ u.o.c. as $j' \to \infty$ for each class (i, k); see Royden [32]. It follows that $Q^i_k(r_{j'}t)/r_{j'}$ converges u.o.c. to a limit $\overline{Q}^i_k(t)$ for each class (i, k) as $j' \to \infty$. Any such limit $(\overline{Q}(\cdot), \overline{T}(\cdot))$ is said to be a *fluid limit*. Each fluid limit is a fluid solution satisfying (1.7)-(1.15). The following proposition follows immediately from the proof of Theorem 4.1 of Chen [8].

Proposition 1.3.1 For a given queueing discipline, the queueing network is pathwise stable if and only if the fluid limit $(\overline{Q}(\cdot), \overline{T}(\cdot))$ is unique and given by $\overline{Q}^i_k(t) = 0$ and $\overline{T}^i_k(t) = m^i_k t$ for each class (i, k).

One immediate corollary is the following result, which first appeared in Chen [8].

Corollary 1.3.2 If the fluid model is globally weakly stable, the corresponding queueing network is globally pathwise stable.

Lemma 1.3.3 Let C be a set of classes and $w = (w_k^i)_{(i,k)\in C}$ weights such that for each fluid limit $(\overline{Q}(\cdot), \overline{T}(\cdot))$:

$$\sum_{(i,k)\in C} w_k^i \overline{T}_k^i(t) \leq t$$

for all $t \geq 0$. If

$$\sum_{(i,k)\in C} w_k^i m_k^i > 1,$$

then the total volume of fluid in the fluid network diverges to infinity as $t \to \infty$. Therefore, the queueing network is not globally weakly stable, and for almost every sample path, the total number of customers in the network diverges to infinity as time $t \to \infty$.

Proof. Let $(\overline{Q}(\cdot), \overline{T}(\cdot))$ be a fluid limit. Because it is also a fluid solution with initial volume zero, it follows from (1.7) that for each class (i,k) we have

$$\sum_{\ell=1}^{k} \overline{Q}_\ell^i(t) = t - \mu_k^i \overline{T}_k^i(t).$$

Therefore,

$$\begin{aligned}
\sum_{(i,k)\in C} w_k^i m_k^i \left(\sum_{\ell=1}^{k} \overline{Q}_\ell^i(t) \right) &= \left(\sum_{(i,k)\in C} w_k^i m_k^i \right) t - \sum_{(i,k)\in C} w_k^i \overline{T}_k^i(t) \\
&> \left(\sum_{(i,k)\in C} w_k^i m_k^i - 1 \right) t.
\end{aligned}$$

Thus, the total volume of fluid in the fluid network diverges to infinity as $t \to \infty$ and by Proposition 1.3.1 the queueing network is not globally weakly stable. Furthermore, It follows from Dai [13, Theorem 3.2] that the total number of customers in the system goes to infinity as $t \to \infty$. □

1.4 Main Results

We show that virtual stations and push starts define necessary and sufficient conditions for the global pathwise stability of any two-station, open multiclass queueing network.

Definition 1.4.1 *Virtual stations* are sets of classes satisfying:

1. No class of a first excursion is in a virtual station, i.e., a class (i,k) in a virtual station must be in one of the excursions numbered 2, 3, $\ldots$, $|E_i|$.

2. If the last class of an excursion is in a virtual station, then every class of that excursion is in the virtual station and if a first class of an excursion is in a virtual station, then every first class of that excursion is in the virtual station. Thus, a virtual station must have either none of the classes, all of the classes, or all but the last class of each excursion.

3. If any class of an excursion is in a virtual station, then the last class of the preceding excursion cannot be in the virtual station.

Definition 1.4.2 A *push start set* is a set F of classes satisfying:

1. If class (i,k) is in F, then each class (i,k'), where $1 \le k' < k$, is also in F, and

2. If a first class of an excursion is in F, then every first class of the excursion is in F.

For a set C of classes, we let C_A denote the classes of C served at station A and C_B, those served at station B.

Theorem 1.4.3 A two-station open multi-class queueing network is globally pathwise stable if and only if for each push start set F and each virtual station C in the subnetwork consisting of classes not in F, we have

$$\frac{\sum_{(i,k)\in C_A} m_k^i}{1-\sum_{(i,k)\in F_A} m_k^i} + \frac{\sum_{(i,k)\in C_B} m_k^i}{1-\sum_{(i,k)\in F_B} m_k^i} \le 1. \qquad (1.16)$$

Furthermore, if (1.16) is violated for some sets F and C, then there is a static buffer priority discipline under which for almost every sample path, the total number of customers in the network diverges to infinity with time.

We also show that these same conditions are both necessary and sufficient to ensure the weak stability of the corresponding fluid model.

Theorem 1.4.4 A two-station fluid network is globally weakly stable if and only if the service times satisfy the conditions (1.16) of Theorem 1.4.3.

We further show that a two-station fluid network is globally (strongly) stable if and only if the service times satisfy the conditions of Theorem 1.4.3 with strict inequality.

Theorem 1.4.5 A two-station fluid network is globally (strongly) stable if and only if

$$\frac{\sum_{(i,k)\in C_A} m_k^i}{1-\sum_{(i,k)\in F_A} m_k^i} + \frac{\sum_{(i,k)\in C_B} m_k^i}{1-\sum_{(i,k)\in F_B} m_k^i} < 1 \tag{1.17}$$

for each F and C in Theorem 1.4.3.

One immediate corollary of Theorem 1.4.3 and Theorem 1.4.4 is:

Corollary 1.4.6 A two-station open multi-class queueing network is globally pathwise stable if and only if the corresponding two-station fluid network is globally weakly stable.

Finally, we show that under additional assumptions on the arrival and service processes, if the mean service times are in the interior of the global stability region, the queueing network is positive Harris recurrent.

Theorem 1.4.7 Under additional assumptions (1.23)–(1.26) in Section 1.6 on interarrival times and service times, if (1.17) holds for each F and C in Theorem 1.4.3, the queueing network is positive Harris recurrent under each non-idling queueing discipline.

We outline the proofs of Theorems 1.4.3, 1.4.4 and 1.4.5 in Section 1.5, and the proof of Theorem 1.4.7 in Section 1.6.

1.5 Virtual Stations and Push Starts

Harrison and Nguyen [24] and independently Dumas [18] observed that under some static priority queueing disciplines, certain classes in the Lu-Kumar [13] and Rybko-Stolyar [33] networks cannot receive service simultaneously even though they are served at different stations. Figure 1 illustrates the simplest queueing network exhibiting this phenomenon. When we give priority to the second buffer at station A and the first buffer at station B, the two high priority classes can never receive service simultaneously if they both start out empty. Lemma 1.5.1 generalizes and formalizes this observation.

For a set C of classes, we call a preempt-resume static buffer priority queueing discipline that gives higher priority at station A to classes in C_A and at station B to classes in C_B a *C-priority discipline*. The following lemma was proved in Dai and Vande Vate [15, Proposition 3.1].

Lemma 1.5.1 Let C be a virtual station in an initially empty two-station queueing network. Under a C-priority discipline,

$$\left(\sum_{(i,k)\in C_A} Q_k^i(t)\right)\left(\sum_{(i,k)\in C_B} Q_k^i(t)\right) = 0$$

for all $t \geq 0$.

Since classes in C_A and C_B can never be served simultaneously, their combined effect on system capacity is like that of classes served by a single station — they are part of a ***virtual station***. Just like real stations, if the system is to have sufficient capacity, the traffic intensity at each virtual station must not exceed 1.

Combining Lemma 1.5.1 with Lemma 1.3.3, gives the following necessary conditions for global stability.

Corollary 1.5.2 Let C be a virtual station in an initially empty two-station queueing network. If the system is globally pathwise stable then

$$\sum_{(i,k)\in C_A} m_k^i + \sum_{(i,k)\in C_B} m_k^i \leq 1. \tag{1.18}$$

Proof. From Lemma 1.5.1 we have

$$\sum_{(i,k)\in C} T_k^i(t) \leq t$$

for almost every sample path, each class (i,k) and each $t \geq 0$. It follows that

$$\sum_{(i,k)\in C} \overline{T}_k^i(t) \leq t$$

for each fluid limit. Because the network is globally pathwise stable, it follows from Lemma 1.3.3 that (1.18) holds. □

Remark 1.5.1 Lemmas 1.5.1 and 1.3.3 generalize the argument used in the proof of Lemma 2.2 of Dumas [18]. Our necessary conditions (1.18) improve, in the two-station case, those obtained by Dumas [19, 20] for d station queueing networks.

In the Lu-Kumar network of Figure 1, the virtual station consisting of classes 2 and 4 is the only virtual station that is not itself a subset of the classes at a station. Thus, if the system is globally pathwise stable, we must have

$$m_2 + m_4 \leq 1. \tag{1.19}$$

The example of Figure 2 illustrates the second phenomenon influencing the capacity of two-station queueing networks. Note that classes 4 and 6 constitute a virtual station in this network. This virtual station, however, imposes a stronger limitation on the capacity of this network than simply requiring that its traffic intensity be at most 1, i.e., stronger than the condition:

$$m_4 + m_6 \leq 1.$$

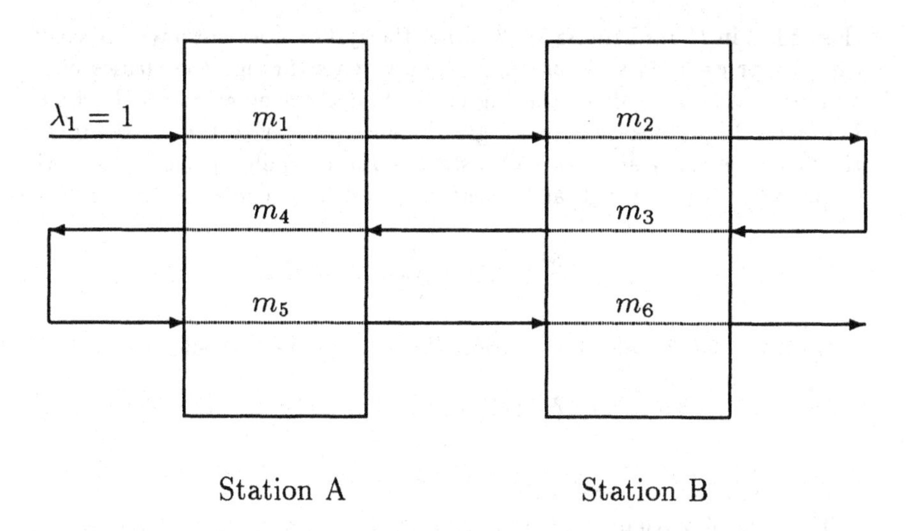

FIGURE 2. A six class network.

If we give highest priority to the first two classes, or *push start* these two classes, customers will tend to pass right through and arrive at class 3 at essentially the same average rate 1. Thus, we can think of the last four classes of this network as a new queueing network with customers arriving at the average rate 1. There is one essential difference, however, between this new queueing network and the network of Figure 1. Namely, over the course of time the server at station A will tend to dedicate the fraction m_1 of his time to class 1 leaving him with only the fraction $1 - m_1$ for classes 4 and 5. Likewise, the server at station B will tend to dedicate the fraction m_2 of his time to class 2 leaving him with only the fraction $1 - m_2$ for classes 3 and 6.

The new network is like the network in Figure 1 insofar as giving next highest priority to class 4 at station A and to class 6 at station B exposes a virtual station — these two classes cannot be served simultaneously. Now, serving class 4 will tend to take the fraction $m_4/(1 - m_1)$ of server A's remaining time and serving class 6 will tend to take the fraction $m_6/(1 - m_2)$ of server B's remaining time. Since these two services cannot occur simultaneously, (1.19) leads to:

$$\frac{m_4}{1 - m_1} + \frac{m_6}{1 - m_2} \leq 1.$$

In other words, the traffic intensity at a virtual stations is magnified by push starting earlier classes because this leaves the servers with less capacity for the classes of the virtual station.

Consider a push start set F. The set F partitions the queueing network into those classes in F and a smaller queueing network consisting of the classes not in F. By Lemma 1.2.2, either the system does not have sufficient capacity or each class of customer tends to pass through the classes of F and arrive at the smaller queueing network at the average rate 1. Further, Lemma 1.2.2 implies that, if we give highest priority to the classes in F, the traffic intensity at each virtual station of the smaller queueing network is magnified by the fact that the server at station A dedicates the fraction

$$M_A(F) = \sum_{(i,k)\in F_A} m_k^i$$

of his time and the server at station B dedicates the fraction

$$M_B(F) = \sum_{(i,k)\in F_B} m_k^i$$

of his time to serving the classes of F.

The role of virtual stations as described in Corollary 1.5.2, combined with this influence of push starting, proves the necessity of the conditions in Theorem 1.4.3 for the global stability of two-station queueing networks.

Proof of Theorem 1.4.3. To establish the necessity of conditions (1.16), suppose the system is globally pathwise stable and consider a push start set F and a virtual station C in the subnetwork consisting of the classes not in F. Consider a preempt-resume static buffer priority queueing discipline that gives highest priority to the classes of F and next highest priority to the classes of C. By the arguments in the proof of Theorem 2.1 in [15, Section 4], under this queueing discipline, almost all sample paths satisfy:

$$\frac{\sum_{(i,k)\in C_A} T_k^i(t)}{1 - M_A(F)} + \frac{\sum_{(i,k)\in C_B} T_k^i(t)}{1 - M_B(F)} \le t + \epsilon(t)$$

for all $t \geq 0$, where $\epsilon(t)/t \to 0$. Hence under this queueing discipline, each fluid limit $(\overline{Q}(\cdot), \overline{T}(\cdot))$ must satisfy

$$\frac{\sum_{(i,k)\in C_A} \overline{T}_k^i(t)}{1 - M_A(F)} + \frac{\sum_{(i,k)\in C_B} \overline{T}_k^i(t)}{1 - M_B(F)} \le t. \tag{1.20}$$

If

$$\frac{\sum_{(i,k)\in C_A} m_k^i}{1 - \sum_{(i,k)\in F_A} m_k^i} + \frac{\sum_{(i,k)\in C_B} m_k^i}{1 - \sum_{(i,k)\in F_B} m_k^i} > 1, \tag{1.21}$$

then it follows from Lemma 1.3.3 that the total volume of fluid in the buffers diverges to infinity with time, and the total number of customers goes to infinity as $t \to \infty$. By Proposition 1.3.1, this contradicts the assumption that the network is globally pathwise stable. Therefore, we have proved the necessity of (1.16).

To prove the sufficiency of conditions (1.16), suppose that (1.16) holds for each push start set F and each virtual station C. It follows from Theorem 1.7.6 that the fluid model is globally weakly stable. Hence by Corollary 1.3.2 the queueing network is globally pathwise stable.

Finally, we prove that if (1.16) is violated for some sets F and C, then there is a static buffer priority discipline under which for almost every sample path, the total number of customers in the network diverges to infinity with time. Theorem 2.1 of [15] shows that under any preempt-resume the same queueing discipline used above the number of customers in the system goes to infinity with time. In other words, whenever a two-station queueing network is not globally pathwise stable, there is a preempt-resume static priority queueing discipline under which it is not stable. Thus, we conclude that the preempt-resume static priority queueing disciplines are "worst" in the class of non-idling queueing disciplines. □

Proof of Theorem 1.4.4. Suppose that (1.16) holds for each push start set F and each virtual station C, it follows from Theorem 1.7.6 that the fluid model is globally weakly stable.

Suppose that there is some push start set F and some virtual station C such that (1.16) is violated or (1.21) holds. Consider the same queueing discipline used in the proof of Theorem 1.4.3. Let $(\overline{Q}(\cdot), \overline{T}(\cdot))$ be a fluid limit. Then $(\overline{Q}(\cdot), \overline{T}(\cdot))$ is a fluid solution. If $\sum_{(i,k)\in F} \overline{Q}_k^i(t) > 0$ for some t, then the fluid model is not globally weakly stable, thus proving the theorem. Now assume that $\sum_{(i,k)\in F} \overline{Q}_k^i(t) = 0$ for all $t \geq 0$. It follows from the proof of Theorem 1.4.3 that (1.20) holds, and hence the total volume of fluid diverges to infinity. Thus, the fluid model is not globally weakly stable. □

In the following proof we offer a very indirect argument showing that if (1.17) is violated for some push start set F and virtual station C, then the fluid model is not globally stable. In Dai and Vande Vate [14], we offer a direct proof of this fact that constructs a particular fluid solution showing the fluid model is not globally stable.

Proof of Theorem 1.4.5. Suppose that (1.17) holds for each push start set and each virtual station C, it follows from Theorem 1.7.4 that the fluid model is globally stable.

Now suppose that (1.17) is violated for some push start set F and some virtual station C. Namely,

$$\frac{\sum_{(i,k)\in C_A} m_k^i}{1 - \sum_{(i,k)\in F_A} m_k^i} + \frac{\sum_{(i,k)\in C_B} m_k^i}{1 - \sum_{(i,k)\in F_B} m_k^i} \geq 1. \tag{1.22}$$

We would like to show that the fluid model is *not* globally stable. Suppose that, to the contrary, the fluid model is globally stable. Let (i_0, k_0) be a fixed class not in F. Consider a sequence of initial starting conditions for the queueing network. In the rth starting condition, $Q_{k_0}^{i_0}(0) = r$

and $Q_k^i(0) = 0$ for all other classes (i,k). Also consider the queueing discipline used in the proof of Theorem 1.4.3. Let $(\overline{Q}(\cdot), \overline{T}(\cdot))$ be a limit point of $\{(Q(r\cdot)/r, T(r\cdot)/r), r \geq 1\}$. Because the fluid model is assumed to be globally stable, the subnetwork consisting of classes in F is also globally stable. By Chen [8], the subnetwork is globally weakly stable. Thus, $\sum_{(i,k)\in F} \overline{Q}_k^i(t) = 0$ for $t \geq 0$. It follows the proof of Theorem 1.4.3 that (1.20) holds for $t \geq 0$, and the arguments used in the proof of Lemma 1.3.3 show that

$$\sum_{(i,k)\in C} w_k^i m_k^i \left(\sum_{\ell=1}^{k} \overline{Q}_\ell^i(t) \right) \geq \sum_{(i,k)\in C} w_k^i m_k^i > 0$$

for each t. Because $(\overline{Q}(\cdot), \overline{T}(\cdot))$ is a fluid solution, the fluid model is not globally (strongly) stable. □

1.6 Positive Harris recurrence

In this section we define *positive Harris recurrence*—a stronger notion of stability used in recent studies of queueing networks including Sigman [35], Kaspi and Mandelbaum [25], Meyn and Down [30] and Dai [11]. To define positive Harris recurrence, we first need to define the *state* of the queueing network and this depends on the queueing discipline used. To be concrete, assume that the queueing discipline is a preempt-resume static buffer priority discipline. Examples of other disciplines can be found in Dai [11]. Under a preempt-resume static buffer priority discipline, the state $X(t)$ at time t is defined by the number of customers in each buffer, the service time remaining for the first customer in each buffer, and the time until the next customer arrives for each type.

We assume all random variables are defined on a probability space $(\Omega, \mathcal{F}, \mathsf{P})$ with expectation operator E. For each type $i \in I$, let $\xi_i = \{\xi_i(r), r \geq 1\}$ be a sequence of interarrival times for type i customers. Similarly, for each class (i,k), let $\eta_k^i = \{\eta_k^i(r), r \geq 1\}$ be a sequence of service times. We assume that

$$\xi_1, \ldots, \xi_n, \eta_1^1, \ldots, \eta_{c_n}^n \quad \text{are mutually independent iid sequences,} \tag{1.23}$$

$$\mathsf{E}[\xi_i(1)] < \infty \text{ and } \mathsf{E}[\eta_k^i(1)] < \infty \quad \text{for } i \in I \text{ and } k \in 1, 2, \ldots, c_i. \tag{1.24}$$

We assume further that, for each $i \in I$, there exists some integer $j_i > 0$ and some function $p_i(x) \geq 0$ on $\mathbb{R}_+$ with $\int_0^\infty p_i(x)\,dx > 0$, such that

$$\mathsf{P}\{\xi_i(1) \geq x\} > 0, \quad \text{for each } x > 0, \tag{1.25}$$

$$\mathsf{P}\Big\{a \leq \sum_{r=1}^{j_i} \xi_i(r) \leq b\Big\} \geq \int_a^b p_i(x)\,dx, \quad \text{for each } 0 \leq a < b. \tag{1.26}$$

Under condition (1.23), the process $X = \{X(t), t \geq 0\}$ is a strong Markov process (see Dai [11]). The Markov process $X = \{X(t), t \geq 0\}$ is said to be *positive Harris recurrent* if it possesses a stationary distribution; see Meyn and Tweedie [31] or Section 3 of Dai [11]. The queueing network is positive Harris recurrent under a queueing discipline if the corresponding state process $X = \{X(t), t \geq 0\}$ is positive Harris recurrent.

Remark 1.6.1 If interarrival times and service times are independent and exponentially distributed, the state process is a continuous time Markov chain taking discrete values. Hence positive Harris recurrence is equivalent to positive recurrence for a Markov chain.

Proof of Theorem 1.4.7. If the stability conditions in (1.17) hold, we prove in Section 1.7 that the corresponding fluid model is strongly stable under any non-idling discipline. Because (1.23)–(1.26) hold, it follows from Theorem 4.2 of Dai [11] that the queueing network is positive Harris recurrent under any non-idling discipline. □

1.7 Stability of The Fluid Model

In this section we outline the arguments used to show that the conditions of Theorem 1.4.5 are sufficient to ensure stability of the fluid network by showing that when they are satisfied there is a potential function or Lyapunov function G proving that the system drains to zero regardless of the initial conditions. Detailed proofs will appear in Dai and Vande Vate [14]. At the end of the section, we comment on the modifications necessary to establish Theorem 1.4.3 for the case of global weak stability.

We define the Lyapunov function G to be the maximum of two linear potential functions — one for each station — and so it is piecewise-linear. The linear potential function at each station is defined in terms of the volume of fluid in the system that must pass through the station. In particular, we let $Z^i_k(t)$ denote the volume of fluid i that has already entered the network by time t, but has not yet received class (i,k) service, i.e.,

$$Z^i_k(t) = t - \mu^i_k \overline{T}^i_k(t) = \sum_{\ell \leq k} \overline{Q}^i_\ell(t).$$

We define the linear potential function at a station to be a weighted combination of the volume of fluid in the system that must pass through the classes the station serves. In particular, given weights $x = (x^i_k)$ for the classes, we define the *generalized workloads* at station A and station B to be:

$$G_A(x,t) \quad = \quad \sum_{i \in I} \sum_{k \in A_i} x^i_k Z^i_k(t) \quad \text{and}$$

$$G_B(x,t) = \sum_{i \in I} \sum_{k \in B_i} x_k^i Z_k^i(t).$$

When the weights x are the service times m, $G_A(m,t)$ and $G_B(m,t)$ are the total workloads at station A and station B, respectively, as used in Wein [36].

Notice that for each fluid solution $(\overline{Q}(\cdot), \overline{T}(\cdot))$, $\overline{T}(t)$ is a Lipschitz continuous function of t. Hence $\overline{T}(\cdot)$ is absolutely continuous and for almost all (in Lebesgue measure) t in $(0,\infty)$, $\overline{T}(\cdot)$ is regular or has derivative at time t. If t is a regular point of $\overline{T}(\cdot)$, it follows from (1.7) that t is a regular point of $\overline{Q}(\cdot)$, and hence of $G_A(x,\cdot)$ and $G_B(x,\cdot)$. Let

$$G(x,t) = \max\{G_A(x,t), G_B(x,t)\},$$

then $G(x,t)$ is Lipschitz continuous in t. Hence for almost all t, $G(x,\cdot)$ and $\overline{T}(\cdot)$ are regular at t. In the following, whenever a derivative is used at time t, it is assumed that t is such a regular point.

If the piecewise linear function $G(x,t)$ is strictly positive whenever some buffer is not empty and there is $\epsilon > 0$ such that whenever some buffer is not empty,

$$\dot{G}(x,t) \equiv \frac{\partial G(x,t)}{\partial t} \leq -\epsilon,$$

then after time $\tau = G(x,0)/\epsilon$ all buffers will have drained to zero; proving that the fluid network is stable. Dai and Weiss [16] showed that G will satisfy these conditions if there is $\epsilon > 0$ and weights $x > 0$ such that:

$$G_A(x,t) \leq G_B(x,t) \quad \text{whenever} \quad V_A(t) = 0, \tag{1.27}$$

$$G_B(x,t) \leq G_A(x,t) \quad \text{whenever} \quad V_B(t) = 0, \tag{1.28}$$

$$\frac{\partial G_A(x,t)}{\partial t} \leq -\epsilon \quad \text{whenever} \quad V_A(t) > 0, \text{ and} \tag{1.29}$$

$$\frac{\partial G_B(x,t)}{\partial t} \leq -\epsilon \quad \text{whenever} \quad V_B(t) > 0. \tag{1.30}$$

Thus we have the following lemma.

Lemma 1.7.1 If there exists $\epsilon > 0$ and $x > 0$ such that (1.27)–(1.30) hold, the fluid network is stable under non-idling dispatching policies.

We transform into a linear program the problem of finding weights $x > 0$ such that $G_A(x,t)$ and $G_B(x,t)$ satisfy (1.27)–(1.30). The linear program has solutions with strictly positive objective values if and only if the desired weights x exist and any solution with strictly positive objective value provides weights satisfying the desired conditions.

Finding the largest possible value of ϵ for which there are weights x satisfying (1.27)–(1.30) reduces to solving the following linear program for

ϵ and x:

$$\text{maximize } \epsilon \tag{1.31}$$

$$\text{subject to:}$$

$$\sum_{k\in A_i, k>\ell(i,e)} x_k^i - \sum_{k\in B_i, k\geq \ell(i,e)} x_k^i \leq 0 \text{ for } i \in I, e \in E_B^i \tag{1.32}$$

$$\sum_{k\in B_i, k>\ell(i,e)} x_k^i - \sum_{k\in A_i, k\geq \ell(i,e)} x_k^i \leq 0 \text{ for } i \in I, e \in E_A^i \tag{1.33}$$

$$\sum_{i\in I}\sum_{k\in A_i} x_k^i - x_\ell^i/m_\ell^i + \epsilon \leq 0 \text{ for } i \in I, \ell \in A_i \tag{1.34}$$

$$\sum_{i\in I}\sum_{k\in B_i} x_k^i - x_\ell^i/m_\ell^i + \epsilon \leq 0 \text{ for } i \in I, \ell \in B_i \tag{1.35}$$

$$x \geq 0 \tag{1.36}$$

The constraints (1.32)–(1.36) define a cone with the single extreme point given by $x = 0$ and $\epsilon = 0$. Thus, we have the following lemma:

Lemma 1.7.2 The linear program (1.31)–(1.36) either:

1. has optimum objective value 0, in which case there are no weights $x > 0$ such that G_A and G_B satisfy (1.27)–(1.30) and so there is no piecewise-linear Lyapunov function of the form of G proving that the fluid network is stable; or

2. has unbounded objective values, in which case each solution (x, ϵ) with $\epsilon > 0$ provides weights $x > 0$ such that $G(x, t)$ is a piecewise-linear Lyapunov function proving that the fluid network is stable.

We translate the linear program (1.31)–(1.36) into the following equivalent *parametric network flow* problem. The parametric network flow problem is equivalent to the linear program in the sense that the linear program has unbounded objective values if and only if there is a value of the parameter β for which the network flow problem has strictly positive objective value.

$$\begin{aligned}
&\text{maximize } \epsilon && (1.37)\\
&\text{subject to:} && \\
-\sum_{\ell\in f(i,e)} x^i_\ell + s^i_e - z^i_e &= 0 \text{ for } i \in I \text{ and } e \in E^i_A && (1.38)\\
\sum_{\ell\in f(i,e)} x^i_\ell - s^i_e + z^i_e &= 0 \text{ for } i \in I \text{ and } e \in E^i_B && (1.39)\\
-x^i_{\ell(i,e)} + z^i_e + s^i_{e+1} &= 0 \text{ for } i \in I \text{ and } e \in E^i_A && (1.40)\\
x^i_{\ell(i,e)} - z^i_e - s^i_{e+1} &= 0 \text{ for } i \in I \text{ and } e \in E^i_B && (1.41)\\
\sum_{i\in I}\sum_{k\in A_i} x^i_k + \epsilon &= 1 && (1.42)\\
-\sum_{i\in I}\sum_{k\in B_i} x^i_k - \epsilon &= -\beta && (1.43)\\
m^i_k &\le x^i_k \text{ for } i \in I \text{ and } k \in A_i && (1.44)\\
\beta m^i_k &\le x^i_k \text{ for } i \in I \text{ and } k \in B_i && (1.45)\\
x, s, z &\ge 0. && (1.46)
\end{aligned}$$

To see the equivalence between these two problems, observe that the linear program (1.31)–(1.36) has unbounded objective values if and only if it admits a solution (x, ϵ) with $\epsilon > 0$. We can scale any such solution to obtain a solution satisfying (1.42) and, by choosing β appropriately, (1.43). Now, using (1.42) we can transform (1.34) into (1.44) and using (1.43), we can transform (1.35) into (1.45). Observe that the last constraint of (1.32) or (1.33) for each excursion implies all the other constraints for the excursion, so we ignore the others. Finally, we obtain (1.38)–(1.41) from (1.32)–(1.33) by adding slack variables s^i_e for the excursions at A and z^i_e for the excursions at B and taking successive differences of the resulting equations.

The linear program (1.37)–(1.46) is a network flow problem with right-hand-sides and lower bounds that depend on the parameter β. In fact, there are weights x satisfying (1.38)–(1.46) if and only if the minimum flow in this network from the node for (1.42) to the node for (1.43) is strictly less than $\max\{1, \beta\}$ — the remaining flow can be assigned to ϵ.

This network has no directed cycles and so induces a partial order on the nodes in which node i precedes node j or $i \preceq j$ if and only if there is a path from i to j. The node for (1.42) is the least element in this partial order since there is no path to it and the node for (1.43) is the greatest element since there is a path to it from every node. An *upper ideal* of a partial order is a subset S of the elements with the property that if $i \in S$ and $i \preceq j$ then $j \in S$. Note that the node for (1.43) is a member of every upper ideal whereas the node for (1.42) is a member only of the upper ideal consisting of all the nodes. Henceforth, we omit these two nodes from all upper ideals.

The following theorem characterizes the minimum flow in an acyclic network in terms of the upper ideals of the corresponding partial order. It is an immediate consequence of the well-known Max Flow–Min Cut Theorem [22]. The *capacity* of an upper ideal S is the sum of the lower bounds on edges from $i \notin S$ to $j \in S$.

Theorem 1.7.3 The minimum volume of flow in an acyclic network with infinite edge capacities is equal to the maximum capacity of an upper ideal.

Theorem 1.7.3 leads to sufficient conditions for global strong stability of the two-station fluid network in terms of β and the service times. Eliminating β via Fourier-Motzkin Elimination and some algebraic manipulation yields the conditions of Theorem 1.4.5. Hence, we have the following theorem.

Theorem 1.7.4 If condition (1.17) holds for each push start set F and each virtual station C, the linear program (1.31)–(1.36) has unbounded objective value, and hence the fluid model is globally (strongly) stable.

In the case of weak stability, Lemma 1.7.1 can be generalized as follows.

Lemma 1.7.5 If there exists $x > 0$ such that (1.27)–(1.30) hold with $\epsilon = 0$, the fluid network is globally weakly stable.

In this case, we need only find a solution to (1.38)–(1.46) with $\epsilon \geq 0$ for some $\beta > 0$. Such a solution will provide weights $x > 0$ satisfying (1.38)–(1.36) with $\epsilon = 0$.

Theorem 1.7.6 If condition (1.16) holds for each push start set F and each virtual station C, the linear program (1.37)–(1.46) has a feasible solution with $x > 0$ and $\epsilon \geq 0$, and hence the fluid model is globally weakly stable.

Acknowledgments: We thank Vincent Dumas for conversations that helped in the development of the virtual station concept. J. G. Dai is supported in part by National Science Foundation grant DMI-94-57336 and by US-Israel Binational Science Foundation grant 94-00196. J. H. Vande Vate is supported in part by Airforce Office of Scientific Research Grant F49620–95–1–0121.

1.8 References

[1] Altman, E., Foss, S. G., Riehl, E., and Stidham, S. Performance bounds and pathwise stability for generalized vacation and polling

systems. (1994). Preprint.

[2] Bertsimas, D., Gamarnik, D., and Tsitsiklis, J. N. Stability conditions for multiclass fluid queueing networks. (1995). Preprint.

[3] Botvich, D. D. and Zamyatin, A. A. Ergodicity of conservative communication networks. Rapport de recherche 1772, INRIA, 1992.

[4] Bramson, M. Instability of FIFO queueing networks. *Annals of Applied Probability* **4**, 414–431 (1994).

[5] Bramson, M. Instability of FIFO queueing networks with quick service times. *Annals of Applied Probability* **4**, 693–718 (1994).

[6] Bramson, M. Convergence to equilibria for fluid models of FIFO queueing networks. *Queueing Systems: Theory and Applications* (1995). To appear.

[7] Bramson, M. Convergence to equilibria for fluid models of processor sharing queueing networks. (1995). Preprint.

[8] Chen, H. Fluid approximations and stability of multiclass queueing networks I: Work-conserving disciplines. *Annals of Applied Probability* **5**, 637–665 (1995).

[9] Chen, H. and Mandelbaum, A. Discrete flow networks: Bottlenecks analysis and fluid approximations. *Mathematics of Operations Research* **16**, 408–446 (1991).

[10] Chen, H. and Zhang, H. Stability of multiclass queueing networks under FIFO service discipline. (1995). Preprint.

[11] Dai, J. G. On positive Harris recurrence of multiclass queueing networks: A unified approach via fluid limit models. *Annals of Applied Probability* **5**, 49–77 (1995).

[12] Dai, J. G. Stability of open multiclass queueing networks via fluid models. In Kelly, F. and Williams, R. J., editors, *Stochastic Networks*, volume 71 of *The IMA volumes in mathematics and its applications*, pages 71–90, New York, 1995. Sringer-Verlag.

[13] Dai, J. G. A fluid-limit model criterion for instability of multiclass queueing networks. *Annals of Applied Probability* (1996). To appear.

[14] Dai, J. G. and VandeVate, J. The stability of two-station fluid networks. (1996). Preprint.

[15] Dai, J. G. and VandeVate, J. Virtual stations and the capacity of two-station queueing networks. (1996). Preprint.

[16] Dai, J. G. and Weiss, G. Stability and instability of fluid models for re-entrant lines. *Mathematics of Operations Research* (1996).

[17] Down, D. and Meyn, S. A survey of Markovian methods for stability of networks. (1994). Preprint.

[18] Dumas, V. A multiclass network with non-linear, non-convex, non-monotonic stability conditions. (1995). Preprint.

[19] Dumas, V. *Approches fluides pour la stabilité et l'instabilité de réseaux de files d'attente stochastiques à plusieurs classes de clients.* PhD thesis, L'école Polytechnique, Paris, France, 1996.

[20] Dumas, V. Stability and irreducibility of multiclass networks with priorities. In preparation, 1996.

[21] El-Taha, M. and Stidham Jr., S. Sample-path stability conditions for multiserver input-output processes. *Journal of Applied Mathematics and Stochastic Analysis* **7**, 437–456 (1994).

[22] Ford, L. and Fulkerson, D. *Flows in networks.* Princeton University Press, Princeton, New Jersey, 1962.

[23] Foss, S. and Rybko, A. Stability of multiclass Jackson-type networks. (1995). Preprint.

[24] Harrison, J. M. and Nguyen, V. Some badly behaved closed queueing networks. In Kelly, F. P. and Williams, R. J., editors, *Stochastic Networks*, volume 71 of *The IMA volumes in mathematics and its applications*, pages 117–124, New York, 1995. Springer-Verlag.

[25] Kaspi, H. and Mandelbaum, A. Regenerative closed queueing networks. *Stochastics and Stochastics Reports* **39**, 239–258 (1992).

[26] Kelly, F. P. Networks of queues with customers of different types. *J. Appl. Probab.* **12**, 542–554 (1975).

[27] Kumar, P. R. and Meyn, S. P. Stability of queueing networks and scheduling policies. *IEEE Transactions on Automatic Control* **40**, 251–260 (1995).

[28] Kumar, P. R. and Seidman, T. I. Dynamic instabilities and stabilization methods in distributed real-time scheduling of manufacturing systems. *IEEE Transactions on Automatic Control* **AC-35**, 289–298 (1990).

[29] Lu, S. H. and Kumar, P. R. Distributed scheduling based on due dates and buffer priorities. *IEEE Transactions on Automatic Control* **36**, 1406–1416 (1991).

[30] Meyn, S. P. and Down, D. Stability of generalized Jackson networks. *Annals of Applied Probability* **4**, 124–148 (1994).

[31] Meyn, S. P. and Tweedie, R. L. Stability of Markovian processes II: Continuous time processes and sample chains. *Adv. Appl. Probab.* **25**, 487–517 (1993).

[32] Royden, H. L. *Real analysis.* MacMillan, New York, 3rd edition, 1988.

[33] Rybko, A. N. and Stolyar, A. L. Ergodicity of stochastic processes describing the operation of open queueing networks. *Problems of Information Transmission* **28**, 199–220 (1992).

[34] Seidman, T. I. 'First come, first served' can be unstable! *IEEE Transactions on Automatic Control* **39**, 2166–2171 (1994).

[35] Sigman, K. The stability of open queueing networks. *Stoch. Proc. Applns.* **35**, 11–25 (1990).

[36] Wein, L. M. Scheduling networks of queues: Heavy traffic analysis of a multistation network with controllable inputs. *Operations Research* **40**, S312–S334 (1992).

[37] Whitt, W. Large fluctuations in a deterministic multiclass network of queues. *Management Sciences* **39**, 1020–1028 (1993).

2
Stable Priority Disciplines for Multiclass Networks

Hong Chen and David D. Yao

ABSTRACT It is well known that a multiclass network may not be stable under the usual traffic condition that the arrived nominal workload is less than one. That this instability may happen in many "favorite" disciplines (such as first-in-first-out, shorted expected service times, and shorted expected remaining service times) has been the focus of many recent studies. In this paper, we address the stability issue from a different angle by asking: whether given any network satisfying the traffic condition, a simple priority discipline can be identified under which the network is guaranteed to be stable. We show this can be done under the condition of *acyclic class transfer*. This covers a wide class of networks, including all networks with a deterministic routing mechanism, such as re-entrant lines and Kelly networks.

2.1 Introduction

Ever since the discovery of several counterexamples by Bramson [4] [5], Rybko and Stolyar [24], and Seidman [25] that a multiclass queueing network may not be stable under the usual traffic condition (that the arrived nominal workload — or, *nominal traffic intensity* — is less one), much research effort has been devoted to the search of sufficient conditions to guarantee that a multiclass queueing network is stable under given service disciplines. The body of literature on this subject is growing; a partial list includes:

- Bertsimas, *et al* [2], Chen [8], Dai and VandeVate [13], Dai and Weiss [14], Down and Meyn [15], and Kumar and Meyn [20], focusing on general work-conserving disciplines;
- Botvich and Zamyatin [3], Chen and Zhang [10], and Dumas [16], focusing on certain priority disciplines;
- Bramson [6], Chen and Zhang [9], Foss and Rybko [17], Rybko and Stolyar [24], and Winograd and Kumar [27], focusing on the first-come-first-served discipline.

Instead of investigating whether a queueing network is stable under a specific service discipline, here our starting point is to ask a different question: given a network and the traffic condition, whether one can identify a "simple" service discipline for which the network is stable. By "simple", we mean those service disciplines that are easy to characterize and to implement; for instance, priority disciplines, or more generally, index policies. The difficulty involved here has been highlighted by several studies, e.g., Banks and Dai [7] and Rybko and Stolyar [24], which demonstrated that some "favorite" disciplines, such as the expected shortest remaining processing time first and the expected shortest processing time first may not guarantee stability in general.

Most of the above references establish stability via a general and powerful theory that links the stability of a (stochastic) queueing network to that of a corresponding (deterministic) fluid network. This theory was first proposed by Rybko and Stolyar [24], and generalized and proved by Dai [11] (also see Chen [8], Dai [12], Meyn [23] and Stolyar [26]). This is also the approach taken in this paper. Specifically, we show that for any multiclass fluid network, with an *acyclic class transfer* mechanism, a simple, static priority discipline can be identified, which guarantees the stability of the fluid network under the usual traffic condition. This class includes virtually all the multiclass networks that have been so far studied in the stability literature; in particular, it includes networks with a deterministic routing mechanism, such as re-entrant lines and (generalized) Kelly networks.

The organization of this paper is as follows. The fluid network model is formulated in §2.2. The main result, stated in terms of fluid networks, is presented in §2.3 and proved in §2.4. Finally, in §2.5, we recast the main result in the context of queueing networks.

2.2 The Network Model

A fluid network consisting of J stations with K classes of fluid can be characterized by a set of four parameters (α, P, μ, C): where

- $\alpha = (\alpha_k)$ is a K-dimensional nonnegative vector, with α_k being the (exogenous) *inflow* rate of class k;
- $P = (p_{ik})$ is a $K \times K$ non-negative matrix, the *flow-transfer* matrix, with p_{ik} being the probability (proportion) of class i fluid switching to class k;
- $\mu = (\mu_k)$ is a K-dimensional positive vector, with μ_k being the processing rate (or, potential *outflow* rate) of class k;
- $C = (c_{jk})$ is a $J \times K$ binary (0-1) matrix, the *constituent* matrix, with $c_{jk} = 1$ if class k fluid is processed at station j, and $c_{jk} = 0$ otherwise. (Note that while each row of C can have multiple 1's, each column has exactly one entry of 1.)

We assume that P is a substochastic matrix, with a spectral radius strictly less than 1. (In this sense, the network is an "open network".) Hence, the matrix $(I - P')$ is invertible, and the inverse is a nonnegative matrix.

Let $\sigma(k)$ denote the station at which class k fluid is processed, i.e., $\sigma(k) = j$ if and only if $c_{jk} = 1$. Since each column k of C has exactly one entry of 1, $\sigma(k)$ is well defined. Let $C(j) = \{k : \sigma(k) = j\}$ be the set of classes that are processed at station j.

The dynamics of the network can be described as follows: For any class k, the fluid flows into the network (from external sources) at rate α_k, and is stored and processed at station $\sigma(k)$; after being processed it is split into a set of other classes: a proportion of p_{ki} transfers to class i, and enters into station $\sigma(i)$, while the proportion $1 - \sum_i p_{ki}$ leaves the network. The maximum rate at which class k fluid can be processed (at station $\sigma(k)$) is μ_k. This happens when the entirety (100%) of the station's capacity is allocated to class k. Otherwise, the outflow rate of class k fluid from station $\sigma(k)$ is less than μ_k.

Let $t \geq 0$ index time. Let $Q_i(t)$ denote the fluid level of class i at time t. Let $T_i(t)$ denote the *total* amount of time that station $\sigma(i)$ has allocated to processing class i fluid over the time interval $[0, t]$. Furthermore, denote

$$Q(t) = (Q_i(t)), \quad Q_i = \{Q_i(t), t \geq 0\}, \quad Q = (Q_i);$$

and similarly,

$$T(t) = (T_i(t)), \quad T_i = \{T_i(t), t \geq 0\}, \quad T = (T_i).$$

We shall refer to Q as the fluid-level process, and T as the allocation process, or a *service discipline.*

Now, given the initial fluid level, $Q(0)$, the fluid-level process $Q = (Q_i)$ can be determined as follows:

$$Q_i(t) = Q_i(0) + \alpha_i t + \sum_{k=1}^{K} \mu_k T_k(t) p_{ki} - \mu_i T_i(t),$$

or, in matrix form,

$$Q(t) = Q(0) + \alpha t - (I - P')DT(t), \tag{2.1}$$

where I is the identity matrix, D is a diagonal matrix with the components of μ at the diagonal, and prime ($'$) denotes matrix transpose.

Clearly,

$$T_k(\cdot) \text{ must be nondecreasing}, \quad k = 1, ..., K. \tag{2.2}$$

Let $U(t) := et - CT(t)$, with e being a vector of all 1's. Then $U_j(t)$ (the jth coordinate of $U(t)$) indicates the *total* idle time of station j over the time interval $[0, t]$. Clearly,

$$U_j(\cdot) \text{ must be nondecreasing}, \quad j = 1, ..., J. \tag{2.3}$$

Any allocation process T satisfying (2.1)-(2.3) is called a *feasible* discipline. If, in addition, T satisfies

$$\sum_{k\in C(j)} \int_0^\infty Q_k(t)dU_j(t) = 0, \qquad j = 1,...,J, \tag{2.4}$$

then it is called a *work-conserving* discipline. Many commonly used disciplines can be modeled by imposing more conditions on T (see, for example, Dai and Weiss [14] for priority disciplines, and Chen and Zhang [9] for the FIFO discipline).

A fluid network is stable under a feasible discipline T if for any initial fluid level $Q(0)$ and any $\epsilon > 0$, there exists a finite time t_0 such that $e'Q(t_0) < \epsilon$, where Q is the fluid-level process corresponding to T.

2.3 The Main Result

Note that no generality is lost by assuming that $p_{kk} = 0$ for all classes $k = 1,...K$. (That is, there no "self-loop", in terms of class transfers.) For we can always write the $(I - P')D$ part of (2.1) as $(I - \hat{P}')\hat{D}$, with $\hat{P}$ having all zero diagonal entries:

$$\hat{p}_{kk} = 0, \quad \hat{p}_{ik} = p_{ik}/(1 - p_{kk}), \ i \neq k;$$

and

$$\hat{\mu}_k = (1 - p_{kk})\mu_k.$$

And then study the network $(\alpha, \hat{P}, \hat{\mu}, C)$. Hence, below we assume there is no self-loop for any class.

Now, for our main result to hold, we need to assume that the class transfer mechanism is *acyclic*. This can be defined precisely as follows. Let

$$I(k) = \{i = 1,...,K : p_{ik} > 0\}$$

denote the set of fluid classes that can *directly* transfer to class k. Denote $i \to k$, if $i \in I(k)$; and extend this to satisfy transitivity, i.e., $i \to k$ and $j \to i$ implies $j \to k$. That is, $j \to k$ means that class j can directly or indirectly transfer to class k. Let $\mathcal{I}(k) = \{j : j \to k\}$ denote the set of all fluid classes that can directly or indirectly transfer to class k. (Note that both $I(k)$ and $\mathcal{I}(k)$ could be empty sets.) If $j \in \mathcal{I}(k)$ (or equivalently, $j \to k$), we say there is a *path* from j to k, When a path starts and ends with the same class, we call it a *loop*. Clearly, k is in a loop if and only if $k \in \mathcal{I}(k)$.

The key condition we need here is the following:

- **Acyclic Class Transfer (ACT)**: $k \notin \mathcal{I}(k)$, for all classes $k = 1,...,K$. That is, there is no loop in class transfers.

Alternative, **(ACT)** can be expressed as $j \rightarrow k$ implies $k \not\rightarrow j$ for all $j \neq k$.

Clearly, under **(ACT)**, there must exist one or several classes to which no other class will transfer. Call these *root* classes, and assign them the smallest class indices (e.g., 1,2,...,i, say). Denote this set of classes as $\mathcal{L}_0$ (level 0). At the next level, identify all the classes that can *only* be transferred into directly from the root classes, and number them as $i+1, \ldots, k$, for instance. In general, the set of classes at level $\ell \geq 1$ can be specified as follows:

$$\mathcal{L}_\ell = \{i \notin \cup_{j=0}^{\ell-1}\mathcal{L}_j : \; I(i) \subseteq \cup_{j=0}^{\ell-1}\mathcal{L}_j\}. \tag{2.5}$$

That is, at level ℓ include a class i (which has not been included in the previous levels) if *all* its "predecessors" (under the relation $\rightarrow$) have already been included (and numbered) in the previous levels. Continue level by level in this fashion until all classes are assigned to a level and numbered.

An example will better illustrate the above mechanism. Suppose there are five classes. Suppose $I(1) = \emptyset$, $I(2) = I(3) = \{1\}$, $I(4) = \{2\}$, $I(5) = \{1,2,3\}$. Then, we have $\mathcal{L}_0 = \{1\}$, $\mathcal{L}_1 = \{2,3\}$, and $\mathcal{L}_2 = \{4,5\}$. Note that at level 1, class 5 cannot be included, since its predecessors, classes 2 and 3, are not in the root level, although another of its predecessors, class 1, is.

Two (related) facts follow immediately from this scheme:

- $i \rightarrow k$ implies $i < k$; and hence,
- P is upper triangular.

Obviously, the upper-triangulation of P following the above mechanism is not unique, since the numbering of classes at the same level can be arbitrary. Also note that the numbering scheme will not in general result in a "tree" or "forest" structure (as evident from the above 5-class example). In other words, the **(ACT)** condition is more general than such structures.

A *priority discipline* T is a numbering of the K classes such that for any t, and any i, k:

$$i < k, \; \sigma(i) = \sigma(k), \; Q_i(t) > 0 \quad \Rightarrow \quad \dot{T}_k(t) = 0. \tag{2.6}$$

That is, the lower class (k) will not be allocated any processing capacity if there is a positive level of a higher class (i) fluid present at the same station.

Theorem 2.3.1 Suppose that the (usual) traffic condition,

$$\rho := CD^{-1}(I - P')^{-1}\alpha < e, \tag{2.7}$$

is in force. (The strict inequality above holds for each component of the two vectors, ρ and e.) Then under condition **(ACT)**, any priority discipline that leads to the upper-triangulation of P is stable.

Remarks

1. Note that under **(ACT)** the priority discipline as specified above can be interpreted as fluid going through (and getting processed at) a set of buffers at different "levels", such that the fluid in a buffer at a higher level (i.e., closer to the "root") has a higher priority. With this view, the priority discipline that leads to the upper-triangulation of P can perhaps be called the *generalized first-buffer-first-served* discipline.

2. Note that a special case of Theorem 2.3.1 is when P is an incidence matrix, i.e., each of its entry is binary: either 0 or 1. This corresponds to a *deterministic* class transfer (or, routing) mechanism. Clearly, in this case, there exists a permutation (or, a rename) of the classes such that P becomes upper triangular. Indeed, this case includes the well known model of so-called re-entrant lines, for which the first-buffer-first-served discipline is known to be stable; refer to Kumar [19], Lu and Kumar [22], and Dai and Weiss [14]. (In fact, by allowing more than one entrance point, this special case is more general than re-entrant lines. This also motivates the above notion of generalized first-buffer-first-served discipline.)

   On the other hand, note that although the last-buffer-first-served discipline, again in the context of re-entrant lines, have also been shown to be stable in the above references, a "generalized last-buffer-first-served" discipline need not be stable, in view of the example created by Rybko and Stolyar [24].

3. One way to model a deterministic class transfer mechanism is the so-called "Kelly network" (Kelly [7]). This is a network in which there are I types of jobs indexed by $i = 1, ..., I$, and each job of type i requires n_i processing stages, indexed by $s = 1, ..., n_i$. Each type of jobs follows its own deterministic routing. Different job characteristics — routing and processing requirements, for instance — can be accounted for via introducing new job types. (Note that here our notion of a Kelly network puts no restriction whatsoever on the arrival and service processes: no distributional assumptions, and no homogeneity among the types is required. This is in contrast with some papers in the literature, where a Kelly network is taken to be one that belongs to the "product-form" class. In view of this, the network here should perhaps be called a "generalized Kelly network".)

   Hence, in view of the remarks in 2 above, for any Kelly network a stable priority discipline can be identified. To do so, simply assign class indices to the job types as follows: start with all types at stage 1, the root level; followed by all types at stage 2, the next level; and so forth. This results in the generalized first-buffer-first-served discipline remarked in 1.

4. On the other hand, our network model here, even under the **(ACT)** condition, may *not* be turned into an equivalent Kelly network. Consider a simple example with a single station and three classes ($J = 1$ and $K = 3$). Class 1 jobs correspond to external arrivals with a rate of α, which upon service completion becomes a class 2 or a class 3 job with equal probability. Both class 2 and class 3 jobs will leave the system upon service completion. Suppose the external arrival process is Poisson, and the service times are exponential with rates $\mu = (\mu_1, \mu_2, \mu_3)'$, and the service discipline is FIFO. Then this network is equivalent to the following Kelly network, also under a FIFO service discipline. It has two types of jobs and each type has two stages; the arrival processes for both types are independent Poisson with rate $\alpha/2$; the service times for both types at stage 1 are exponential with rate μ_1, the service times for the two types at stage 2 are exponential with rates μ_2 and μ_3, respectively. Furthermore, when the external arrivals follow a renewal process and the service times are non-exponential, the above equivalence may still hold with the added complication that arrival processes in the Kelly network will be non-renewal.

   However, if the FIFO discipline is replaced by a priority discipline, then the above transformation will not work. Note that under FIFO, in the original network, after serving each class 1 job, the server continues serving either a class 2 or a class 3 job, with equal probability; and this is equivalent to, in the Kelly network, the server serving either a type 1 job or a type 2 job, with equal probability, for both stages (consecutively). Now suppose the priority is to serve classes 1, 2 and 3, in this order, in the original network. Then, in the Kelly network, there is no equivalent priority assignment.

   This lack of equivalence between the two models, in a way, points to a limitation of our result. For if there were such an equivalence, then given any network with a matrix P, we could always turn it into an equivalent Kelly network (possibly with an infinite number of types), and with the remarks in 2 and 3 above, we could always identify a stable priority discipline. In other words, a stable priority discipline would exist, without requiring *any* condition such as **(ACT)**. At this point in time, however, whether such a result is true at all remains an open problem.

2.4 Proof of the Theorem

Before we prove Theorem 2.3.1, let us introduce another piece of notation:

$$\beta := D^{-1}(I - P')^{-1}\alpha. \tag{2.8}$$

Note that β is a K-dimensional vector, with β_k being the nominal traffic intensity of class k fluid. In contrast, ρ_j is the nominal traffic intensity of station j, i.e., the total traffic of all classes that are processed at station j. (Hence, if $j = \sigma(k)$, then β_k is one of the terms that sum to ρ_j.) Clearly, (2.7) implies $\beta < e$.

Now, consider the priority discipline as specified in the upper-triangulation of P preceding Theorem 2.3.1; i.e., class 1 has the highest priority and class K the lowest priority (rename classes if necessary). Let $\dot{Q}_i(t)$ and $\dot{T}_i(t)$ denote the derivatives, with respect to t, of $Q_i(t)$ and $T_i(t)$, respectively.

Since class 1 has the highest priority (at station $\sigma(1)$), we must have $\dot{T}_1(t) = 1$; and since $p_{k1} = 0$ for all k, we have

$$\dot{Q}_1(t) = \alpha_1 + \sum_k p_{k1}\mu_k \dot{T}_k(t) - \mu_1 \dot{T}_1(t) = \alpha_1 - \mu_1 < 0,$$

where the inequality is due to (2.8): $\beta_1 = \alpha_1/\mu_1 < 1$. Hence, Q_1 will reach zero in a finite time. After that, to maintain it at zero, simply set $\dot{T}_1(t) = \alpha_1/\mu_1$.

Now suppose that the fluid levels of all classes in $H := \{1, ..., i-1\}$ have been driven down to zero, after a finite time and have been maintained at zero. Let $L := \{i, ..., K\}$ denote the complement of H. Below, we use (H, L) to partition all matrices and vectors under discussion, and the partitioned objects will be indexed by the subscripts H and L accordingly, with the exception of I and e, which carry no such subscripts, since their dimensions are obvious from the context. (Note that all operations, such as transpose, inverse and derivative, apply *after* the partition. Hence, P'_{LH} denotes the transpose of the LH block of P, *not* the LH block of P'.)

To maintain Q_H at the zero level, we have (applying the partition to (2.1) and taking derivatives on both sides):

$$\begin{aligned} \dot{Q}_H(t) &= \alpha_H + P'_{LH} D_L \dot{T}_L(t) - (I - P'_H) D_H \dot{T}_H(t) \\ &= \alpha_H - (I - P'_H) D_H \dot{T}_H(t) = 0, \end{aligned}$$

since $P'_{LH} = 0$. Hence,

$$\dot{T}_H(t) = D_H^{-1}(I - P'_H)^{-1}\alpha_H. \tag{2.9}$$

Similarly,

$$\dot{Q}_L(t) = \alpha_L + P'_{HL} D_H \dot{T}_H(t) - (I - P'_L) D_L \dot{T}_L(t).$$

Substituting (2.9) into the above, we have

$$\dot{Q}_L(t) = [\alpha_L + P'_{HL}(I - P'_H)^{-1}\alpha_H] - [I - P'_L] D_L \dot{T}_L(t).$$

On the other hand, applying the partition to (2.8), we have

$$\begin{aligned}\beta_H &= D_H^{-1}(I-P'_H)^{-1}(\alpha_H + P'_{LH}D_L\beta_L)\\ &= D_H^{-1}(I-P'_H)^{-1}\alpha_H, \qquad (2.10)\\ \beta_L &= D_L^{-1}(I-P'_L)^{-1}(\alpha_L + P'_{HL}D_H\beta_H)\\ &= D_L^{-1}(I-P'_L)^{-1}[\alpha_L + P'_{HL}(I-P'_H)^{-1}\alpha_H]. \qquad (2.11)\end{aligned}$$

Hence, writing

$$\tilde{\alpha}_L := \alpha_L + P'_{HL}(I-P'_H)^{-1}\alpha_H = (I-P'_L)D_L\beta_L, \qquad (2.12)$$

we have

$$\dot{Q}_L(t) = \tilde{\alpha}_L - [I-P'_L]D_L\dot{T}_L(t).$$

Next, we show how to handle class i, the highest priority class in L. Let us first examine the feasibility of the allocation, in terms of $(\dot{T}_H, \dot{T}_L)$. Let

$$\sigma(H) := \{\sigma(k), k\in H\} \quad \text{and} \quad \sigma(L) := \{\sigma(k), k\in L\}$$

denote, respectively, the set of stations that process fluid classes H and L. (In general, the two sets will overlap.) To ensure that $U(t)$ is nondecreasing, we insist that $C\dot{T}(t)\le e$; that is,

$$C_{\sigma(H),H}\dot{T}_H + C_{\sigma(H),L}\dot{T}_L \le e, \quad C_{\sigma(L),H}\dot{T}_H + C_{\sigma(L),L}\dot{T}_L \le e.$$

(When $\sigma(H)$ and $\sigma(L)$ overlap, there will be some redundant inequalities above, an innocuous situation here and below.) Comparing (2.9) and (2.10), we can write the above as

$$C_{\sigma(H),L}\dot{T}_L \le \tilde{e}_{\sigma(H)}, \quad C_{\sigma(L),L}\dot{T}_L \le \tilde{e}_{\sigma(L)},$$

where

$$\tilde{e}_{\sigma(H)} = e - C_{\sigma(H),H}\beta_H, \quad \tilde{e}_{\sigma(L)} = e - C_{\sigma(L),H}\beta_H.$$

Note that $\tilde{e}_{\sigma(H)}$ and $\tilde{e}_{\sigma(L)}$ represent the remaining capacity (after maintaining the classes in H at zero) at the stations. And, $\rho = C\beta < e$ implies that all components in $\tilde{e}_{\sigma(H)}$ and $\tilde{e}_{\sigma(L)}$ are positive, making it possible to specify a feasible T_L, which we do next. (Note that since $\dot{T}_H(t) = \beta_H$, T_H is feasible.)

Now, we treat i, the highest priority class in L, in the same fashion as we did earlier with class 1. In particular, if $Q_i(t) > 0$, then the entirety of the *remaining* capacity at station $\sigma(i)$ — after processing the classes in H, if any, at that station — is allocated to i, i.e., $\dot{T}_i(t) = \tilde{e}_{\sigma(i)}$. This way, we have

$$\dot{Q}_i(t) = \tilde{\alpha}_i + \sum_{k\in L}\mu_k\dot{T}_k(t)p_{ki} - \mu_i\dot{T}_i(t) = \tilde{\alpha}_i - \mu_i\tilde{e}_{\sigma(i)},$$

since $p_{ki} = 0$, $k\in L$. To be able to drive Q_i to zero in a finite time, we need the right hand side above to be negative.

From (2.12), we have $\tilde{\alpha}_i = \mu_i \beta_i$; hence we need $\beta_i < \tilde{e}_{\sigma(i)}$. However, partitioning $\rho = C\beta < e$ in the same fashion as we did to $C\dot{T}(t) \le e$ above yields

$$\begin{aligned} C_{\sigma(H),L}\beta_L &< e - C_{\sigma(H),H}\beta_H = \tilde{e}_{\sigma(H)} \\ C_{\sigma(L),L}\beta_L &< e - C_{\sigma(L),H}\beta_H = \tilde{e}_{\sigma(L)}. \end{aligned}$$

Since $\sigma(i) \in \sigma(L)$, we do have the desired $\beta_i < \tilde{e}_{\sigma(i)}$.

Hence, we can remove i from L and add it into H, and repeat the above procedure until the fluid levels of all classes are driven down to zero, in finite time.

2.5 Stable Priority Disciplines for Multiclass Queueing Networks

Now we extend the main result in §2.3 to multiclass queueing networks that correspond to the multiclass fluid network described in §2.2. The queueing network consists of J single-server stations indexed by $j = 1, ..., J$, and K job classes indexed by $k = 1, ..., K$. Let (α, P, μ, C) be the same set of parameters in §2.2. Exogenous arrivals of class k jobs follow a renewal process with a mean interarrival time $1/\alpha_k$, $k = 1, ..., K$. When $\alpha_k = 0$, it is assumed that there are no exogenous arrivals for class k jobs. Class k jobs are served at station $\sigma(k)$, and their service times form an i.i.d. sequence with a mean $1/\mu_k$, where $\sigma(\cdot)$ is defined via matrix C in the same way as in §2.2. Assume that the arrival and service processes are mutually independent and independent among different classes. Upon its service completion, a class k job may be transferred into a class i job with probability p_{ki}, and leave the network with probability $1 - \sum_{i=1}^{K} p_{ki}$, independent of anything else.

Let $\mathcal{E} = \{k : \alpha_k > 0\}$, and let $u_k(n)$ be the interarrival time between the $(n-1)$st and the nth jobs of class k, $k \in \mathcal{E}$ and $n = 1, 2, ...$ (The 0th job arrives at time 0.) Assume that for $k \in \mathcal{E}$, $P\{u_k(1) \ge x\} > 0$, for $x \ge 0$, and that there exists an n and a function $p(x) \ge 0$ on $[0, \infty)$ with $\int_0^\infty p(x)dx > 0$ such that

$$P\left\{a \le \sum_{\ell=1}^{n} u_k(\ell) \le b\right\} \ge \int_a^b p(x)dx, \qquad \text{for any } 0 \le a < b, \quad k \in \mathcal{E}.$$

Following Remark 1 of §2.3, a generalized first-buffer-first-served discipline refers to any permutation (possible renumbering) of the class indices, $\{1, ..., K\}$, that leads to the upper-triangulation of the matrix P, such that class i has (preemptive) priority over class k whenever $i < k$ and $\sigma(i) = \sigma(k)$.

The queueing network described above is stable if its underlying Markov process is positive Harris recurrent. (Referred to Dai [11] for more details.) Therefore, our Theorem 2.3.1, along with Theorem 4.1 and Lemma 5.3 of Dai [11], yields the following result.

Theorem 2.5.1 Suppose that the (usual) traffic condition,

$$\rho := CD^{-1}(I - P')^{-1}\alpha < e,$$

is in force. Then under condition **(ACT)**, the multiclass queueing network is stable under any generalized first-buffer-first-served priority discipline.

Acknowledgments: This research was initiated when David Yao was visiting the University of British Columbia in August, 1995, and was partially supported by a Canadian NSERC grant. David Yao has also been supported in part by NSF Grant MSS-92-16490 and a matching grant from EPRI.

2.6 References

[1] Berman, A. and R.J. Plemmons. (1979). *Nonnegative Matrices in the Mathematical Science.* Academic Press.

[2] Bertsimas, D., D. Gamarnik, and Tsitsiklis, Stability conditions for multiclass fluid networks (preprint), 1994.

[3] Botvich, D.D. and A.A. Zamyatin, Ergodicity of conservative communication networks, Rapport de recherche 1772, INTRA, October 1992.

[4] Bramson, M., Instability of FIFO queueing networks, *Annals of Applied Probability*, **4**, 414–431, 1994.

[5] Bramson, M., Instability of FIFO queueing networks with quick service times, *Annals of Applied Probability*, **4**, 693–718, 1994.

[6] Bramson, M., Convergence to equilibria for fluid models of FIFO queueing networks (preprint), 1994.

[7] Banks, J., and J.G. Dai, Simulation studies of multiclass queueing networks (preprint), 1995.

[8] Chen, H., Fluid approximations and stability of multiclass queueing networks: Work-conserving disciplines, *Annals of Applied Probability*, **5**, 1995.

[9] CHEN, H., and H. ZHANG, Stability of multiclass queueing networks under FIFO service discipline (preprint), 1994.

[10] CHEN, H., and H. ZHANG, Linear Lyapunov functions for stability of queueing networks under priority service disciplines (in preparation), 1996.

[11] DAI, J.G., On positive Harris recurrence of multiclass queueing networks: a unified approach via fluid limit models, *Annals of Applied Probability*, **5**, 49–77, 1995.

[12] DAI, J.G., Instability of multiclass queueing networks (preprint), 1995.

[13] DAI, J.G., and J. VANDEVATE, The stability regions for two-station queueing networks (in preparation), 1996.

[14] DAI, J.G., and G. WEISS, Stability and instability of fluid models for certain reentrant lines, *Mathematics of Operations Research* (to appear), 1994.

[15] DOWN, D., and S. MEYN, Piecewise linear test functions for stability of queueing networks, *Proceedings of the 33rd Conference on Decision and Control*, 1994.

[16] DUMAS, V., A multiclass network with non-linear, non-convex, non-monotonic stability conditions (preprint), 1995.

[17] FOSS, S., and RYBKO, A., Stability of multiclass Jackson–type networks, preprint, 1995.

[18] KELLY, F.P., *Reversibility and Stochastic Networks,* Wiley, New York, 1979.

[19] KUMAR, P.R., Re-entrant Lines, *Queueing Systems,* **13**, 87-110, 1993.

[20] KUMAR, P.R., and S.P. MEYN, Stability of queueing networks and scheduling policies, *Tech. rep., C.S.L.,* University of Illinois, 1993.

[21] KUMAR, P.R., and T.I. SEIDMAN, Dynamic instabilities and stabilization methods in distributed real–time scheduling of manufacturing systems, *IEEE Transactions on Automatic Control*, **35**, 289–298, 1990.

[22] LU, S.H., and P.R. KUMAR, Distributed scheduling based on due dates and buffer priorities, *IEEE Transactions on Automatic Control*, **36**, 1406–1416, 1991.

[23] MEYN, S.P., Transience of multiclass queueing networks via fluid limit models (preprint), 1994.

[24] RYBKO, A.N., and A.L. STOLYAR, Ergodicity of stochastic processes describing the operations of open queueing networks, *Problemy Peredachi Informatsii*, 28, 2–26, 1992.

[25] SEIDMAN, T.I., 'First come first serve' is unstable (preprint), 1993.

[26] STOLYAR, A.L., On the stability of multiclass queueing networks (preprint).

[27] WINOGRAD, G.I., and P.R. KUMAR., The FCFS service discipline: stable network topologies, bounds on traffic burstiness and delay, and control by regulators, *Mathematical and Computer Modelling: Special Issue on Recent Advances in Discrete Event Systems* (to appear), 1995.

[2[illegible]] Rybko, [illegible] and [illegible] Stolyar. Ergodicity of stochastic [illegible] using the [illegible] of open queueing networks. [illegible] [illegible] Information [illegible]

[24] Serfozo, R. [illegible] (in preparation).

[2[illegible]] S[illegible], A. [illegible] On the stability of multiclass queueing [illegible] (preprint).

[2[illegible]] [illegible] and P. R. Kumar. [illegible] [illegible] and [illegible] [illegible]

3
Closed Queueing Networks in Heavy Traffic: Fluid Limits and Efficiency

Sunil Kumar and P.R. Kumar

ABSTRACT We address the behavior of stochastic Markovian closed queueing networks in heavy traffic, i.e., when the population trapped in the network increases to infinity. All service time distributions are assumed to be exponential. We show that the fluid limits of the network can be used to study the asymptotic throughput in the infinite population limit. As applications of this technique, we show the efficiency of all policies in the class of Fluctuation Smoothing Policies for Mean Cycle Time (FSMCT), including in particular the Last Buffer First Serve (LBFS) policy for all reentrant lines, and the Harrison–Wein balanced policy for two station reentrant lines. By "efficiency" we mean that they attain bottleneck throughput in the infinite population limit.

3.1 Introduction

Consider a closed queueing network with a population size N. Typically, the objective is to schedule such networks to maximize the throughput. In this paper we are interested in the heavy traffic behavior, i.e., as $N \to \infty$, of the throughput. We assume that all service times are exponentially distributed.

Let $\lambda^u(x_n)$ denote the throughput when the closed network is started with the initial condition x_n, and scheduling policy u is employed. The population size is $|x_n|$. Let λ^* denote the maximum throughput sustainable by the network. We say that the scheduling policy u is *efficient* if $\lim_{|x_n| \to \infty} \lambda^u(x_n) = \lambda^*$. Our goal in this paper is to address the efficiency of scheduling policies.

Our approach is via the "fluid limits" of the closed network. These fluid limits (see Dai [4] and Chen and Mandelbaum [2]) are obtained by suitably scaling the queue lengths in the network and applying a functional strong law of large numbers to this scaled network to obtain a limit process. This deterministic limit process, called the fluid limit, is not unique. However, every such limit must necessarily obey a set of integral equations.

It is known that by studying such fluid limits one can study the stability, i.e., positive Harris recurrence, of *open* queueing networks, see Dai [4]. It has been shown by Lu and Kumar [13] that the First Buffer First Serve (FBFS) policy, the Last Buffer First Serve (LBFS) policy and the Least Slack (LS) scheduling policy are all stable for open reentrant lines, under a bursty deterministic model. Using the fluid limit approach, Dai [4] has established that the FBFS policy is stable for stochastic reentrant lines, while Dai and Weiss [3] and Kumar and Kumar [11] have established the stability of the LBFS policy. Finally, the stability of all Fluctuation Smoothing Policies for Mean Cycle Time (FSMCT), a special subset of LS policies has been established in Kumar and Kumar [12]. Note that the LBFS policy is a special case of such an FSMCT policy. Such policies have been shown to perform well in the simulation studies of [14] for closed reentrant lines.

In this paper we employ the fluid limit approach for *closed* queueing networks. By studying the integral equations, we can determine a superset of (i.e., a larger set containing) the fluid limits. We relate the throughputs of the fluid limits to the asymptotic throughput of the reentrant line. We thereby obtain sufficient conditions for the reentrant line to be efficient, i.e., to achieve the maximal achievable throughput in the infinite population limit.

We utilize this approach to establish the efficiency of certain specific scheduling policies for reentrant lines. Even though there is no notion of "first" or "last buffer" in a closed reentrant line, let us designate some arbitrary buffer as the "last buffer." Consider any resulting FSMCT policy. We establish that all such FSMCT policies are efficient. Lest it be regarded that all policies which are stable for open networks are also efficient for closed networks, we note that the FBFS policy is inefficient for a closed network, as shown in Harrison and Nguyen [7], even though it is stable for open networks.

For *two-station* closed networks, by studying the reflected Brownian motion approximation of the network, Harrison and Wein [8] have devised a buffer priority policy which they conjecture is asymptotically optimal in heavy traffic. We establish the efficiency of this Harrison–Wein policy for two-station closed reentrant lines. Such a result for general two-station closed networks has been shown earlier by Jin, Ou and Kumar [2] using the very different method of functional bounding by linear programs.

In the next section we describe the model of the closed queueing network, and also describe the special case of a closed reentrant line. In Section 3 we discuss the existence of fluid limits, and specify the set of integral equations that these limits satisfy. In Section 4 we define the notion of efficiency, and obtain conditions on the fluid limits which guarantee it. In Section 5 we refine the limit for buffer priority policies. In Section 6 we establish the efficiency of all FSMCT policies, regardless of network topology, for closed reentrant lines. In Section 7 we establish the efficiency of the Harrison–Wein policy for two-station reentrant lines. Finally, we provide some examples of networks with inefficient fluid flows in Section 8.

3.2 The Stochastic Closed Network Model

The network we consider consists of S servers labeled $\{1, 2, \ldots, S\}$. There are L buffers labelled $b_1, b_2, \ldots, b_L$. Customers at buffer b_i require exponentially distributed service time with mean $m_i = \frac{1}{\mu_i}$ from server $\sigma(i) \in \{1, 2, \ldots, S\}$. On completing service at buffer b_i, customers move to buffer b_j with probability p_{ij}. The routing matrix $P = [p_{ij}]$ is stochastic and irreducible. Hence if the initial population size is $|x_n| = N$, then these N customers are trapped in the system. Let $\pi = \pi P$ be the unique invariant probability measure associated with P.

A special case of the above model is the *closed reentrant line*, which may be described as follows. Customers begin processing at buffer b_1, located at server $\sigma(1) \in \{1, \ldots, S\}$. Upon completing service, they proceed to buffer b_2 located at server $\sigma(2) \in \{1, \ldots, S\}$. Let b_L at server $\sigma(L)$ be the last buffer visited. The sequence $\{\sigma(1), \ldots, \sigma(L)\}$ is the *route* of the customer. At the same time that a customer completes service at buffer b_L, a new customer is released into buffer b_1. This is what is termed as a "closed loop" release policy in manufacturing [18, 16], and a "window" based admission policy in communication networks [17, 1]. The end result is a closed network with routing matrix given by $P = [p_{ij}]$, where $p_{ij} = 1$ if $j = (i+1) \bmod L$ and $= 0$ otherwise. Note that in this case, $\pi_i = 1/L$, for all i.

In a closed network, one cannot really talk of an exit or entry point, but, for convenience, we will continue to call b_L as the "last buffer." The goal is to maximize the "throughput" of the system. For reentrant lines it is immediate that the throughput can be taken to be the rate of departures from buffer b_L. In the case of the more general network, following [2] we define a "normalized" throughput as follows. Consider a scheduling policy u. Under u, let

$$\begin{aligned} w_i(t) &:= 1 \text{ if } \sigma(i) \text{ is working on customer in } b_i \text{ at } t, \\ &:= 0 \text{ otherwise.} \end{aligned}$$

Let $\beta_i := E[w_i(t)]$. Now

$$\sum_j \mu_i \beta_i p_{ij} = \sum_j \mu_j \beta_j p_{ji}.$$

But π is the unique invariant measure of P. Therefore, there exists an $\alpha > 0$ such that

$$\alpha = \frac{\beta_i \mu_i}{\pi_i}.$$

We call α the *normalized throughput*, and define the (unnormalized) *throughput* as $\lambda^u(x_n) := \pi_L \alpha = \mu_L \beta_L$.

For each server $\sigma \in \{1, 2, \ldots, S\}$, define

$$\rho_\sigma := \left(\sum_{\{1 \le k \le L, \ \sigma(k) = \sigma\}} \pi_k m_k \right), \tag{3.1}$$

and

$$\alpha^* := \min_{\sigma \in \{1,2,\ldots,S\}} \left(\sum_{\{1 \le k \le L, \ \sigma(k) = \sigma\}} \pi_k m_k \right)^{-1}. \tag{3.2}$$

We call ρ_σ the *nominal normalized load* on server σ, and α^* the *normalized bottleneck throughput*. We define the (unnormalized) bottleneck throughput of the network as

$$\lambda^* := \pi_L * \alpha^*. \tag{3.3}$$

We note that the long term rate of departures from b_L can never exceed λ^*. Any server σ achieving the "min" on the RHS of (3.2) is thus a bottleneck.

We assume that the scheduling policy employed is *non–idling*; that is, a server cannot be idle when any of its buffers is non–empty, and *stationary*; that is, the policy depends only on the current buffer lengths. In the following, we will mainly restrict attention to buffer priority policies which are nonidling and stationary. If a priority policy is used, we assume that it is preemptive resume. By this we mean that the service of a customer is interrupted to serve a higher priority customer whenever any such customer is present at the station, and is resumed to work on the remaining portion of the service time whenever there is no such higher priority customer.

3.3 The Fluid Limits

In this section, we define the fluid limits and describe the integral equations which they must satisfy. With the sole exception that the network is closed rather than open, all the results in this section are the same as in [4]. First we introduce the notation and some preliminaries. Let $Q_k(t)$ denote the queue length, and $v_k(t)$ the residual service time at buffer b_k at time t. Let $D_k(t)$ denote the number of departures from buffer b_k in $[0, t]$. Let $\Phi_l^k(D_k(t))$ denote the numbers of these departures in $[0, t]$ which were routed from b_k to b_l. Then $A_k(t)$, the total number of arrivals to buffer b_k in $[0, t]$, including the initial condition, is given by:

$$A_k(t) = Q_k(0) + \sum_{l=1}^{L} \Phi_k^l(D_l(t)), \ k = 1, \ldots, L.$$

Also, we have

$$Q_k(t) = A_k(t) - D_k(t).$$

Let $T_k(t)$ be the amount of time in $[0,t]$ that server $\sigma(k)$ spends working on b_k, and let $B_\sigma(t) := \sum_{\{j:1\leq j\leq L, \text{ and } \sigma(j)=\sigma\}} T_j(t)$ be the amount of time server σ was busy in $[0,t]$. Let $I_\sigma(t) := t - B_\sigma(t)$ denote the idle time of σ. Because of the non–idling assumption, we have

$$\int_0^\infty \left[\left(\sum_{\{j:\,1\leq j\leq L \text{ and } \sigma(j)=\sigma\}} Q_j(t)\right)\wedge 1\right] dI_\sigma(t) = 0. \tag{3.4}$$

The state of the system is given by

$$X(t) := (Q_1(t),\ldots,Q_L(t),v_1(t),\ldots,v_L(t)).$$

For $x = (q_1,...,q_L,v_1,...,v_L)$ we define $|x| := \sum_{k=1}^L (q_k + v_k)$. In the sequel, we shall denote explicit dependence on the initial condition x of $X(t)$ by adding a superscript x to the variable of interest. For any function f, let

$$\bar{f}^x(t) := \frac{1}{|x|} f^x(|x|t), \text{ for } t \geq 0 \text{ and } x > 0,$$

denote its scaled version. This is the so–called "fluid scaling."

For simplicity, we make the additional assumption that all the service time distributions are exponential. This allows us to ignore the effects of v_k. The state $x(t)$ is then redefined to consists only of the queue lengths $(Q_1(t),\ldots,Q_L(t))$, and $|Q(t)| := \sum_{i=1}^L Q_i(t) \equiv N$.

For each initial condition x_n, we consider the resulting vector process $(\bar{D}^{x_n},\bar{A}^{x_n},\bar{Q}^{x_n},\bar{T}^{x_n},\bar{I}^{x_n})$ as an element of $\mathcal{D}^d_{\mathcal{R}}[0,\infty)$, the space of $\mathcal{R}^d$-valued right continuous paths with left limits, endowed with the Skorohod topology. Thus we can define the countable collection of $\mathcal{D}^d_{\mathcal{R}}$-valued variables (one for each x_n) on a common probability space. This is the probability space with respect to which all the results in the sequel are stated.

First, we state a result which is about the same as that of Dai [4]. It is the closed network version of Theorem 4.4 given in [4] for open networks.

Theorem 3.3.1 *Almost surely*, the following holds. For every sequence of initial conditions with $|x_n| \to \infty$, there exists a further subsequence x_{n_l} such that along this subsequence, as $l \to \infty$,

$$(\bar{D}^{x_{n_l}},\bar{A}^{x_{n_l}},\bar{Q}^{x_{n_l}},\bar{T}^{x_{n_l}},\bar{I}^{x_{n_l}}) \Rightarrow (\bar{D},\bar{A},\bar{Q},\bar{T},\bar{I}), \tag{3.5}$$

where "$\Rightarrow$" denotes uniform convergence on compacts. Furthermore, the limit processes, called the *fluid limits*, satisfy:

$$\bar{D}_k(t) = \mu_k \bar{T}(t), \tag{3.6}$$

$$\bar{A}_k(t) = q_k + \sum_{l=1}^L p_{lk}\bar{D}_l(t), \tag{3.7}$$

$$\bar{Q}_k(t) = \bar{A}_k(t) - \bar{D}_k(t), \tag{3.8}$$

$$\bar{I}_\sigma(t) = t - \sum_{\{j:1\le j\le L \text{ and } \sigma(j)=\sigma\}} \bar{T}_j(t), \tag{3.9}$$

$$0 \le \bar{T}_k(t_1) \le \bar{T}_k(t_2) \text{ and } 0 \le \bar{I}_\sigma(t_1) \le \bar{I}_\sigma(t_2) \text{ if } t_1 \le t_2, \tag{3.10}$$

$$|\bar{I}_\sigma(t) - \bar{I}_\sigma(s)| \le |t-s|,\ |\bar{T}_k(t) - \bar{T}_k(s)| \le |t-s|, \text{ and } \bar{Q}_k(t) \text{ is also Lipschitz}, \tag{3.11}$$

$$\int_0^\infty \left(\sum_{\{j:1\le j\le L \text{ and } \sigma(j)=\sigma\}} \bar{Q}_j(t) \wedge 1 \right) d\bar{I}_\sigma(t) = 0, \tag{3.12}$$

$$\text{and } \sum_{k=1}^{L} q_k = 1. \tag{3.13}$$

Proof. The equality (3.6) follows from the renewal theorem while (3.7) follows from the classical functional strong law of large numbers, see also Lemma 4.2 of [4]. We use Lemma 4.1 of [4] to conclude uniform convergence on compacts. Since

$$|\bar{T}_k^{x_n}(t) - \bar{T}_k^{x_n}(s)| \le |t-s|, \text{ for all } x_n,$$

$\{\bar{T}_k^{x_n}\}$ is relatively compact in $\mathcal{D}_{\mathcal{R}}[0,\infty)$; see Ethier and Kurtz [6], Theorem 6.3, p.123. Therefore there is a convergent subsequence in the Skorohod topology. Also, the Lipschitz property guarantees the continuity of the limit. Since the limit is continuous, it follows that the convergence is uniform on compacts, see [6]. The other quantities, namely $\bar{A}_k^{x_n}(t)$, $\bar{Q}_k^{x_n}(t)$ and $\bar{I}_\sigma^{x_n}(t)$, depend on $\bar{D}_k^{x_n}(t)$ and $\bar{T}_k^{x_n}(t)$ in an affine fashion, and hence contain convergent subsequences by the continuous mapping theorem.

The equations (3.8–3.10) follow from the fact that they hold for every sample path of the scaled processes. The result (3.11) is also immediate. The result (3.13) follows from the scaling by $|x_n|$. Only (3.12) needs justification. For this, note that $\left(\sum_{\{j:1\le j\le L \text{ and } \sigma(j)=\sigma\}} \bar{Q}_j^{x_n}(t) \wedge 1\right)$ is continuous in $\bar{Q}_j^{x_n}(t)$, and $\bar{I}_\sigma^{x_n}(t)$ is continuous and nondecreasing. So by Lemma 2.4 of Dai and Williams [5],

$$\int_0^t \left(\sum_{\{j:1\le j\le L \text{ and } \sigma(j)=\sigma\}} \bar{Q}_j^{x_{n_l}}(s) \wedge 1 \right) d\bar{I}_\sigma^{x_{n_l}}(s)$$

$$\to \int_0^t \left(\sum_{\{j:1\le j\le L \text{ and } \sigma(j)=\sigma\}} \bar{Q}_j(s) \wedge 1 \right) d\bar{I}_\sigma(s),$$

for all t. This and (3.4) yield (3.12). □.

It should be noted that the derivatives of the processes $\bar{A}_i$, $\bar{Q}_i$, $\bar{D}_i$ and $\bar{I}_\sigma$ exist at a.e. t. Such times of differentiablity are called "regular times," as in Dai and Weiss [3].

3.4 Fluid Limit Sufficient Condition for Efficiency

In this section we define the notion of "efficiency" and provide sufficient conditions for it in terms of the throughput of the fluid limit.

Given a stationary scheduling policy u and an initial state x, we note that the throughput $\lambda^u(x)$ can be expressed as

$$\lambda^u(x) := \lim_{T\to\infty} \frac{E\left[D_L^x(T)\right]}{T}.$$

Definition The stationary scheduling policy u is said to be *efficient* if for every sequence of initial conditions $\{x_n\}$ with $|x_n| \uparrow \infty$, we have

$$\lim_n \lambda^u(x_n) = \lambda^* \tag{3.14}$$

where λ^* is defined in (3.3). (More precisely, there is a sequence of stationary, non-idling scheduling policies, one for each $|x_n| = N$.)

The next lemma provides a sufficient condition for efficiency based on calculating the limits in the opposite order to (3.14).

Lemma 3.4.1 If for every sequence x_n with $|x_n| \uparrow \infty$,

$$\limsup_{T\to\infty} \limsup_n \frac{D_L^{x_n}(|x_n|T)}{|x_n|T} \geq \lambda^* \text{ a.s.,} \tag{3.15}$$

then (3.14) also holds for every such sequence.

Proof. Consider a sequence x_n with $|x_n| \uparrow \infty$. For every x_n, let x'_n be a state with the same population size, and lying in a closed communicating class with the least throughput, i.e.,

$$|x_n| = |x'_n| \text{ and } \lambda^u(x'_n) = \min_{\{y_n:\ |y_n|=|x_n|\}} \lambda^u(y_n). \tag{3.16}$$

Since x'_n is in a closed communicating class,

$$\lim_{T\to\infty} \frac{D_L^{x'_n}(T)}{T} = \lambda^u(x'_n) \text{ a.s.} \tag{3.17}$$

Fix ω a sample point in our probability space, outside of a set of zero measure excluded in (3.15) above. Let $D_L^{x'_n}$ be the sequence of $\mathcal{D}_{\mathcal{R}}$-valued variables corresponding to ω. Then, from (3.15) above applied to the sequence x'_n, given any x'_n, T and $\epsilon > 0$, we can find x'_{n_l} with $|x'_{n_l}| \geq |x'_n|$ and $T' \geq T$ such that,

$$\frac{D_L^{x'_{n_l}}(|x'_{n_l}|T')}{|x'_{n_l}|T'} \geq \lambda^* - \epsilon.$$

So we have

$$\limsup_{x'_n} \limsup_{T\to\infty} \frac{D_L^{x'_n}(T)}{T} \geq \lambda^* \text{ a.s.}$$

From (3.17), we thus obtain

$$\limsup_{x'_n} \lambda^u(x'_n) \geq \lambda^* \text{ a.s.}$$

Then from (3.16) we have, for *every* sequence x_n with $|x_n| \uparrow \infty$,

$$\limsup_{x_n} \lambda^u(x_n) \geq \lambda^*. \tag{3.18}$$

Hence

$$\liminf_{x_n} \lambda^u(x_n) \geq \lambda^*, \tag{3.19}$$

and the result (3.14) follows since $\lambda^u(x) \leq \lambda^*$ for every x. □

We are now ready to prove the main sufficient condition for efficiency, which is that all the fluid limits have maximal throughput.

Theorem 3.4.2 Suppose that for every fluid limit $\bar{D}_L(t)$

$$\limsup_{t\to\infty} \frac{\bar{D}_L(t)}{t} \geq \lambda^* \text{ a.s.} \tag{3.20}$$

Then (3.15) holds, and hence the stationary non-idling scheduling policy u is efficient.

Proof. Fix ω a sample point in our probability space, outside of the set of zero measure excluded in Lemma 3.4.1. Let $D_L^{x_n}$, $|x_n| \uparrow \infty$, be the sequence of $\mathcal{D}_{\mathcal{R}}$-valued variables corresponding to ω. For brevity we will omit the explicit dependence on ω. Let $\bar{D}_L^{x_{n_k}}$ be any convergent subsequence, converging to $\bar{D}_L$. Fix $\epsilon > 0$ arbitrary. Then from (3.20) for every T, $\exists\, T < t_1 < \infty$ such that,

$$\frac{\bar{D}_L(t_1)}{t_1} \geq \lambda^* - \epsilon.$$

Moreover, from Theorem 3.3.1 for every t_1, $\exists K(t_1, \epsilon)$ such that for all $k > K$,

$$\left| \frac{D_L^{x_{n_k}}(|x_{n_k}|t_1)}{|x_{n_k}|} - \bar{D}_L(t_1) \right| < \epsilon.$$

Hence,

$$\frac{D_L^{x_{n_k}}(|x_{n_k}|t_1)}{|x_{n_k}|t_1} \geq \lambda^* - \epsilon(1 + \frac{1}{t_1}) \text{ for all } k > K(t_1, \epsilon). \tag{3.21}$$

So,

$$\liminf_{x_{n_k}} \frac{D_L^{x_{n_k}}(|x_{n_k}|t_1)}{|x_{n_k}|t_1} \geq \lambda^* - \epsilon(1 + \frac{1}{t_1}).$$

Since for every T, $\exists\, t_1 > T$ for which this holds, we have

$$\limsup_{t\to\infty} \liminf_{x_{n_k}} \frac{D_L^{x_{n_k}}(|x_{n_k}|t)}{|x_{n_k}|t} \geq \lambda^* - \epsilon.$$

Since ϵ was arbitrary, we have

$$\limsup_{t\to\infty} \liminf_{x_{n_k}} \frac{D_L^{x_{n_k}}(|x_{n_k}|t)}{|x_{n_k}|t} \geq \lambda^*.$$

The above holds for almost all ω, and hence the result follows. □

3.5 Buffer Priority Policies

Consider a buffer priority policy u. Note that it is stationary and non-idling, and also, as shown in Jin, Ou and Kumar [2], there is a single communicating class.

We use the notation "$b_k \succ b_l$" if $\sigma(k) = \sigma(l)$, $k \neq l$, and b_k has higher priority than b_l. Let $H_k := \{1 \leq j \leq L \mid b_j \succ b_k\} \cup \{k\}$ denote the set of buffers having at least as high a priority as b_k, and let

$$U_k(t) := t - \sum_{j \in H_k} T_j(t),$$

denote the time *not* spent working on them respectively. The following lemma identifies additional constraints satisfied by the fluid limits of the buffer priority policy u.

Lemma 3.5.1 For every sequence of initial conditions x_n with $|x_n| \uparrow \infty$, there exists a subsequence x_{n_l} such that $\bar{U}_k^{x_{n_l}}(t) \to \bar{U}_k(t)$ in the sense of Theorem 3.3.1, i.e., almost sure convergence uniformly on compacts. Furthermore, the fluid limits satisfy the additional constraints:

$$\bar{U}_k(t) = t - \sum_{j \in H_k} \bar{T}_j(t), \tag{3.22}$$

$$0 \leq \bar{T}_k(t) \leq \bar{U}_j(t) \leq t \text{ for all } j \in H_k,\ j \neq k \tag{3.23}$$

$$0 \leq \bar{U}_k(t_1) \leq \bar{U}_k(t_2) \text{ if } t_1 \leq t_2, \tag{3.24}$$

$$\int_0^\infty \left(\sum_{j \in H_k} \bar{Q}_j(t) \wedge 1 \right) d\bar{U}_k(t) = 0, \tag{3.25}$$

Proof. Identical to that of Theorem 3.3.1. □

Now we can provide a sufficient condition based on Theorem 3.4.2 which depends only on the solutions to the constraints (3.6–3.13) and (3.22–3.25).

Theorem 3.5.2 (i) Suppose every solution $\bar{D}_L(t)$ of (3.6–3.13, 3.22–3.25) satisfies

$$\limsup_{t\to\infty} \frac{\bar{D}_L(t)}{t} \geq \lambda^*. \tag{3.26}$$

Then the buffer priority scheduling policy u is efficient.

(ii) If for every solution $\bar{D}_L(t)$ of (3.6–3.13, 3.22–3.25), there exists a time $T < \infty$ such that

$$\frac{d}{dt}\bar{D}_L(t) \geq \lambda^* \text{ for almost every } t > T, \tag{3.27}$$

then the buffer priority scheduling policy u is efficient.

Proof. The result (i) is immediate from Theorem 3.4.2 using the fact that the fluid limits almost surely satisfy (3.6–3.13, 3.22–3.25). The result (ii) follows by noting that (3.27) implies (3.26). □

We now provide a sufficient condition for (3.26), in terms of the idleness process at the bottleneck machine.

Lemma 3.5.3 If

$$\bar{I}_\sigma(t) = 0 \text{ for all } t, \text{ for some } \sigma \in \{1, 2, ...S\}, \tag{3.28}$$

for a solution of (3.6–3.13, 3.22–3.25), i.e., if one of the servers never idles in the fluid limit, then the solution satisfies (3.26).

Proof. Suppose $\bar{I}_1(t) = 0$ for all t. Then

$$\sum_{\{k:\sigma(k)=1\}} m_k \bar{D}_k(T) = T \text{ for all } T \geq 0. \tag{3.29}$$

Now assume (3.26) does not hold, that is,

$$\liminf_{t\to\infty} \frac{\bar{D}_L(t)}{t} < \lambda^*.$$

Suppose that $\{t_l\}$ is a subsequence of times at which

$$\lim_{l\to\infty} \frac{\bar{D}_L(t_l)}{t_l} = \liminf_{t\to\infty} \frac{\bar{D}_L(t)}{t}.$$

We can assume (by taking further subsequences if necessary) that all the limits $\lim_{l\to\infty} \frac{\bar{D}_k(t_l)}{t_l}$ for $k = 1, 2, ..., L$ exist along the subsequence $\{t_l\}$.

Note that $\sum_{j=1}^{L} \bar{Q}_j(t) \leq 1$ for all $t \geq 0$. From (3.7) we have

$$\left| \sum_{k=1}^{L} \bar{D}_k(t_l) p_{kj} - \sum_{k=1}^{L} \bar{D}_j(t_l) p_{jk} \right| \leq 1.$$

So we have

$$\sum_{k=1}^{L} p_{kj} \lim_{l \to \infty} \frac{\bar{D}_k(t_l)}{t_l} = \lim_{l \to \infty} \frac{\bar{D}_j(t_l)}{t_l} \sum_{k=1}^{L} p_{jk}.$$

But π is the unique invariant probability measure associated with P. So we must have

$$\lim_{l \to \infty} \frac{\bar{D}_k(t_l)}{t_l} = \pi_k \alpha \text{ for some constant } \alpha > 0, \text{ for all } k. \tag{3.30}$$

Since $\pi_L \alpha < \lambda^*$, we have $\alpha < \alpha^*$. From (3.30), for a fixed $\delta > 0$ sufficiently small, for any T, we can find a $t_l \geq T$ such that

$$\bar{D}_k(t_l) < (\pi_k \alpha^* - \delta) t_l, \text{ for all } k = 1, 2, ..., L.$$

Thus, from (3.29),

$$t_l = \sum_{\{k:\sigma(k)=1\}} m_k \bar{D}_k(t_l) \leq \left(\sum_{\{k:\sigma(k)=1\}} m_k \pi_k \right) [(\alpha^* - L\delta) t_l].$$

But $\sum_{\{k:\sigma(k)=1\}} m_k \pi_k \alpha^* \leq 1$, leading to a contradiction. Thus (3.26) must hold. □

3.6 Efficiency of All FSMCT Policies

In this section, we utilize Theorem 3.5.2(ii) to establish that all policies in the class of Fluctuation Smoothing Policies for Mean Cycle Time (FSMCT) are efficient for all closed reentrant lines, regardless of network topology. Since the Last Buffer First Serve (LBFS) policy is a member of this class of FSMCT policies, its efficiency is also thus established.

First, as described in Section 2, we designate some buffer b_L as the last buffer. The FSMCT policies can be described as follows, see [15]. At any given instant t, machine σ should work on the first part in that non-empty buffer b_k for which

$$k = \arg\min_j \sum_{i=j+1}^{L} (Q_i(t) - \xi_i). \tag{3.31}$$

Here, $\xi_i \in \mathcal{R}$, for $i = 1, 2, ..., L$, is intended to be an estimate of the mean number of parts in b_i, usually chosen by some empirical method. This policy attempts to regulate the downstream shortfall and by doing so, attempts to reduce fluctuations. It has been shown to perform well in simulation studies of reentrant lines[14].

However, we shall let $\{\xi_i\}$ be arbitrary real numbers, subject only to the assumption that the partials sums $\sum_{i=j+1}^{L} \xi_i$ are unique with unique fractional parts for each j. This can always be done by changing the fractional parts of the ξ_i's by arbitrarily small amounts. This also means that the minimizer in (3.31) above is unique. One should note that if a different buffer had been chosen as the "last" buffer, then one obtains a different policy. The following result applies to *all* the resulting FSMCT policies from any choice of the last buffer b_L, and any choice of ξ_i's.

The LBFS scheduling policy functions as follows. If buffers b_j and b_k share the same server, i.e., $\sigma(j) = \sigma(k)$, and $j < k$, then priority is given to b_k. Therefore if $k < L$, b_k can be worked on by $\sigma(k)$ only if $b_j = 0$ for all $j = k+1, \ldots, L$, with $\sigma(j) = \sigma(k)$. Note that b_L is never preempted. Note that this corresponds to the case when all the ξ_i above have been chosen negative in the FSMCT policy. Hence the LBFS policy is a special case of FSMCT policies.

Theorem 3.6.1 Every FSMCT policy is efficient.

Proof. We show in Lemma 3.6.2 below that the integral equations describing the fluid limits of the queueing network under any FSMCT policy are identical to those describing the LBFS policy, along the lines of Kumar and Kumar [12]. Then we establish in Lemma 3.6.3 that (3.27) holds for LBFS policies. We thus conclude the validity of the result using Theorem 3.5.2(ii). □

Lemma 3.6.2 The fluid limits of any FSMCT policy obey the integral equations (3.6-3.13) and (3.22-3.25) with the H_k denoting the set of buffers which are downstream of b_k, i.e.,

$$H_k := \{b_j \mid k \leq j \leq L\}.$$

This means that the integral equations describing the fluid limits of the queueing network under any FSCMCT policy are identical to those describing the LBFS policy.

Proof. The proof is based on Theorem 2 of Kumar and Kumar [12]. Note that we need only prove (3.25). This is the nontrivial assertion which proves that the fluid limits of all FSMCT policies are identical those of LBFS. Define $Q(t) := (Q_1(t), \ldots, Q_L(t))$, $\xi := (\xi_1, \ldots, \xi_L)$, and

$$f_k(Q(t), \xi) := \min(1, g_k(Q(t), \xi))$$

where

$$g_k(Q(t),\xi) := \sum_{\{j:\sigma(j)=\sigma(k)\}} Q_j(t) \text{ if } k = \min\{j : \sigma(j) = \sigma(k)\}$$

$$:= \sum_{\substack{\{j\geq k \ \& \\ \sigma(j)=\sigma(k)\}}} Q_j(t) \prod_{\substack{\{n<j \ \& \\ \sigma(n)=\sigma(k)\}}} \left(\sum_{i=n+1}^{j} Q_i(t) - \xi_i \right)^+ \quad \text{otherwise.}$$

Now it follows from (3.31) that $f_k(Q(t),\xi) > 0$ implies that the highest priority non-empty buffer at $\sigma(k)$ at time t, say b_j, is such that $j \geq k$. Thus the policy and the non–idling assumption require that machine $\sigma(k)$ work on some buffer in the set $\{b_j : j \geq k \text{ and } \sigma(j) = \sigma(k)\}$. So if we define $I_k(t) := t - \sum_{j\geq k} T_j(t)$, the amount of time in $[0,t]$ not spent on buffers $\{b_j,\ j \geq k\}$, we must have

$$\int_0^\infty f_k(Q(s),\xi)\, dI_k(s) = 0 \text{ for } k = 1, ..., L. \tag{3.32}$$

Note that from (3.32), we have for trajectories starting from initial conditions x_{n_l},

$$\int_0^\infty f_k(\bar{Q}^{x_{n_l}}(s), \bar{\xi}^{x_{n_l}})\, d\bar{I}_k^{x_{n_l}}(s) = 0, \text{ for all } x_{n_l}, \tag{3.33}$$

where, by $\bar{\xi}^x$, we mean $\frac{\xi}{|x|}$. Now, f_k is a continuous bounded function on $\mathcal{R}^{2L}$. Also, $(\bar{Q}^{x_{n_l}}, \bar{\xi}^{x_{n_l}}) \Rightarrow (\bar{Q}, 0)$ in $D_{\mathcal{R}^{2L}}[0,\infty)$, $\bar{I}_k^{x_{n_l}} \Rightarrow \bar{I}_k$ in $C_{\mathcal{R}}[0,\infty)$, and $\bar{I}_k^{x_{n_l}}$ is non–decreasing for each l. We can use Lemma 2.4 of Dai and Williams [5] to conclude that uniformly for all t in any compact subset of $\mathcal{R}$,

$$\int_0^t f_k(\bar{Q}^{x_{n_l}}(s), \bar{\xi}^{x_{n_l}})\, d\bar{I}_k^{x_{n_l}}(s) \rightarrow \int_0^t f_k(\bar{Q}(s), 0)\, d\bar{I}_k(s).$$

From the above and (3.33), we obtain

$$\int_0^\infty f_k(\bar{Q}(s), 0)\, d\bar{I}_k(s) = 0. \tag{3.34}$$

Note that (3.34) does not involve ξ's. So it must be the same as the integral equation obtained under the LBFS policy. Alternately, we can see that

$$f_k(\bar{Q}(t), 0) > 0 \Leftrightarrow \sum_{\{k\leq j\leq L,\, \sigma(j)=\sigma(k)\}} \bar{Q}_j(t) > 0.$$

So (3.34) becomes

$$\int_0^\infty \sum_{\{k\leq j\leq L,\, \sigma(j)=\sigma(k)\}} \bar{Q}_j(s)\, d\bar{I}_k(s) = 0, \tag{3.35}$$

thus yielding (3.25) and completing the proof. □

Lemma 3.6.3 *(3.27) holds for any LBFS policy.*

Proof. Let σ^* be a bottleneck server, i.e.,

$$\sum_{\{j:\, 1\leq j\leq L \text{ and } \sigma(j)=\sigma^*\}} m_j = \frac{1}{\lambda^*}.$$

First, let us assume that σ^* is unique. We will relax this assumption later. Let b_{k^*} be the lowest indexed buffer at σ^*, i.e.,

$$b_{k^*} = \min\{j : 1 \leq j \leq L \text{ and } \sigma(j) = \sigma^*\}.$$

Then we claim that $\exists T < \infty$ such that for all $t > T$, $\bar{Q}_k(t) = 0$ for all $k \neq k^*$, and $\bar{Q}_{k^*}(t) = 1$. By Dai and Weiss [3], Prop. 4.2[1], this in turn implies that for all $k = 1, 2, ..., L$, and almost all $t > T$,

$$\frac{d}{dt}\bar{D}_k(t) = \lambda^*,$$

thus establishing (3.27).

We now prove the claim that there exists $T < \infty$ such that $\bar{Q}_k(t) = 0$ for all $k \neq k^*$ and $\bar{Q}_{k^*}(t) = 1$, for all $t > T$. The arguments in the proof will be very similar to those in Kumar and Kumar [11] and Dai and Weiss [3].

Suppose $k^* = L$. This implies in particular that b_L is the only buffer served by $\sigma(L)$. Consider $W(t) := \sum_{k=1}^{L-1} \bar{Q}_k(t)$ at a regular time t. Note that b_{L-1} has highest priority at $\sigma(L-1)$ since $\sigma(L-1) \neq \sigma(L)$ by assumption. Then, since $\sigma(L)$ is the bottleneck and the only buffer it serves is b_L, for almost all t such that $\bar{Q}_{L-1}(t) > 0$, we have

$$\frac{d}{dt}W(t) \leq \mu_L - \mu_{L-1} < 0.$$

So by Lemma 2.2 of [3], $\exists t_{L-1} \leq \frac{1}{\mu_{L-1}-\mu_L}$ such that $Q_{L-1}(t_{L-1}) = 0$. Arguing as in Lemmas 2 and 3 of [11], define

$$k := \max\{i < L-1 \mid \sum_{\{j:\, i\leq j\leq L \text{ and } \sigma(j)=\sigma(i)\}} m_j \geq m_{L-1}\}$$

if the set on the right hand side is non–empty, and $k := 0$ otherwise. As shown in [11], it follows that

$$\bar{Q}_{k+1}(t_{L-1} + \delta) = \cdots = \bar{Q}_{L-1}(t_{L-1} + \delta) = 0, \text{ for all } \delta \geq 0.$$

This argument can be iterated backwards from b_k. First we observe that there exists a time t_k such that $\bar{Q}_k(t_k) = 0$, noting that the input rate to

[1] Though [3] deals with fluid limits of open networks, their proposition can be adapted to closed networks with the modification that $d_0(t) = d_L(t)$.

the section $\{b_1, ..., b_k\}$ is no larger than λ^* while the output rate from b_k is $\frac{1}{\sum_{\{i:k\leq i\leq L \text{ and } \sigma(i)=\sigma(k)\}} m_i} > \lambda^*$. Then at time t_k we identify a k' such that

$$k' = \max\{j < k | \sum_{\{i:j\leq i\leq L \text{ and } \sigma(i)=\sigma(j)\}} m_i \geq \sum_{\{n:k\leq n\leq L \text{ and } \sigma(n)=\sigma(k)\}} m_n\}.$$

We argue that $b_{k'+1}, ..., b_k$ are empty at t_k and remain empty thereafter. The arguments are identical to those in [11] and are omitted. Iterating backwards from $b_{k'}$ completes the proof, when we reach b_1, leading to

$$\bar{Q}_1(t_1 + \delta) = \cdots = \bar{Q}_{L-1}(t_1 + \delta) = 0 \text{ for all } \delta \geq 0. \tag{3.36}$$

Suppose $k^* < L$. Then it is enough to show that there is a $t_{k^*+1} < \infty$ such that

$$\bar{Q}_{k^*+1}(t) = \cdots = \bar{Q}_L(t) = 0 \text{ for all } t \geq t_{k^*+1}. \tag{3.37}$$

All we would then have left to do for $t \geq t_{k^*+1}$ is to apply the case when $k^* = L$ to the section $b_1, b_2, ..., b_{k^*}$ with μ_j for $j = 1, 2, ..., k^*$ being replaced by

$$\tilde{\mu}_j := \mu_j \left(1 - \sum_{\{i:\, i>j \text{ and } \sigma(i)=\sigma(j)\}} m_i \lambda^*\right).$$

See (16) of [11] for the idea behind this argument. Now we prove (3.37). The key idea is the following. Suppose at a regular time τ, for some $j > k^*$, $\bar{Q}_j(\tau) > 0$ and $\bar{Q}_{j+1}(\tau) = \bar{Q}_{j+2}(\tau) = \cdots = \bar{Q}_L(\tau) = 0$. Then under LBFS,

$$\frac{d}{dt}\bar{D}_L(\tau) \geq \frac{1}{\sum_{\{i:j\leq i\leq L \text{ and } \sigma(i)=\sigma(j)\}} m_i} > \lambda^*.$$

Since this exit rate cannot be sustained for almost all $\tau \geq t$, it follows that at some t_{k^*+1},

$$\bar{Q}_{k^*+1}(t_{k^*+1}) = \cdots = \bar{Q}_L(t_{k^*+1}) = 0.$$

Moreover as shown in Lemma 3 of [11], this imples (3.37).

Let us now relax the assumption that the bottleneck server is unique. Suppose that there are exactly two bottleneck servers. Suppose k_1^* and k_2^* are the lowest priority buffers at bottleneck servers σ_1^* and σ_2^* respectively, and $k_1^* > k_2^*$. Then we proceed as follows, imitating the arguments above. For brevity, we will not repeat the arguments and we will confine ourselves to indicating the sections which empty. The exact same argument as leading to (3.37) shows that there exists a $t_{k_1^*+1} < \infty$ such that

$$\bar{Q}_{k_1^*+1}(t_{k_1^*+1} + \delta) = \cdots = \bar{Q}_L(t_{k_1^*+1}\delta) = 0 \text{ for all } \delta \geq 0.$$

Now consider the section $b_{k_2^*+1}, \ldots, b_{k_1^*-1}$. Using the same argument as that leading to (3.36), we see that there exists a $T_1 < \infty$ such that

$$\bar{Q}_{k_2^*+1}(T_1 + \delta) = \cdots = \bar{Q}_{k_1^*-1}(T_1 + \delta) = 0 \text{ for all } \delta \geq 0.$$

Finally, we consider the section $b_1, \ldots, b_{k_2^*-1}$. The input rate to this section cannot exceed λ^* after $t_{k_1^*+1}$. Also note that there is no buffer of higher priority than buffers $b_1, \ldots, b_{k_2^*-1}$ which is nonempty after T_1. Once again, using the arguments leading to (3.36), we see that there exists a $T_2 < \infty$ such that

$$\bar{Q}_1(T_2 + \delta) = \cdots = \bar{Q}_{k_2^*-1}(T_2 + \delta) = 0 \text{ for all } \delta \geq 0.$$

Thus we have shown that there exists a $T' < \infty$ such that

$$\sum_{\{k : k \neq k_1^* \text{ and } k \neq k_2^*\}} \bar{Q}_k(T' + \delta) = 0 \text{ for all } \delta \geq 0,$$

and so for all $\delta \geq 0$,

$$\bar{Q}_{k_1^*}(T' + \delta) + \bar{Q}_{k_2^*}(T' + \delta) = 1.$$

By applying Dai and Weiss [3], Prop. 4.2, we obtain the result. The extension to any number of bottleneck servers is immediate. □

3.7 Efficiency of the Harrison–Wein Policy

For closed queueing networks with *two* servers, Harrison and Wein [8] have examined the reflected Brownian motion approximation associated with the heavy traffic scenario, and conjectured that a particular buffer priority policy provides maximal throughput in the infinite population limit. In this section we prove this conjecture for two-station closed reentrant lines.

To describe their policy, it is convenient to imagine that when a customer leaves b_L, it exits from the system only to be replaced by a new customer in b_1. (As mentioned in Section 2, closed queueing networks arise from such window based admission control strategies). Let P_L be the same matrix as P except that all the elements of the L-th row are set to zero. Then $V := (I - P_L)^{-1} = I + P_L + P_L^2 + \cdots$ exists, and its ij-th element v_{ij} = expected number of visits to buffer b_j, before exit from a customer starting in b_i. Then

$$M_{\sigma,i} := \sum_{\{j : i \leq j \leq L \text{ and } \sigma(j) = \sigma\}} m_j v_{ij}$$

is the mean amount of work on a customer in b_i still remaining to be done by server σ prior to the customer's exit from b_L. Also let ρ_σ be the relative utilization of server σ as in (3.1). At the first server, $\sigma = 1$, rank the buffers to give higher priority to buffers with *smaller* values of the index

$$\eta_j := \rho_2 M_{1,j} - \rho_1 M_{2,j}.$$

At the second server ($\sigma = 2$) rank the buffers to give higher priority to buffers with *larger* values of the index η_j. The resulting buffer priority policy is enforced in a preemptive resume fashion and will be called the *Harrison–Wein policy* hereafter. The following result has been established in Jin, Ou and Kumar [2] using a different approach based on functional bounding by linear programs.

Theorem 3.7.1 The Harrison–Wein policy for two-station closed reentrant lines is efficient.

Proof. This proof uses Theorem 3.5.2(i). In the sequel we will use the notation

$$d_k(t) := \frac{d}{dt}\bar{D}_k(t),$$

to denote the derivative at the regular time t when the derivative exists.

Note that for reentrant lines

$$M_{\sigma,j} = \sum_{\{j \le k \le L \text{ and } \sigma(k)=\sigma\}} m_k,$$

and we can take ρ_σ to be $M_{\sigma,1}$. The next result identifies some properties of the priority ordering used in the Harrison–Wein policy.

Lemma 3.7.2 Under the Harrison–Wein policy, the following are true.

(i) If $\sigma(k) = \sigma((k+1) \bmod L)$ then $b_{(k+1)\bmod L} \succ b_k$.

(ii) If $j \le i-1$, $\sigma(j) = 1$ and $b_j \succ b_i$, then

$$\sum_{\{k: j \le k \le i-1; \sigma(k)=2\}} m_k \ge \frac{\rho_2}{\rho_1} \sum_{\{k: j \le k \le i-1; \sigma(k)=1\}} m_k.$$

(iii) If $j \ge i+1$, $\sigma(j) = 1$ and $b_j \succ b_i$, then

$$\sum_{\{k: j \le k \le L; \sigma(k)=2\}} m_k + \sum_{\{k: 1 \le k \le i-1; \sigma(k)=2\}} m_k$$

$$\ge \frac{\rho_2}{\rho_1}\left[\sum_{\{k: j \le k \le L; \sigma(k)=1\}} m_k + \sum_{\{k: 1 \le k \le i-1; \sigma(k)=1\}} m_k\right].$$

Above, empty sums are taken to be zero.

Proof. (i) If $\sigma(k) = \sigma(k+1) = 1$ for some $k \in \{1, 2, ..., L-1\}$, then $\eta_k = \eta_{k+1} + \rho_2 m_k > \eta_{k+1}$ and so b_{k+1} always has higher priority than b_k. Similarly, if $\sigma(k) = \sigma(k+1) = 2$, for some $k \in \{1, 2, ..., L-1\}$, then $\eta_k = \eta_{k+1} - \rho_1 m_k < \eta_{k+1}$ and so b_{k+1} again has higher priority than b_k. If $\sigma(1) = \sigma(L) = 1$, then $0 = \eta_1 < \eta_L = \rho_2 m_L$ and so b_1 has higher priority than b_L. The result for the case when $\sigma(1) = \sigma(L) = 2$ is obtained similarly.

(ii) $\sigma(j) = \sigma(i) = 1$ and $b_j \succ b_i$ imply $\eta_j \leq \eta_i$, and so we have

$$\rho_2 \sum_{\{k:j \leq k \leq L; \sigma(k)=1\}} m_k - \rho_1 \sum_{\{k:j \leq k \leq L; \sigma(k)=2\}} m_k$$
$$\leq \rho_2 \sum_{\{k:i \leq k \leq L; \sigma(k)=1\}} m_k - \rho_1 \sum_{\{k:i \leq k \leq L; \sigma(k)=2\}} m_k,$$

from which we get

$$\rho_2 \sum_{\{k:j \leq k \leq i-1; \sigma(k)=1\}} m_k \leq \rho_1 \sum_{\{k:j \leq k \leq i-1; \sigma(k)=2\}} m_k.$$

Then (ii) is immediate.

(iii) As before, $\sigma(j) = \sigma(i) = 1$ and $b_j \succ b_i$ imply $\eta_i \geq \eta_j$, and so

$$\rho_2 \sum_{\{k:i \leq k \leq j-1; \sigma(k)=1\}} m_k \geq \rho_1 \sum_{\{k:i \leq k \leq j-1; \sigma(k)=2\}} m_k.$$

Thus,

$$\rho_1\rho_2 - \rho_2 \sum_{\{k:i \leq k \leq j-1; \sigma(k)=1\}} m_k \leq \rho_1\rho_2 - \rho_1 \sum_{\{k:i \leq k \leq j-1; \sigma(k)=2\}} m_k,$$

from which (iii) follows. □

Now we are ready to prove (3.28). Let us assume without loss of generality that $\rho_1 \leq \rho_2$, i.e, server 1 is faster. Then we claim that $\bar{I}_2(t) = 0$ for all t, for all solutions to (3.6–3.13, 3.22–3.25).

Suppose not. Then for some regular t, we must have

$$\frac{d}{dt}\bar{I}_2(t) > 0.$$

For this to happen, from (3.22–3.25) and Prop. 4.2 of Dai and Weiss [3], we must have $\sum_{\{k:1 \leq k \leq L; \sigma(k)=2\}} \bar{Q}_k(t) = 0$ and $d_k(t) = d_{(k-1)\text{mod}L}(t)$ for all $1 \leq k \leq L$ such that $\sigma(k) = 2$. Necessarily, it must also be true that

$$\sum_{\{k:1 \leq k \leq L; \sigma(k)=2\}} m_k d_k(t) < 1.$$

Now, $\sum_{\{k:1 \leq k \leq L; \sigma(k)=1\}} \bar{Q}_k(t) = 1$, and so

$$\sum_{\{k:1 \leq k \leq L; \sigma(k)=1\}} m_k d_k(t) = 1.$$

Thus,

$$\sum_{\{k:1\le k\le L;\sigma(k)=2\}} m_k d_k(t) - \sum_{\{k:1\le k\le L;\sigma(k)=1\}} m_k d_k(t) < 0. \tag{3.38}$$

At the regular time t, let b_{k^*} be the highest priority nonempty buffer at server 1.

If b_{k^*} is the lowest priority buffer at server 1, then we must have $d_j(t) = d_{(j-1)\text{mod}L}(t)$ for all $j \neq k^*$, since every other buffer in the network must be empty (recall that server 2 is empty). Thus, $d_j(t) = d_{k^*}(t)$ for all j. Hence (3.38) yields

$$\left(\sum_{\{k:1\le k\le L;\sigma(k)=2\}} m_j - \sum_{\{k:1\le k\le L;\sigma(k)=1\}} m_j\right) d_{k^*}(t) = (\rho_2 - \rho_1) d_{k^*}(t) < 0,$$

which however cannot hold because $\rho_1 \le \rho_2$.

Thus b_{k^*} is *not* the lowest priority buffer at server 1. Let b_l be the first buffer at server 1 which is of lower priority than b_{k^*}, encountered as we traverse the network in the order $b_{k^*}, b_{k^*+1}, \ldots$ wrapping around at b_L if need be. Now there are only two possibilities because of Lemma 3.7.2(i), either $l > k^* + 1$ or $l \le k^* - 1$. In the first case, $l > k^* + 1$, we must have $d_{k^*}(t) = d_{k^*+1}(t) = \cdots = d_{l-1}(t)$; $d_l(t) = d_{l+1}(t) = \cdots = d_L(t) = 0$ and $d_1(t) = d_2(t) = \cdots = d_{k^*-1}(t) = 0$. Thus (3.38) becomes

$$\left(\sum_{\{j:k^*\le j\le l-1;\sigma(j)=2\}} m_j - \sum_{\{j:k^*\le j\le l-1;\sigma(j)=1\}} m_j\right) d_{k^*}(t) < 0.$$

But from Lemma 3.7.2(ii), since $l > k^* + 1$,

$$\sum_{\{j:k^*\le j\le l-1;\sigma(j)=2\}} m_j \ge \frac{\rho_2}{\rho_1} \sum_{\{j:k^*\le j\le l-1;\sigma(j)=1\}} m_j \ge \sum_{\{j:k^*\le j\le l-1;\sigma(j)=1\}} m_j,$$

and therefore (3.38) cannot hold.

Last, we consider the case when $l \le k^* - 1$. Then $d_{k^*}(t) = d_{k^*+1}(t) = \cdots = d_L(t)$; $d_1(t) = d_2(t) = \cdots = d_{l-1}(t) = d_{k^*}(t)$ (this second set of equations is, of course, irrelevant when $l = 1$), and $d_l(t) = d_{l+1}(t) = \cdots = d_{k^*-1}(t) = 0$. Thus (3.38) becomes

$$\left(\sum_{\{j:k^*\le j\le L;\sigma(j)=2\}} m_j + \sum_{\{j:1\le j\le l-1;\sigma(j)=2\}} m_j\right) d_{k^*}(t)$$
$$- \left(\sum_{\{j:k^*\le j\le L;\sigma(j)=1\}} m_j + \sum_{\{j:1\le j\le l-1;\sigma(j)=1\}} m_j\right) d_{k^*}(t) < 0 \tag{3.39}$$

But from Lemma 3.7.2(iii) we have

$$\sum_{\{j:k^* \leq j \leq L;\sigma(j)=2\}} m_j + \sum_{\{j:1 \leq j \leq l-1;\sigma(j)=2\}} m_j$$

$$\geq \frac{\rho_2}{\rho_1} \left[\sum_{\{j:k^* \leq j \leq L;\sigma(j)=1\}} m_j + \sum_{\{j:1 \leq j \leq l-1;\sigma(j)=1\}} m_j \right]$$

$$\geq \sum_{\{j:k^* \leq j \leq L;\sigma(j)=1\}} m_j + \sum_{\{j:1 \leq j \leq l-1;\sigma(j)=1\}} m_j,$$

and so (3.39) cannot hold.

Thus, in all cases (3.38) cannot hold, and so

$$\frac{d}{dt}\bar{I}_2(t) = 0 \text{ for all regular } t,$$

thus completing the proof by Lemma 3.5.3. □

3.8 Inefficient Reentrant Lines

In this section, we give two examples of reentrant lines for which the integral equations (3.6–3.13) and (3.22–3.25) admit solutions which do not satisfy (3.26). This does not rule out the possibility that the actual fluid limits still satisfy (3.26).

The first example is similar to that of Harrison and Nguyen [7], which is the closed version of the examples in Kumar and Seidman [10] and Lu and Kumar [13], and is a counterexample to the conjecture that if all fluid limits of an open reentrant line under a buffer priority policy with Poisson arrivals empty in finite time, i.e, are stable, then all fluid limits of the corresponding closed reentrant line operated under the same buffer priority policy satisfy (3.26). The second example shows that the topology constraint that there are no self-loops, i.e., $\sigma(k) \neq \sigma(k+1)$, is not enough to guarantee (3.26) in two-server lines. This demonstrates that even in the two-server case when self loops are not the problem, alternate blocking and starvation leading to inefficient utilization could well be a problem. The second example also shows that even if there is only one bottleneck in a system, the fluid flows still need not satisfy (3.26).

Example 3.8.1 The FBFS buffer priority policy can admit an inefficient solution to the fluid limit integral equations.

Consider the closed reentrant line shown in Figure 1. Here $m_1 = m_3 = 1$ and $m_2 = m_4 = 0.2$. So $\lambda^* = 1/1.2$. The scheduling policy used is the FBFS policy which was shown to be stable in the open stochastic case by Dai [4]. So $b_1 \succ b_2$ and $b_3 \succ b_4$. Consider the initial condition $q = (0, 1, 0, 0)$. Then

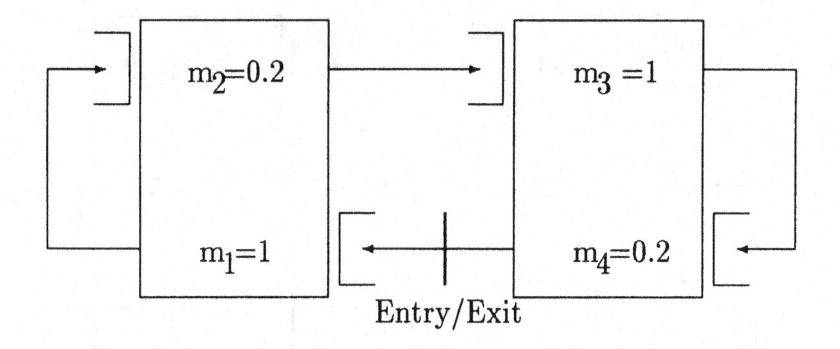

FIGURE 1. System of Example 1 under FBFS

Time	State
0	$(0,1,0,0)$
0.2	$(0,0,0.8,0.2)$
1.0	$(0,0,0,1)$
1.2	$(0.8,0.2,0,0)$
2.0	$(0,1,0,0)$

TABLE 3.1. Inefficient trajectory for Example 1

we can construct a piecewise linear trajectory which evolves as shown in Table 1.

Thus we see that $\frac{\bar{D}_L(t)}{t} \to 1/2 < \lambda^* = 1/1.2$, proving that (3.26) does not hold for this trajectory. Thus, in the closed case, with suitable choice of the "first" buffer, the FBFS policy can lead to inefficient fluid flows.

Example 3.8.2 Two-server reentrant lines without self loops can admit an inefficient solution to the fluid limit integral equations.

Consider the system shown in Figure 2, with the buffer priority policy

Time	State
0	$(0,0,1,0,0,0)$
0.0125	$(0,0,0,0.975,0,0.025)$
0.5	$(0,0,0,0,0,1)$
$0.5+1/49$	$(1-1/49,0,1/49,0,0,0)$
1.5	$(0,0,1,0,0,0)$

TABLE 3.2. Inefficient trajectory for Example 2

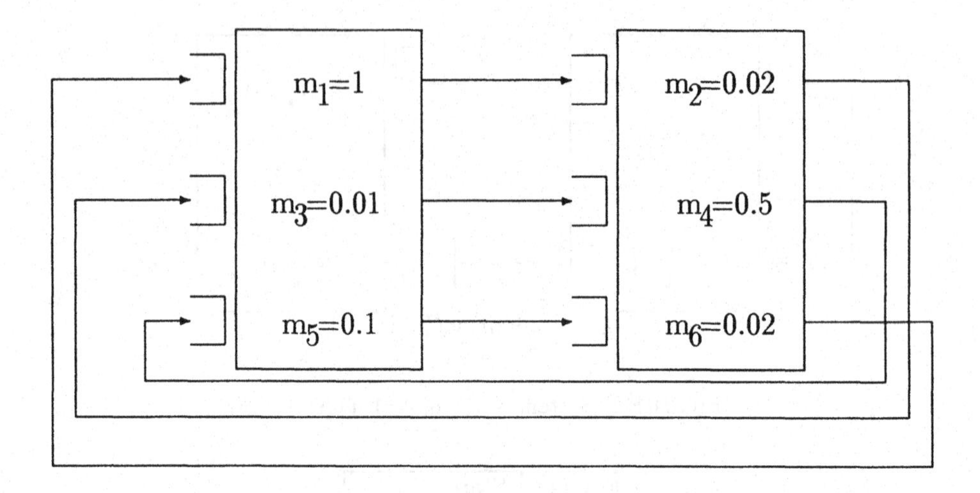

FIGURE 2. System of Example 2 without self-loops

$b_1 \succ b_5 \succ b_3$ and $b_4 \succ b_2 \succ b_6$. The processing times are $m_1 = 1, m_2 = 0.02$, $m_3 = 0.01$, $m_4 = 0.5$, $m_5 = 0.1$, and $m_6 = 0.02$. So $\lambda^* = 1/1.11$. An inefficient piecewise linear trajectory which satisfies the integral equations is shown in Table 2 (with its breakpoints).

Thus we see that $\frac{\bar{D}_L(t)}{t} \to 1/1.5 < 1/1.11$ and hence (3.26) cannot be satisfied.

3.9 Concluding Remarks

We have investigated the fluid limit approach for analyzing the heavy traffic behavior of closed queueing networks. We have shown that the efficiency of such networks can be established by studying the throughput of the fluid limits. Using this approach, we have proved the efficiency of the class of all FSMCT policies, including the Last Buffer First Serve policy, for general closed reentrant lines, and the Harrison–Wein buffer priority policy for two-station closed reentrant lines. We have also presented examples which illustrate how inefficiency can arise in closed reentrant lines and shown that buffer priority policies which lead to stable fluid flows in an open network may still inefficiently utilize servers in the corresponding closed fluid flow network.

Acknowledgments: The authors are grateful to an anonymous reviewer for suggesting a slight extension to the original Lemma 4.1 and Theorem 4.2 which featured “lim infs” rather than “lim sups” in the statements of the results. The research reported here has been supported in part by the U. S. Army Research Office under Grant No. DAAH 04-95-1-0090, the National Science Foundation under Grant No. ECS-94-05371, and the Joint Services Electronics Program under Contract No. N00014-90-J-1270.

3.10 References

[1] D. Bertsekas and R. Gallager. *Data Networks*. Prentice-Hall, Englewood Cliffs, NJ, 1987.

[2] H. Chen and A. Mandelbaum. Discrete flow networks: Bottleneck analysis and fluid approximations. *Math. Oper. Res.*, 16:408–446, 1991.

[3] J. Dai and G. Weiss. Stability and instability of fluid models for certain re-entrant lines. Preprint, February 1994. To appear in Mathematics of Operations Research.

[4] J. G. Dai. Stability of open multiclass queueing networks via fluid models. Technical report, Georgia Institute of Technology, 1994.

[5] J. G. Dai and R. J. Williams. Existence and uniqueness of semimartingale reflecting Brownian motions in convex polyhedrons. Submitted to *Theory of Probability and its Applications*, 1994.

[6] S. N. Ethier and T. G. Kurtz. *Markov Processes: Characterization and Convergence*. Wiley, New York, NY, 1985.

[7] J. M. Harrison and V. Nguyen. Some badly behaved closed queueing networks. Technical Report 3691-94-MSA, Stanford and MIT, May 1994.

[8] J. M. Harrison and L. M. Wein. Scheduling networks of queues: Heavy traffic analysis of a two-station closed network. *Operations Research*, 38(6):1052–1064, 1990.

[9] H. Jin, J. Ou, and P. R. Kumar. The throughput of closed queueing networks–functional bounds, asymptotic loss, efficiency, and the Harrison-Wein conjectures. Submitted to *Mathematics of Operations Research*, October 1994.

[10] P. R. Kumar and T. I. Seidman. Dynamic instabilities and stabilization methods in distributed real-time scheduling of manufacturing systems. *IEEE Transactions on Automatic Control*, AC-35(3):289–298, March 1990.

[11] S. Kumar and P. R. Kumar. The last buffer first policy is stable for stochastic re–entrant lines. Technical report, Coordinated Science Laboratory, University of Illinois, Urbana, IL, 1994.

[12] S. Kumar and P. R. Kumar. Fluctuation smoothing policies are stable for stochastic re-entrant lines. To appear in *Discrete Event Dynamical Systems*, 1996.

[13] S. H. Lu and P. R. Kumar. Distributed scheduling based on due dates and buffer priorities. *IEEE Transactions on Automatic Control*, AC-36(12):1406–1416, December 1991.

[14] Steve C. H. Lu, Deepa Ramaswamy, and P. R. Kumar. Efficient scheduling policies to reduce mean and variance of cycle–time in semiconductor manufacturing plants. *IEEE Transactions on Semiconductor Manufacturing*, 7(3):374–385, August 1994.

[15] Steve H. Lu and P. R. Kumar. Fluctuation smoothing scheduling policies for queueing systems. In *Proceedings of the Silver Jubilee Workshop on Computing and Intelligent Systems*, Indian Institute of Science, Bangalore, December 1993.

[16] D. J. Miller. Simulation of a semiconductor manufacturing line. *Communications of the ACM*, 33(10):98–108, 1990.

[17] J. Walrand. *Communication Networks–A First Course*. Aksen Associates, Boston, 1991.

[18] L. M. Wein. Scheduling semiconductor wafer fabrication. *IEEE Transactions on Semiconductor Manufacturing*, 1(3):115–130, August 1988.

4
Asymptotics and Uniform Bounds for Multiclass Queueing Networks

P. R. Kumar

ABSTRACT Recently it has been shown in [2] that one can obtain linear programs which provide bounds on the coefficients α, ν in the expression $\frac{\alpha N}{N+\nu}$, which then bounds the throughput for every population size N in a closed irreducible network. Also, it has been shown in [1] that one can obtain linear programs which provide bounds on the coefficients $\{c_i\}$ in the expansion $\sum_{i=0}^{M} \frac{c_i \rho}{(1-\rho)^i}$, which then bounds the mean number in an open system for every nominal load ρ in the system. We provide a brief account of these results.

4.1 Introduction

We provide here a brief account of some recent results from [2] and [1] concerning uniform bounds and asymptotics for multiclass closed and open networks.

Briefly, if N is the population size in a closed irreducible network, then one can provide lower and upper bounds on the throughput, which are of the form $\frac{\alpha N}{N+\nu}$. The coefficients α and ν can be obtained through linear programs. These bounds hold for all N, and are therefore referred to as *uniform bounds*. If α above is the maximal throughput capacity of the bottleneck servers in the network, then the constant ν is a bound on the *asymptotic loss*.

For an open network, if ρ is the nominal load on the bottleneck servers, then one can obtain lower or upper bounds on the mean number of customers in the network, which are of the form $\sum_{i=0}^{M} \frac{c_i \rho}{(1-\rho)^i}$. The coefficients $\{c_i\}$ and M can be obtained through the use of linear programs. These bounds are also uniform in ρ. The integer M is a bound on the *pole multiplicity* in heavy traffic.

4.2 Closed Networks

Consider a system with L buffers $\{b_1, b_2, \ldots, b_L\}$ (which can also be regarded as "classes"), and S servers $\{1, 2, \ldots, S\}$. Customers in buffer b_i are served by the server $\sigma(i) \in \{1, 2, \ldots, S\}$, and their service time is exponentially distributed with mean $\frac{1}{\mu_i}$. After such a service, they join buffer with probability p_{ij}. Each server can only serve one customer at a time. (Also, throughout this paper, each buffer can be catered to by only a specific server; thus we do not consider $M/M/c$ type situations). We assume that the routing matrix $p = [p_{ij}]$ is irreducible, and refer to this as the "single route" case. (The "multiple route" case corresponds to the situation where P can be decomposed into several closed communicating classes.) Let N denote the size of the population of customers trapped in this closed network.

Let $\pi = \pi P$ be the invariant probability measure of P. Denote by "$i \in \sigma$", the situation where buffer b_i is served by server σ, i.e., $\sigma(i) = \sigma$.

Using uniformization, we pretend that every buffer has either a real or fictitious customer in service, and sample the system at all such real or fictitious service completion times. Let $x(n) = (x_1(n), x_2(n), \ldots, x_L(n))^T$ denote the vector of buffer lengths at the n-th such sampling time, and let $w_i(n)$ be the indicator variable of whether a customer in b_i is really in service.

We will consider a stationary (in the sense used in Markov decision processes) non-idling scheduling policy. By "non-idling" we mean that a server is idle only when no customers are waiting for it. By "stationary" we mean that the decision variables $(w_1(n), w_2(n), \ldots, w_L(n))$ are chosen as a function only of the state $x(n)$. Without loss of generality we suppose that the system is in steady-state.

We define the throughput $\alpha^u(N)$ under the scheduling policy u when the population size is N by

$$\alpha^u(N) = \frac{\mu_i}{\pi_i} E(w_i).$$

The expression on the right above does not depend on the choice of $i \in \{1, 2, \ldots, L\}$. To see this, note that by balancing the rates at which customers leave from and arrive at buffer b_i, one has $\sum_j E(w_j)\mu_j p_{ji} = \sum_j E(w_i)\mu_i p_{ij}$. Let $\delta_i := E(w_i)\mu_i$. Now note that the only solution $\delta = (\delta_1, \delta_2, \ldots, \delta_L)$ of the equation $\sum_j \delta_j p_{ji} = \sum_j \delta_i p_{ij}$ is $\delta = \alpha\pi$ for some α, due to the uniqueness of the invariant probability measure for the routing matrix P. Hence $\delta_i = E(w_i)\mu_i = \alpha\pi_i$. Thus, as claimed, $\frac{\mu_i}{\pi_i} E(w_i)$ does not depend on i.

The throughput capacity of the system is

$$\alpha^* = \text{Min}_\sigma \left(\sum_{i \in \sigma} \frac{\pi_i}{\mu_i} \right)^{-1}.$$

The throughput $\alpha^u(N)$ can never be greater than α^* for any scheduling policy u and any population size N.

Theorem 4.2.1 **(i)** Consider the linear program (with decision variables $\{q_{ij} : 1 \leq i, j \leq L\}$ and $\{s_i : 1 \leq i \leq L\}$):

$$\text{Max} \sum_{ij} \pi_i p_{ij} (q_{ji} - q_{ii})$$

subject to:

$$\text{Max}_{i \in \sigma(j)} \mu_i \left[\sum_k p_{ik} \left(q_{kj} - q_{ij} - \frac{s_i}{\pi_i} \right) \right]$$

$$+ \sum_{\sigma \neq \sigma(j)} \text{Max}_{i \in \sigma} \mu_i \left[\sum_k p_{ik} \left(q_{kj} - q_{ij} - \frac{s_i}{\pi_i} \right) \right]^+ \leq \alpha^* \quad \forall j,$$

$$\sum_i s_i \leq -1, \text{ and}$$

$$q_{ij} = q_{ji} \quad \forall i, j.$$

Let $\underline{\nu}$ denote its value. Then

$$\alpha^u(N) \leq \alpha^* \frac{N}{N + \underline{\nu}} \text{ for all } u, N.$$

(ii) Consider the linear program:

$$\text{Max} \sum_{ij} \pi_i p_{ij} (q_{ji} - q_{ii})$$

subject to

$$\text{Max}_{i \in \sigma(j)} \mu_i \left[\sum_k p_{ik} \left(q_{kj} - q_{ij} - \frac{s_i}{\pi_i} \right) \right]$$

$$+ \sum_{\sigma \neq \sigma(j)} \text{Max}_{i \in \sigma(j)} \left[\sum_k p_{ik} \left(q_{kj} - q_{ij} - \frac{s_i}{\pi_i} \right) \right]^+ \leq -\alpha^* \quad \forall j,$$

$$\sum_i s_i \leq 1, \text{ and}$$

$$q_{ij} = q_{ji} \quad \forall i, j.$$

If this linear program is feasible, let $\overline{\nu}$ denote its value (it is nonpositive). Then

$$\alpha^u(N) \geq \alpha^* \frac{N}{N - \overline{\nu}} \text{ for all } u, N.$$

Proof.

(i) If $Q = [q_{ij}]$ is the optimal solution, consider the quadratic form $x^T(n)Qx(n)$. Denote by $e_k = (0,0,\ldots,0,1,0,\ldots,0)^T$ the k-th coordinate vector. Given $x(n)$, $x(n+1)$ is equal to $x(n) - e_i + e_k$ with probability $\mu_i w_i(n) p_{ik}$. Hence,

$$E[x^T(n+1)Qx(n+1) \mid x(n)] = x^T(n)Qx(n)$$
$$+2\sum_{ik} \mu_i w_i(n) p_{ik}(e_k - e_i)^T Q x(n)$$
$$+\sum_{ik} \mu_i w_i(n) p_{ik}(e_k - e_i)^T Q(e_k - e_i).$$

Since $E[x^T(n+1)Qx(n+1)] = E[x^T(n)Q(n)]$ in steady-state, it follows that

$$E\left[\sum_{ik} \mu_i w_i(n) p_{ik}(e_k - e_i)^T Q\left\{2x(n) + (e_k - e_i)\right\}\right] = 0.$$

Substituting $Ew_i(n) = \frac{\pi_i \alpha^u(N)}{\mu_i}$, using $\pi = \pi P$, and adding and subtracting $E\sum_{ij} \frac{\mu_i w_i(n) s_i x_j(n)}{\pi_i} = N\alpha^u(N)\sum_i s_i$ with $\{s_i\}$ obtained from the optimal solution, we obtain the identity

$$E\sum_{ij} \mu_i w_i(n) \sum_k p_{ik}\left(q_{kj} - q_{ij} - \frac{s_i}{\pi_i}\right) x_j(n)$$
$$+\alpha^u(N)N\sum_i s_i - \alpha^u(N)\sum_{ij} \pi_i p_{ij}(q_{ij} - q_{ii}) = 0.$$

Now note that by the non-idling nature of the policy, whenever $x_j(n) \geq 1$, one of the variables in the set $\{w_i(n) : i \in \sigma(j)\}$ is equal to one, while all the others are zero. Hence, for any constants $\{\gamma_i\}$, $\sum_{i\in\sigma(j)} \gamma_i w_i(n) x_j(n) \leq \text{Max}_{i\in\sigma(j)}\gamma_i x_j(n)$. On the other hand, if we consider a $\sigma \neq \sigma(j)$, it is possible for all the variables in the set $\{w_i(n) : i \in \sigma\}$ to be equal to zero. Hence, for $\sigma \neq \sigma(j)$, $\sum_{i\in\sigma} \gamma_i w_i(n) x_j(n) \leq \text{Max}\{0, \text{Max}_{i\in\sigma}\gamma_i x_j(n)\}$. Thus, for any constants $\{\gamma_i\}$, we have the upper bounds,

$$\sum_{i\in\sigma} \gamma_i w_i(n) x_j(n) \begin{cases} \leq & \text{Max}_{i\in\sigma}\gamma_i x_j(n) \text{ if } \sigma = \sigma(j) \\ \leq & \text{Max}_{i\in\sigma}\gamma_i^+ x_j(n) \text{ if } \sigma \neq \sigma(j) \end{cases},$$

where $\gamma_i^+ := \text{Max}\{\gamma_i, 0\}$. Now we break up the summation $\sum_i$ in the earlier derived identity as $\sum_{i\in\sigma(j)} + \sum_{\sigma\notin\sigma(j)}\sum_{i\in\sigma}$, and substitute these upper bounds. This gives the inequality,

$$E\sum_j \left[\text{Max}_{i\in\sigma(j)}\mu_i[\sum_k p_{ik}(q_{kj} - q_{ij} - \frac{s_i}{\pi_i})]\right.$$

$$+\sum_{\sigma\neq\sigma(j)} \text{Max}_{i\in\sigma}\mu_i[\sum_k p_{ik}(q_{kj}-q_{ij}-\frac{s_i}{\pi_i})]^+\Bigg] x_j(n)$$

$$+\alpha^u(N)N\sum_i s_i - \alpha^u(N)\sum_{i,j}\pi_i p_{ij}(q_{ij}-q_{ii}) \geq 0.$$

Now we use the inequality constraints in the linear program, which give us,

$$E[\sum_j \alpha^* x_j(n)] + \alpha^u(N)N\sum_i s_i - \alpha^u(N)\sum_{i,j}\pi_i p_{ij}(q_{ij}-q_{ii}) \geq 0.$$

Using $\sum_j x_j(n) = N$, and $\sum_i s_i \leq -1$ gives us,

$$\alpha^* N - N\alpha^u(N) - \alpha^u(N)\sum_{i,j}\pi_i p_{ij}(q_{ij}-q_{ii}) \geq 0.$$

Since $\sum_{i,j}\pi_i p_{ij}(q_{ij}-q_{ii}) = \underline{\nu}$ for the optimal solution $\{q_{ij}, s_i\}$, the result follows.

(ii) Similar.

This result is from [2], to which we refer the reader for more details as well as several consequential results.

There are also extensions of these results to the multiple route case which the author has developed together with his student M. Ginsberg.

4.3 Open Networks

The model for open networks consists of a set of buffers $\{b_1, \ldots, b_2\}$ and servers $\{1, 2, \ldots, S\}$ where customers in buffer b_i are served by the server $\sigma(i) \in \{1, 2, \ldots, S\}$ with an exponentially distributed service time with mean μ_i, as in Section 2. The difference is that there is a Poisson stream of customers of rate λ from the outside, and each such arriving customer joins buffer b_i with probability p_{*i}. Likewise there are also departures to the outside. After completion of service at b_i, a customer leaves the system with probability p_{i*}, or joins a buffer j with probability p_{ij}.

We assume that there exists a unique solution $\{\rho_i : 1 \leq i \leq L\}$ to the traffic equations

$$\lambda p_{*i} + \sum_k \mu_k \rho_k p_{ki} \quad = \quad \mu_i \rho_i \quad \forall i.$$

The quantity ρ_i is then the nominal load on the server $\sigma(i)$ imposed by customers from buffer b_i. Let

$$\rho = \text{Max}_\sigma \sum_{i\in\sigma} \rho_i$$

denote the nominal load on the system's bottleneck servers. The arrival rate $\lambda^* = \frac{\lambda}{\rho}$ is the *capacity* of the system; it corresponds to a nominal load of unity.

A uniform upper bound on the mean number in the system (= $\lambda\times$ mean delay) is provided by the following theorem. Below, a symmetric matrix Q is said to be *copositive* if $x^T Qx \geq 0$ for all x in the positive orthant, i.e., for all vectors $x = (x_1, x_2, \ldots, x_L)^T$ where each $x_i \geq 0$.

Theorem 4.3.1 Let $\{Q^{(m)} : 0 \leq m \leq M\}$ be a family of symmetric matrices such that:

$$\lambda^* \sum_i p_{*i} \left(q_{ij}^{(m)} - 1(m \geq 1) q_{ij}^{(m-1)} \right)$$

$$+ \mathrm{Max}_{i \in \sigma(j)} \mu_i \left(\sum_k p_{ik} q_{kj}^{(m)} - q_{ij}^{(m)} \right)$$

$$+ \sum_{\sigma \neq \sigma(j)} \mathrm{Max}_{i \in \sigma} \mu_i \left[\sum_k p_{ik} q_{kj}^{(m)} - q_{ij}^{(m)} \right]^+ \leq -1(m = M)$$

for $1 \leq j \leq L$ and $0 \leq m \leq M$,

$$q_{ii}^{(m)} - 2 \sum_j p_{ij} q_{ij}^{(m)} + \sum_j p_{ij} q_{jj}^{(m)} \geq 0 \text{ for } 1 \leq j \leq L,$$

$$\sum_i p_{*i} q_{ij}^{(m)} \geq 0 \text{ for } 1 \leq j \leq L,$$

$$\sum_{m=0}^{M} \frac{1}{(1-\rho)^m} Q^{(M-m)} \text{ is copositive.}$$

Then if

$$c_m = \sum_i \mu_i \frac{\rho_i}{\rho} \left[q_{ii}^{(M-m)} - \sum_j p_{ij} q_{ij}^{(M-m)} \right],$$

we have

$$\limsup_{n \to \infty} E \left[\frac{1}{n} \sum_{t=0}^{n-1} |x(n)| \right] \leq \sum_{i=0}^{M} \frac{\rho c_i}{(1-\rho)^i} \text{ for all } 0 \leq \rho < 1.$$

Above, $|x(n)| = \sum_{i=1}^{L} x_i(n)$ is the total number in the system.

Proof.
For any symmetric matrix Q, by proceeding as in Theorem 2.1, we obtain

$$E\left[\frac{1}{2}x^T(t+1)Qx(t+1)\right] = E\left[\frac{1}{2}x^T(t)Qx(t)\right]$$
$$+E\sum_{j=1}^{L} x_j(t)\left[\lambda\sum_i p_{*i}q_{ij} + \sum_i \mu_i w_i(t)\left(\sum_k p_{ik}q_{kj} - q_{ij}\right)\right]$$
$$+\frac{1}{2}\sum_i \lambda p_{*i}q_{ii} + \frac{1}{2}\sum_{i=1}^{L}\mu_i w_i(t)\left(q_{ii} - 2\sum_j p_{ij}q_{ij} + \sum_j p_{ij}q_{jj}\right).$$

With the choice $Q = Q^{(M)}$, this identity yields the inequality,

$$E\left[\frac{1}{2}x^T(t+1)Q^{(M)}x(t+1)\right] \le E\left[\frac{1}{2}x^T(t)Q^{(M)}x(t)\right] - E|x(t)|$$
$$+(\lambda-\lambda^*)E\left[\sum_{j=1}^{L} x_j(t)\sum_{i=1}^{L} p_{*i}q_{ij}^{(M)}\right] + \lambda^* E\left[\sum_{j=1}^{L} x_j(t)\sum_{i=1}^{L} p_{*i}q_{ij}^{(M-1)}\right]$$
$$+\frac{1}{2}\sum_i \lambda p_{*i}q_{ii}^{(M)}$$
$$+\left[\frac{1}{2}\sum_{i=1}^{L}\mu_i w_i(t)\left(q_{ii}^{(M)} - 2\sum_j p_{ij}q_{ij}^{(M)} + \sum_j p_{ij}q_{jj}^{(M)}\right)\right],$$

where we have written the term $\lambda\sum_i p_{*i}q_{ij}^{(M)}$ as
$\lambda^*\sum_i p_{*i}q_{ij}^{(M)} - \lambda^*\sum_i p_{*i}q_{ij}^{(M-1)} + (\lambda-\lambda^*)\sum_i p_{*i}q_{ij}^{(M)} + \lambda^*\sum_i p_{*i}q_{ij}^{(M-1)}$,
and also used the upper bound on $\sum_{i\in\sigma}\gamma_i w_i(t)x_j(t)$ from the proof of Theorem 2.1. Telescoping this inequality, using the bound

$$\frac{1}{n}\sum_{t=0}^{n-1} Ew_i(t) \le \rho_i + o(1).$$

on the rate at which work arrives for buffer b_i, gives us

$$E[\frac{1}{n}\sum_{t=0}^{n-1} x_j(t)] - (\lambda-\lambda^*)E\sum_{j=1}^{L}[\frac{1}{n}\sum_{t=0}^{n-1} x_j(t)]\sum_{i=1}^{L} p_{*i}q_{ij}^{(M)}$$
$$-\lambda^* E\sum_{j=1}^{L}[\frac{1}{n}\sum_{t=0}^{n-1} x_j(t)]\sum_{i=1}^{L} p_{*i}q_{ij}^{(M-1)}$$
$$\le \sum_{i=1}^{L}\mu_i\rho_i\left(q_{ii}^{(M)} - \sum_j p_{ij}q_{ij}^{(M)}\right)$$
$$+\frac{1}{2}\frac{x^T(0)Q^{(M)}x(0)}{n} - \frac{1}{2}\frac{x^T(n)Q^{(M)}x(n)}{n} + o(1).$$

Now, beginning with $m = M-1$, and continuing with $m = M-2, M-1, \ldots, 0$, we repeatedly substitute for $E\sum_{j=1}^{L}[\frac{1}{n}\sum_{t=0}^{n-1} x_j(t)]\sum_{i=1}^{L} p_{*i}q_{ij}^{(m)}$ from

$$-(\lambda-\lambda^*)E\sum_{j=1}^{L}[\frac{1}{n}\sum_{t=0}^{n-1} x_j(t)]\sum_{i=1}^{L} p_{*i}q_{ij}^{(m)}$$

$$-1(m \geq 1)\lambda^* E\sum_{j=1}^{L}[\frac{1}{n}\sum_{t=0}^{n-1} x_j(t)]$$

$$\leq \sum_{i=1}^{L}\mu_i\rho_i\left(q_{ii}^{(m)} - \sum_j p_{ij}q_{ij}^{(m)}\right)$$

$$+\frac{1}{2}\frac{x^T(0)Q^{(m)}x(0)}{n} - \frac{1}{2}\frac{x^T(n)Q^{(m)}x(n)}{n} + o(1),$$

which is obtained with the choice $Q = Q^{(m)}$. This yields the result.

Note that the copositivity condition in the Theorem statement can be replaced by a stronger nonnegativity requirement on that matrix. (Or one can follow the procedure below to obtain the best matrices $\{Q^{(m)}\}$ and subsequently check the copositivity condition).

The upper bound on the mean number of the system is ripe for optimization by linear programming to determine the best c_i's. For example, if the best heavy traffic bound is desired, then one can determine the lowest value of the integer M at which such a feasible $\{Q^{(m)} : 0 \leq m \in M\}$ exists, and then minimize c_M. For the best light traffic bound one would minimize $\sum_{i=0}^{M} c_i$. There is further freedom in the coefficients which can also be utilized.

A uniform lower bound on the mean number in the system is provided below. The proof of this result is similar to that of Theorem 3.1.

Theorem 4.3.2 Suppose $Q^{(0)}$ and $Q^{(1)}$ are symmetric matrices such that:

$$\lambda^*\sum_i p_{*i}q_{ij}^{(0)} + \mathrm{Max}_{i\in\sigma(j)}\mu_i\left[\sum_k p_{ik}q_{kj}^{(0)} - q_{ij}^{(0)}\right]$$

$$+\sum_{\sigma\neq\sigma(j)}\mathrm{Max}_{i\in\sigma}\mu_i\left[\sum_k p_{ik}q_{kj}^{(0)} - q_{ij}^{(0)}\right]^+ \leq 0 \text{ for} 1 \leq j \leq L,$$

$$\lambda^*\sum_i p_{*i}\left(q_{ij}^{(1)} - q_{ij}^{(0)}\right) + \mathrm{Max}_{i\in\sigma(j)}\mu_i\left[\sum_k p_{ik}q_{kj}^{(1)} - q_{ij}^{(0)}\right]$$

$$\sum_{\sigma\neq\sigma(j)}\mathrm{Max}_{i\in\sigma}\mu_i\left[\sum_k p_{ik}q_{kj}^{(1)} - q_{ij}^{(1)}\right]^+ \leq 1 \text{ for} 1 \leq j \leq L,$$

$$\sum_i p_{*i} q_{ij}^{(1)} \geq 0 \text{ for } 1 \leq j \leq L.$$

Define

$$c_m = \sum_i \mu_i \frac{\rho_i}{\rho} \left[\sum_j p_{ij} q_{ij}^{(1-m)} - q_{ii}^{(1-m)} \right].$$

Then

$$\limsup_{n \to \infty} \frac{1}{n} E \sum_{t=0}^{n-1} |x(n)| \leq \frac{\rho c_i}{(1-\rho)} + \rho c_0 \text{ for all } 0 \leq \rho < 1.$$

This result is ripe for linear programming to determine the best constants c_0 and c_1.

We refer the reader to [1] for the details of the proofs of these results as well as several consequential ones.

4.4 References

[1] C. Humes, Jr., J. Ou, and P. R. Kumar. Delay and throughput bounds for open and closed queueing networks: Uniform functional expansions, heavy traffic pole multiplicities, stability and efficiency. Coordinated Science Laboratory, University of Illinois, July 1995. Submitted for publication.

[2] H. Jin, J. Ou, and P. R. Kumar. The throughput of closed queueing networks–functional bounds, asymptotic loss, efficiency, and the Harrison-Wein conjectures. Submitted to *Mathematics of Operations Research*, October 1994.

[illegible]

REFERENCES

[illegible]

5
Stability of Discrete-Time Jackson Networks with Batch Movements

Masakiyo Miyazawa

ABSTRACT We introduce a discrete-time Jackson network with batch movements. Not more than one node simultaneously completes service, but arbitrary sizes of batch arrivals, departures and transfers are allowed under a Markovian routing of batches including changes of their sizes. This model corresponds with the continuous-time network work with batch movements studied by Miyazawa and Taylor [11], but needs a care for the discrete-time setting. It is shown that the stationary joint distribution of queue length vector is stochastically bounded by a product of geometric distributions under the stability condition. Its improvements and tightness are discussed. We also provide an algorithm to calculate the decay rates of the geometric distributions, which answers the stability of the network as well.

5.1 Introduction

Discrete-time queueing models have been recently paid much attention due to its applicability to digital and high speed communication systems (see e.g. [2]). Batch and/or concurrent movements of customers are essentials features of these models. Another important feature is the dependency among arrival traffics. However, the latter complicates analysis, and one usually has to simplify the models like tree networks (see e.g. [3]). We here consider a rather general network structure, but restrict those models in which population dynamics over the networks are described by Markov chains. Under certain local balance conditions, those queueing networks exhibit to have stationary distributions in closed form (see e.g. [1], [6], [10], [12]). However, such networks require very specific rules for the batch size distributions for arriving customers as well as those of departing ones. On the other hand, the Jackson network of continuous-time, which is the standard model in continuous-time, is easily discretized. But no concurrent feature is essentially included in such a discretization.

In this paper, we modify a discrete-time Jackson network so to include batch and concurrent movements, which does not have local balance anymore. We assume a rather general Markovian routing of batches, which allows to change the transferred batch sizes. Arriving batch sizes at each node are *i.i.d.* with a general distribution depending on the node. At each time slot, one node is randomly selected with a given probability, and requests for a random number of customers to depart from the node according to another general distribution. If there are not sufficiently many customers at the node, all customers there leave the system. Otherwise those customers are transferred to other single node or to the outside according to the routing function. During these transfers, the batch sizes may be increased or decreased, which means that new customers from the outside are added or some of customers in the batches leave the system. So far, the model allows certain batch and concurrent movements of customers, though simultaneous completion of service at different nodes are still prohibited.

This model is motivated not only by the fact that the batch arriving feature is important for discrete-time queueing networks but also by its tractability for analysis. In fact, the model can be considered as a discrete-time version of the open queueing network with batch movements, studied by Miyazawa and Taylor [11]. In the latter model, Poisson arrivals, exponentially distributed service times and Markovian routing are assumed. However, we have to incorporate specific features due to discrete-time. Not only for this but also for the paper to be self contained, we give a full proofs for our results.

The above discrete-time model does not have the so called product form for the stationary distribution of the queue length vector, whose components are the numbers of customers at nodes. Furthermore, its marginal distribution for each node is not geometric, since the isolated node behaves like the $GI/G/1$ queue. So the model is intractable for exact analysis, but it is shown that, if we modify the network by adding extra arrivals in a certain way, the modified network has a product form geometric distribution for the queue length vector, which gives a stochastic bound for the original network. This bound is tight for the single node networks, and expected to do so in general. We here extend the corresponding results, obtained for the continuous-time case in [11]. Thus we can evaluate tail probabilities concerning customer populations over nodes. This is particularly important in performance evaluation of high seed switching systems, e.g. for considering the bandwidth (see e.g. [8]).

The main result of the paper is to get this stochastic bound. We then consider to improve the bound, which includes a direct proof for the tightness in the single node networks with and without a feedback loop. We also provide an algorithm to calculate the decay rates of the geometric distributions, which simultaneously answers the stability of the network. Those results would be applied to the corresponding continuous-time model as well.

This paper is split into six sections. We first introduce the discrete-time Jackson network with batch movements. The stochastic bound is obtained in Section 3. Section 4 discusses an improvement of the stochastic bound and its tightness. An algorithm for calculating the decay rates is given in Section 5. Finally, Section 6 discusses related theoretical issues on the stochastic bound. We are there concerned with quasi-reversibility and large deviation.

5.2 Jackson Networks with Batch Movements

We first introduce a discrete-time Jackson network, and then will include batch movement features. Suppose a network is composed of N nodes, numbered from 1 to N, and time is slotted. Let $J = \{1, 2, \cdots, N\}$, $J_0 = \{0\} \cup J$, $Z_+ = \{1, 2, \cdots\}$ and $Z_0 = \{0\} \cup Z_+$. In each time slot, either one of the following events occurs, which is independent of the past history and the present network state. With probability p_0, an exogenous customer arrives, and is routed to node j with probability $r(0, j)$ for $j \in J$. With probability p_i ($i \in J$), if there are any customers in node i, one customer departs from node i, and is routed to node j with probability $r(i, j)$ for $j \in J_0$, where we include $j = 0$ as the outside. Otherwise, nothing occurs at all nodes. It is assumed that $\sum_{i \in J_0} p_i \leq 1$, $\sum_{j \in J_0} r(i, j) = 1$ for $i \in J_0$, where $r(0, 0) = 0$, and $\{r(i, j)\}_{i,j \in J_0}$ is irreducible. We refer to this network as *the discrete-time Jackson network*.

Obviously, the discrete-time Jackson network is obtained by observing the queue length vector, just before its jump instants in the corresponding continuous-time Jackson network. Denote this embedded process by $\{X_t\}_{t \in Z_0}$. Hence, the stationary distribution π_0 of the queue length vector X_t at a time slot boundary is given by

$$\pi_0(\boldsymbol{n}) = \prod_{i=1}^{N} (1 - \rho_i) \rho_i^{n_i} \qquad (\boldsymbol{n} = (n_1, \cdots, n_N) \in Z_0^N), \tag{5.1}$$

if $\rho_i < 1$ for all $i \in J$, where ρ_i is a unique solution of the following traffic equation.

$$\rho_i p_i = \sum_{j \in J_0} \rho_j p_j r(j, i) \qquad (i \in J), \tag{5.2}$$

where $\rho_0 = 1$. Note that (5.2) also holds for $i = 0$, since $r(0, 0) = 0$.

Time slots of the discrete-time Jackson network are classified into three types, arrival slots for exogenous customers, departure slots for customers completed service and empty slots for no events, where the departure slots include routing. Observe the network at the end of each departure slot or of each empty slot if the network is empty. Then, we have the model in which one node completes service at each time slot unless the network is empty. We refer to this model as *the departure slotted Jackson network*. Note that, in each time slot of this model, batch of exogenous customers first arrive with probability p_0, and then one customer completes service if any. Here, the batch size of exogenous customers has the geometric distribution with mean $\frac{p_0}{1-p_0}$, and those customers are individually and independently routed to node j with probability $r(0,j)$.

Denote the queue length vector process of the departure slotted Jackson network by $\{\tilde{X}_t\}_{t\in Z_0}$. Suppose this process and $\{X_t\}$ are stationary. Since $\tilde{X}_t$ results when service is completed or no event occurs under the network being empty, we have, for a normalizing constant C_0,

$$C_0 P(\tilde{X}_t = \boldsymbol{n}) = \sum_{i\in J} \sum_{j\in S_+(\boldsymbol{n})\cup\{0\}} P(X_{t-1} = \boldsymbol{n} + \boldsymbol{e}_i - \boldsymbol{e}_j) p_i r(i,j)$$
$$+ (1-p_0) P(X_{t-1} = \mathbf{o}) 1_{\{\mathbf{o}\}}(\boldsymbol{n}), \tag{5.3}$$

where $S_+(\boldsymbol{n}) = \{i \in J | n_i \geq 1\}$, 1_A is the indicator function of a set A, $\boldsymbol{e}_i$ is the unit vector such that its i-th component is one, and the other components are zero for $i \in J$, and $\boldsymbol{e}_0 = \mathbf{o}$. By using (5.1) and (5.2), the right-hand side of (5.3) becomes:

$$\sum_{j\in S_+(\boldsymbol{n})\cup\{0\}} \left(p_j - p_0 r(0,j)\frac{1}{\rho_j}\right) \pi_0(\boldsymbol{n}) + (1-p_0)\,\pi_0(\mathbf{o}) 1_{\{\mathbf{o}\}}(\boldsymbol{n}).$$

Thus, the stationary distribution π_0' of $\tilde{X}_t$ is given by

$$\pi_0'(\boldsymbol{n}) = C_0^{-1} \sum_{j\in S_+(\boldsymbol{n})\cup\{0\}} \left(p_j - p_0 r(0,j)\frac{1}{\rho_j}\right) \prod_{i=1}^{N} \rho_i^{n_i}, \tag{5.4}$$

for $\boldsymbol{n} = (n_1, \cdots, n_N) \in Z_0^N \setminus \{\mathbf{o}\}$. For applications, the departure slotted model might be more interesting, since service should not be stopped during exogenous arrivals. However, since it can be converted from the standard slot model, we are mainly concerned with the standard slot model.

Let us generalize the discrete-time Jackson network in the following way. First we assume that exogenous customers arrive in batches, and those batch sizes are mutually independent. The batch enters node j as size m batch with probability $r(0,(j,m))$. The batch size distribution arriving at node j is denoted by A_j, i.e. the probability mass $a_j(m)$ of A_j equals $\frac{r(0,(j,m))}{\sum_{m\in Z_0} r(0,(j,m))}$. Those batch arrivals occur with probability p_0 in the same way as exogenous customer arrivals in the discrete-time Jackson network. Similarly, node i is selected for service completion with probability p_i, and requests a random number of customers to depart, which are independent of every thing else and has distribution B_i. If there are sufficiently many customers at node i, those customers depart from the node. Let k be the size of this batch, then the batch is routed to node j as size m with probability $r((i,k),(j,m))$ for $j \in J$ and $r((i,k),0)$ for $j = 0$. On the other hand, if there are less customers than the requested batch size, the node is just emptied. We referred to this model as *the discrete-time Jackson network with batch movements.*

The above model may be used for the performance evaluation of high speed switching systems, in which departure batch sizes may be deterministic, but arriving batch sizes are random. Here, p_i corresponds with the processing speed at node i. For computer network applications, the change of batch sizes during their transfers might describe additional task or partial completion of the original task. There seem many applications of the batch arrival and batch transfer features of the model. Of course, the model itself is hard to analyze, since the single node network is equivalent to the waiting time process of $GI/G/1$ queue in which interarrival and service times of customers are correlated. So we are only concerned with a stochastic bound and stability of the model.

We now describe the queue length vector process of the discrete-time Jackson network with batch movements as a Markov chain. Denote this Markov chain by $\{Y_t\}_{t\in Z_0}$. Define $c_{i,j}(k,m) = b_i(k)r((i,k),(j,m))$ for $i,j \in J$, $c_{i,0}(k) = b_i(k)r((i,k),0)$ for $i \in J$, $c_{0,j}(m) = r(0,(j,m))$ for $j \in J$, where $k,m \in Z_0$, and b_i is the mass probability function of B_i. Then the transition probability p of the Markov chain $\{Y_t\}$ is given by

$$
\begin{aligned}
p(\boldsymbol{n}, \boldsymbol{n} - k\boldsymbol{e}_i + m\boldsymbol{e}_j) &= p_i c_{i,j}(k,m) && (1 \le k \le n_i, m \ge 1)\\
p(\boldsymbol{n}, \boldsymbol{n} - k\boldsymbol{e}_i) &= p_i(c_{i,0}(k) + \overline{B}_i(n_i+1)1_{\{n_i\}}(k)) && (1 \le k \le n_i)\\
p(\boldsymbol{n}, \boldsymbol{n} + m\boldsymbol{e}_j) &= p_0 c_{0,j}(m) && (m \ge 1)\\
p(\boldsymbol{n}, \boldsymbol{n}) &= 1 - \left(p_0 + \sum_{i \in S_+(\boldsymbol{n})} p_i\right),
\end{aligned}
$$

for $i,j \in J$, where $\overline{B}_i(n) = \sum_{l=n}^{\infty} b_i(l)$. Other transition probabilities are zero.

Suppose the transition function p has the stationary distribution π. Then the π is determined by the global balance equation:

$$
\begin{aligned}
&\left(p_0 + \sum_{i \in S_+(\boldsymbol{n})} p_i\right) \pi(\boldsymbol{n})\\
&\quad = \sum_{i \in J} \sum_{j \in S_+(\boldsymbol{n})} \sum_{k=1}^{\infty} \sum_{m=1}^{n_j} \pi(\boldsymbol{n} + k\boldsymbol{e}_i - m\boldsymbol{e}_j) p_i c_{i,j}(k,m)\\
&\qquad + \sum_{i \in J} \sum_{k=1}^{\infty} \pi(\boldsymbol{n} + k\boldsymbol{e}_i) p_i (c_{i,0}(k) + \overline{B}_i(k+1)1_{\{n_i\}}(0))\\
&\qquad + \sum_{j \in S_+(\boldsymbol{n})} \sum_{m=1}^{n_j} \pi(\boldsymbol{n} - m\boldsymbol{e}_j) p_0 c_{0,j}(m)\,. \qquad (5.5)
\end{aligned}
$$

Suppose $\{Y_t\}$ be stationary under π, and denote the tail probability concerning π by $\overline{\Pi}(\boldsymbol{n}) = P(Y_t \geq \boldsymbol{n})$ ($\boldsymbol{n} \in Z_0^N$). By summing up (5.5), we have, for $\boldsymbol{n} \in Z_0^N$,

$$\begin{aligned} \overline{p}\,\overline{\Pi}(\boldsymbol{n}) &= \sum_{i\in S_0(\boldsymbol{n})} p_i \sum_{m_1=n_1}^{u(n_1)} \sum_{m_2=n_2}^{u(n_2)} \cdots \sum_{m_N=n_N}^{u(n_N)} \pi((m_1, m_2, \cdots, m_N)) \\ &\quad + \sum_{i\in J} \sum_{j\in S_+(\boldsymbol{n})} \sum_{k=1}^{\infty} \sum_{m=1}^{\infty} \overline{\Pi}((\boldsymbol{n} + k\boldsymbol{e}_i - m\boldsymbol{e}_j)^+) p_i c_{i,j}(k,m) \\ &\quad + \sum_{i\in J} \sum_{k=1}^{\infty} \overline{\Pi}(\boldsymbol{n} + k\boldsymbol{e}_i) p_i (c_{i,0}(k) + \overline{B}_i(k+1) 1_{\{0\}}(n_i)) \\ &\quad + \sum_{j\in S_+(\boldsymbol{n})} \sum_{m=1}^{\infty} \overline{\Pi}((\boldsymbol{n} - m\boldsymbol{e}_j)^+) p_0 c_{0,j}(m)\,, \end{aligned} \tag{5.6}$$

where $\overline{p} = \sum_{i\in J_0} p_i$, $S_0(\boldsymbol{n}) = \{i \in J | n_i = 0\} (\equiv J \setminus S_+(\boldsymbol{n}))$, $u(n) = 0$ if $n = 0$ and $u(n) = \infty$ otherwise, and $\boldsymbol{n}^+ = \max(\boldsymbol{n}, \boldsymbol{o})$, where vectors are ordered in component-wise. The following remark is important to get a bound for $\overline{\Pi}$.

Remark 5.2.1 From (5.5), the stationary distribution π and hence $\overline{\Pi}$ are unchanged, if, for a positive constant η, all p_i's are changed to ηp_i for $i \in J_0$, and if the value of $p(\boldsymbol{n}, \boldsymbol{n})$ is changed to $1 - \eta\overline{p} + \sum_{i\in S_0(\boldsymbol{n})} \eta p_i$.

Remark 5.2.2 Similarly to the case of the discrete-time Jackson network, we can construct the corresponding departure slotted model for the Jackson network with batch movements. Denote its stationary distribution by π'. Similarly to (5.3), we have, for $\boldsymbol{n} \neq \boldsymbol{o}$ and for a normalizing constant C,

$$\begin{aligned} C\pi'(\boldsymbol{n}) &= \sum_{i\in J} \sum_{k=1}^{\infty} \Big(\sum_{j\in S_+(\boldsymbol{n})} \sum_{m=1}^{n_j} \pi(\boldsymbol{n} + k\boldsymbol{e}_i - m\boldsymbol{e}_j) p_i c_{i,j}(k,m) \\ &\qquad + \pi(\boldsymbol{n} + k\boldsymbol{e}_i) p_i (c_{i,0}(k) + \overline{B}_i(k+1) 1_{\{0\}}(n_j)) \Big) \\ &= (p_0 + \sum_{i\in S_+(\boldsymbol{n})} p_i)\pi(\boldsymbol{n}) - \sum_{j\in S_+(\boldsymbol{n})} \sum_{m=1}^{n_j} \pi(\boldsymbol{n} - m\boldsymbol{e}_j) p_0 c_{0,j}(m), \end{aligned}$$

where the second equality is obtained by (5.5). It is easy to see that $C = \sum_{\boldsymbol{n}\in Z_0^N} \sum_{i\in S_+(\boldsymbol{n})} p_i \pi(\boldsymbol{n}) \geq (\sum_{i\in J} p_i)\overline{\Pi}(\boldsymbol{1})$, where $\boldsymbol{1}$ is the vector in Z_+^N whose components all equal to 1. Hence, the tail probability function $\overline{\Pi}'$ for π' is given by

$$\overline{\Pi}'(\boldsymbol{n}) = C^{-1} \left(\overline{p}\,\overline{\Pi}(\boldsymbol{n}) - \sum_{j\in S_+(\boldsymbol{n})} \sum_{m=1}^{\infty} \overline{\Pi}((\boldsymbol{n} - m\boldsymbol{e}_j)^+) p_0 c_{0,j}(m) \right)$$

$$\leq C^{-1} \sum_{i \in J} p_i \overline{\Pi}(\boldsymbol{n}) \leq \frac{\overline{\Pi}(\boldsymbol{n})}{\overline{\Pi}(\boldsymbol{1})}, \tag{5.7}$$

for $\boldsymbol{n} \in Z_+^N$. Thus a bound for $\overline{\Pi}'$ can be obtained through the one for $\overline{\Pi}$.

5.3 Stochastic Bound

Similar to Miyazawa and Taylor [11], we modify the discrete-time Jackson network with batch movements by adding extra batch arrivals to each node when it is empty. For node i, this extra arrivals occur with probability p_i^+, and its batch size distribution is denoted by A_i^+ with the probability mass function a_i^+. These p_i^+ and A_i^+ are determined later. As in Miyazawa and Taylor [11], we assume the following finiteness conditions concerning batch size distributions.

(a) B_i has a finite first moment m_{B_i} for all $i \in J$.

(b) For each $i, j \in J_0, k \in Z_+$, there exists a $z > 1$ such that $\sum_{m=1}^{\infty} c_{i,j}(k,m) z^m < \infty$.

The queue length vector process of the above modified network is a Markov chain, and has the following transition probability function p^*.

$$\begin{aligned}
p^*(\boldsymbol{n}, \boldsymbol{n} - k\boldsymbol{e}_i + m\boldsymbol{e}_j) &= p_i c_{i,j}(k,m) && (1 \leq k \leq n_i, m \geq 1) \\
p^*(\boldsymbol{n}, \boldsymbol{n} - k\boldsymbol{e}_i) &= p_i (c_{i,0}(k) + \overline{B}_i(n_i + 1) 1_{\{n_i\}}(k)) && (1 \leq k \leq n_i) \\
p^*(\boldsymbol{n}, \boldsymbol{n} + m\boldsymbol{e}_j) &= p_0 c_{0,j}(m) + p_j^+ a_j^+(m) 1_{\{n_j\}}(0) && (m \geq 1) \\
p^*(\boldsymbol{n}, \boldsymbol{n}) &= 1 - \overline{p} + \sum_{i \in S_0(\boldsymbol{n})} (p_i - p_i^+) .
\end{aligned}$$

From the last equation of the above formulas, p_i^+ should be sufficiently small so that $p^*(\boldsymbol{n}, \boldsymbol{n})$ is nonnegative. This feature is different from the continuous-time setting. We will verify this later by rescaling probabilities p_i according to Remark 2.1, but we just assume it in the following lemma.

Lemma 5.3.1 Suppose there exist nonnegative numbers $\gamma_j(m)$ for $j \in J$ and $m \in Z_0$ and $z_i < 1$ for $i \in J$ satisfying the following two sets of equations:

$$\gamma_j(m) = p_0 c_{0,j}(m) + \sum_{i \in J} p_i \sum_{k=1}^{\infty} z_i^k c_{i,j}(k,m) \tag{5.8}$$

$$\tilde{\Gamma}_j(z_j^{-1}) - \tilde{\Gamma}_j(1) = p_j(1 - \tilde{B}_j(z_j)), \tag{5.9}$$

where $\tilde{\Gamma}_j(z) = \sum_{m=1}^{\infty} \gamma_j(m) z^m$ and $\tilde{B}_j(z) = \sum_{m=1}^{\infty} b_j(m) z^m$. If we choose p_i^+ and a_i^+ $(i \in J)$ such that

$$p_i^+ a_i^+(n) = \sum_{m=n+1}^{\infty} \gamma_i(m) z_i^{n-m} , \tag{5.10}$$

and if $p^*(\boldsymbol{n}, \boldsymbol{n}) \geq 0$ for all $\boldsymbol{n} \in Z_0^N$, then the stationary distribution π^* of p^* is given by

$$\pi^*(\boldsymbol{n}) = \prod_{i=1}^{N} (1 - z_i) z_i^{n_i} . \tag{5.11}$$

Remark 5.3.2 (5.8) is interpreted as the traffic equation, while (5.9) determines the decay rate of the queue length distribution at node i. It can be seen that (5.9) has a solution z_j such that $0 < z_j < 1$ for a fixed $\overline{\Gamma}_j$ if and only if $\overline{\Gamma}_j(1) < p_j \tilde{B}_j'(1)$.

Proof. This lemma can be proved in the same way as Theorem 3.1 of Miyazawa and Taylor [11], which uses the time-reversed arguments. We here prove the lemma by directly deriving the balance of probability flow. Denote the probability flow out of state $\boldsymbol{n}$ and into it by $\alpha_{\text{out}}(\boldsymbol{n})$ and $\alpha_{\text{in}}(\boldsymbol{n})$, respectively, under the distribution π^*. We only need to prove $\alpha_{\text{out}}(\boldsymbol{n}) = \alpha_{\text{in}}(\boldsymbol{n})$ for all $\boldsymbol{n} \in Z_0^N$. From the definition of p^* and (5.10), we get

$$\begin{aligned}
\frac{\alpha_{\text{out}}(\boldsymbol{n})}{\pi^*(\boldsymbol{n})} &= \sum_{i \in S_+(\boldsymbol{n})} \sum_{j \in J} \sum_{k=1}^{n_i} \sum_{m=1}^{\infty} p_i c_{i,j}(k,m) \\
&\quad + \sum_{i \in S_+(\boldsymbol{n})} \sum_{k=1}^{n_i} p_i (c_{i,0}(k) + \overline{B}_i(n_i+1) 1_{\{n_i\}}(k)) \\
&\quad + \sum_{j \in J} \sum_{m=1}^{\infty} p_0 (c_{0,j}(m) + p_j^+ a_j^+(m) 1_{\{n_j\}}(0)) \\
&= p_0 + \sum_{i \in S_+(\boldsymbol{n})} p_i + \sum_{i \in S_0(\boldsymbol{n})} \left(\frac{z_i \tilde{\Gamma}_i(z_i^{-1}) - \tilde{\Gamma}_i(1)}{1 - z_i} \right) .
\end{aligned} \tag{5.12}$$

On the other hand, from (5.11) and referring the right-hand side of (5.5), we get

$$\begin{aligned}
\frac{\alpha_{\text{in}}(\boldsymbol{n})}{\pi^*(\boldsymbol{n})} &= \sum_{i \in J} \sum_{j \in S_+(\boldsymbol{n})} p_i \sum_{k=1}^{\infty} \sum_{m=1}^{n_j} c_{i,j}(k,m) z_i^k z_j^{-m} \\
&\quad + \sum_{i \in J} p_i \sum_{k=1}^{\infty} c_{i,0}(k) z_i^k + \sum_{i \in S_0(\boldsymbol{n})} p_i \sum_{k=1}^{\infty} \overline{B}_i(k+1) z_i^k
\end{aligned}$$

$$+ \sum_{j \in S_+(\boldsymbol{n})} \left(p_0 \sum_{m=1}^{n_j} c_{0,j}(m) z_j^{-m} + p_j^+ a_j^+(n_j) z_j^{-n_j} \right). \quad (5.13)$$

From (5.8), the first term of the right-hand side of (5.13) becomes;

$$\sum_{j \in S_+(\boldsymbol{n})} \sum_{m=1}^{\infty} (\gamma_j(m) - p_0 c_{0,j}(m)) z_j^{-m} = \sum_{j \in S_+(\boldsymbol{n})} \left(\tilde{\Gamma}_j(z_j^{-1}) - p_0 \tilde{c}_{0,j}(z_j^{-1}) \right).$$

where $\tilde{c}_{0,j}(z) = \sum_{m=1}^{\infty} c_{0,j}(m) z^m$. Its second, third and fourth terms become, respectively:

$$\sum_{i \in J} p_i \sum_{k=1}^{\infty} c_{i,0}(k) z^k = \sum_{i \in J} p_i \sum_{k=1}^{\infty} z^k \left(b_i(k) - \sum_{j \in J} \sum_{m=1}^{\infty} c_{i,j}(k,m) \right),$$

$$\sum_{i \in S_0(\boldsymbol{n})} p_i \sum_{k=1}^{\infty} \tilde{B}_i(k+1) z_i^k = \sum_{i \in S_0(\boldsymbol{n})} p_i \left(z_i \frac{1 - \tilde{B}_i(z_i)}{1 - z_i} - \tilde{B}_i(z_i) \right),$$

$$\sum_{j \in S_+(\boldsymbol{n})} \left(p_0 \sum_{m=1}^{n_j} c_{0,j}(m) z_j^{-m} + p_j^+ a_j^+(n_j) z_j^{-n_j} \right) = \sum_{j \in S_+(\boldsymbol{n})} p_0 \tilde{c}_{0,j}(z_j^{-1}).$$

Thus, using (5.8) again, the right-hand side of (5.13) is calculated to:

$$\sum_{j \in S_+(\boldsymbol{n})} \left(\tilde{\Gamma}_j(z_j^{-1}) - \tilde{\Gamma}_j(1) \right) - \sum_{j \in S_0(\boldsymbol{n})} \left(p_i z_i \frac{1 - \tilde{B}_i(z_i)}{1 - z_i} - \tilde{\Gamma}_i(1) \right)$$
$$+ \sum_{j \in S_+(\boldsymbol{n})} p_i \tilde{B}_i(z_i) + p_0$$
$$= \sum_{j \in S_+(\boldsymbol{n})} \left(\tilde{\Gamma}_j(z_j^{-1}) - \tilde{\Gamma}_j(1) - \left(p_j 1 - \tilde{B}_j(z_j) \right) \right)$$
$$+ p_0 + \sum_{i \in S_+(\boldsymbol{n})} p_i - \sum_{i \in S_0(\boldsymbol{n})} p_i \left(z_i \frac{1 - \tilde{B}_i(z_i)}{1 - z_i} - \tilde{\Gamma}_i(1) \right).$$

Hence, (5.9) leads to $\alpha_{\text{out}}(\boldsymbol{n}) = \alpha_{\text{in}}(\boldsymbol{n})$. This completes the proof.

Remark 5.3.3 Since (5.9) is used only at the final stage of the proof, it is not hard to see that (5.9) is also necessary for the stationary distribution to have a geometric product, provided that (5.8) holds. On the other hand, $p_j^+ a_j^+(n)$ does not necessarily have the form (5.10). See Section 3 of Miyazawa and Taylor [11] for the details concerning this.

Lemma 5.3.4 If the values of p_i is changed to ηp_i for all $i \in J_0$ for a constant η such that $0 < \eta < 1$, and if the transition function p is changed for this η as in Remark 5.2.1, then p_i^+ is changed to ηp_i^+ for all $i \in J$ in Lemma 3.1.

Proof. From (5.9) and (5.10), it is easy to see z_i does not depend on η, so (5.9) and (5.10) implies that p_i^+ is changed to ηp_i^+.

After modifying the transition function p as in Lemma 3.4 for a sufficiently small η, we clearly have $p^*(\boldsymbol{n}, \boldsymbol{n}) \geq 0$. So the nonnegativity condition in Lemma 3.1 is indeed satisfied.

Remark 5.3.5 One might be interested in how large p_i^+ is. This is evaluated by using (5.9) and (5.10). That is, we have

$$p_i^+ = \frac{z_i\tilde{\Gamma}_i(z_i^{-1}) - \tilde{\Gamma}_i(1)}{1 - z_i} = \frac{p_i z_i\left(1 - \tilde{B}_i(z_i)\right)}{1 - z_i} - \tilde{\Gamma}_i(1) .$$

Hence, easy bounds are given by

$$p_i^+ \leq p_i z_i m_{B_i} - \tilde{\Gamma}_i(1) \leq p_i m_{B_i} - p_0 \sum_{m=1}^{\infty} c_{0,i}(m) .$$

From the middle term, one can see that the extra arrival rates given the system being empty are small for both of light traffic and heavy traffic situations.

Since the extra arrivals increase the customer populations in all nodes, π^* is stochastically not less than π under the modification by η. Furthermore, from Remark 5.2.1 and Lemma 3.4, the stationary distribution π and π^* are unchanged under the modification of p and p^*, respectively. Hence we have obtained the following theorem.

Theorem 5.3.6 Under assumptions (a) and (b), if there exist nonnegative numbers $\gamma_j(m)$ for $j \in J$ and $m \in Z_0$ and $z_i < 1$ for $i \in J$ satisfying (5.8) and (5.9), then π is stochastically not greater than π^* of (5.11), i.e.

$$\overline{\Pi}(\boldsymbol{n}) \leq \overline{\Pi}^*(\boldsymbol{n}) \left(\equiv \prod_{i=1}^{N} z_i^{n_i} \right) \qquad (\boldsymbol{n} \in Z_0^N) . \tag{5.14}$$

Remark 5.3.7 From (5.7), we have, for the departure slotted model,

$$\overline{\Pi}'(\boldsymbol{n}) \leq \frac{\overline{\Pi}^*(\boldsymbol{n})}{\overline{\Pi}(\boldsymbol{1})} \qquad (\boldsymbol{n} \in Z_0^N) . \tag{5.15}$$

The sharper bound for $\overline{\Pi}'$ can be obtained through the observation that the stationary distribution of the queue length vector in the departure slotted model is stochastically not greater than the corresponding distribution of the modified network (see (5.4)).

5.4 Improvement and Tightness

In this section, we consider to improve the upper bound obtained in Theorem 3.6. Let us define the difference between the exact value and the bound by

$$\delta(\boldsymbol{n}) = \overline{\Pi}^*(\boldsymbol{n}) - \overline{\Pi}(\boldsymbol{n}) .$$

We first note that $\overline{\Pi}^*$ satisfies, for $\boldsymbol{n} \in Z_0^N$,

$$\begin{aligned}
\overline{p}\,\overline{\Pi}^*(\boldsymbol{n}) &= \left(\sum_{i\in S_0(\boldsymbol{n})} (p_i - p_i^+)\right) \prod_{i\in S_0(\boldsymbol{n})} (1-z_i) \prod_{j\in S_+(\boldsymbol{n})} z_j^{n_j} \\
&\quad + \sum_{i\in J} \sum_{j\in S_+(\boldsymbol{n})} \sum_{k=1}^{\infty} \sum_{m=1}^{\infty} \overline{\Pi}^*((\boldsymbol{n} + k\boldsymbol{e}_i - m\boldsymbol{e}_j)^+) p_i c_{i,j}(k,m) \\
&\quad + \sum_{i\in J} \sum_{k=1}^{\infty} \overline{\Pi}^*(\boldsymbol{n} + k\boldsymbol{e}_i) p_i (c_{i,0}(k) + \overline{B}_i(k+1) 1_{\{n_i\}}(0)) \\
&\quad + \sum_{j\in S_+(\boldsymbol{n})} \sum_{m=1}^{\infty} \overline{\Pi}^*((\boldsymbol{n} - m\boldsymbol{e}_j)^+) p_0 c_{0,j}(m) \\
&\quad + \sum_{j\in S_+(\boldsymbol{n})} \overline{\Pi}^*((\boldsymbol{n} - n_j\boldsymbol{e}_j)^+)(1-z_j) \sum_{m=n_j}^{\infty} p_j^+ a_j^+(m), \qquad (5.16)
\end{aligned}$$

similarly to (5.6). Hence, subtracting both sides of (5.6) from those of (5.16), we have, for $\boldsymbol{n} \in Z_+^N$,

$$\begin{aligned}
\overline{p}\,\delta(\boldsymbol{n}) &= \sum_{i\in J} \sum_{j\in J} \sum_{k=1}^{\infty} \sum_{m=1}^{\infty} \delta((\boldsymbol{n} + k\boldsymbol{e}_i - m\boldsymbol{e}_j)^+) p_i c_{i,j}(k,m) \\
&\quad + \sum_{i\in J} \sum_{k=1}^{\infty} \delta(\boldsymbol{n} + k\boldsymbol{e}_i) p_i c_{i,0}(k) \\
&\quad + \sum_{j\in J} \sum_{m=1}^{\infty} \delta((\boldsymbol{n} - m\boldsymbol{e}_j)^+) p_0 c_{0,j}(m) \\
&\quad + \sum_{j\in J} \overline{\Pi}^*((\boldsymbol{n} - n_j\boldsymbol{e}_j)^+)(1-z_j) \sum_{m=n_j}^{\infty} p_j^+ a_j^+(m). \qquad (5.17)
\end{aligned}$$

Remark 5.4.1 Equation (5.17) is similar to (5.16) for $\boldsymbol{n} \in Z_+^N$, but δ and $\overline{\Pi}^*$ are different. In particular, $\delta(\mathbf{o}) = 0$, while $\overline{\Pi}^*(\mathbf{o}) = 1$.

We improve the upper bound $\overline{\Pi}^*(\boldsymbol{n})$ by deriving a lower bound of $\delta(\boldsymbol{n})$. From (5.17), we easily see, for $\boldsymbol{n} \in Z_+^N$,

$$\begin{aligned} \delta(\boldsymbol{n}) &\geq \frac{1}{\overline{p}} \overline{\Pi}^*(\boldsymbol{n}) \sum_{j \in J} p_j^+ (1 - z_j) \sum_{m=n_j}^{\infty} a_j^+(m) z_j^{-n_j} \\ &= \frac{1}{\overline{p}} \overline{\Pi}^*(\boldsymbol{n}) \sum_{j \in J} \sum_{m=n_j}^{\infty} \left(z_j^{-m} - z_j^{-n_j} \right) \gamma_j(m). \end{aligned} \tag{5.18}$$

Hence we get:

$$\Pi(\boldsymbol{n}) \leq \left(1 - \frac{1}{\overline{p}} \sum_{j \in J} \sum_{m=n_j}^{\infty} \left(z_j^{-m} - z_j^{-n_j} \right) \gamma_j(m) \right) \overline{\Pi}^*(\boldsymbol{n}). \tag{5.19}$$

This bound does not improve the limiting coefficient of $\overline{\Pi}^*(\boldsymbol{n})$, i.e. the coefficient of $\overline{\Pi}^*(\boldsymbol{n})$ in (5.19) converges to 1 as $\boldsymbol{n}$ tends to infinity. Note that the bound of (5.19) can be further improved by repeatedly substituting the obtained bounds into δ's in the right-hand side of (5.17), in which we substitute 0 for $\delta(\boldsymbol{n})$ for which there exists a $j \in J$ such that $n_j = 0$. These bounds for δ increases by iteration, so it converges to a constant value, which gives the best obtainable bound for $\overline{\Pi}$ by this method. However, it seems hard to evaluate this limit in closed form.

From the view of large deviation, it is interesting to see whether the geometric product form bound $\overline{\Pi}^*$ is tight, i.e. there is no better geometric bound. We conjecture this because the extra arrivals are only added when nodes are empty. However, except for the single node case, it seems very hard to verify it. The single node network without feedback loop can be reduced to the waiting time process in the $GI/G/1$ queue in which the interarrival time is correlated to the service time in a certain way, and we can see that the decay rate is the best (cf. Corollary 2.1 of [11]). In the following, we prove this fact by evaluating the limiting coefficient of $\overline{\Pi}^*(\boldsymbol{n})$. In the view of Lemma 3.4, we can assume $\overline{p} = 1$ without loss of generality, which is used in the following arguments.

For the single node network without feedback loop, (5.17) becomes for $n > 0$,

$$\begin{aligned} \delta(n) &= p_1 \sum_{m=1}^{\infty} \delta(n+m) b(m) + p_0 \sum_{m=1}^{\infty} \delta((n-m)^+) a(m) \\ &\quad + p_1^+ (1 - z) \sum_{m=n}^{\infty} a^+(m), \end{aligned} \tag{5.20}$$

where $z = z_1$, $b(m) = c_{1,0}(m)$, $a(m) = c_{0,1}(m)$ and $a^+(m) = a_1^+(m)$. Define $\tilde{A}(x) = \sum_{m=1}^{\infty} a(m) x^m$, $\tilde{B}(x) = \sum_{m=1}^{\infty} b(m) x^m$ and $\tilde{\Delta}(x) = \sum_{m=1}^{\infty} \delta(m) x^m$.

Multiplying x^n to both sides of (5.20), we have

$$\tilde{\Delta}(x) = p_1\tilde{\Delta}(x)\tilde{B}(x^{-1}) - p_1\sum_{m=1}^{\infty}\sum_{n=1}^{m}\delta(n)x^n b(m)x^{-m} + p_0\tilde{\Delta}(x)\tilde{A}(x)$$
$$+p_0\frac{(1-z)x}{1-x}\left(\frac{z\tilde{A}(z^{-1})-1}{1-z} - \frac{zx\tilde{A}(z^{-1})-\tilde{A}(x)}{1-zx}\right). \quad (5.21)$$

Hence, we get

$$\tilde{\Delta}(x) \le \frac{p_0\frac{(1-z)x}{1-x}\left(\frac{z\tilde{A}(z^{-1})-1}{1-z} - \frac{zx\tilde{A}(z^{-1})-\tilde{A}(x)}{1-zx}\right)}{1-p_0\tilde{A}(x)-p_1\tilde{B}(x^{-1})},$$

which yields, by applying Lopital's rule,

$$\limsup_{x\uparrow z^{-1}}(1-xz)\tilde{\Delta}(x) \le \frac{p_0\left(\tilde{A}'(z^{-1}) - \frac{z}{1-z}\left(\tilde{A}(z^{-1})-1\right)\right)}{p_0\tilde{A}'(z^{-1})-p_1z^2\tilde{B}'(z)}. \quad (5.22)$$

On the other hand, since

$$\tilde{B}'(z) \le \frac{1-\tilde{B}(z)}{1-z},$$

and $z \in (0,1)$ is the solution of the equation (5.9):

$$p_0\left(\tilde{A}(z^{-1})-1\right) = p_1\left(1-\tilde{B}(z)\right),$$

we have

$$p_1z^2\tilde{B}'(z) < zp_1\tilde{B}'(z) \le p_1\frac{z\left(1-\tilde{B}(z)\right)}{1-z} = p_0\frac{z\left(\tilde{A}(z^{-1})-1\right)}{1-z}.$$

Note that $p_0\tilde{A}'(z^{-1}) - p_1z^2\tilde{B}'(z) > 0$, because $p_0\tilde{A}(x)+p_1\tilde{B}(x^{-1})$ is a convex function for $x \in ((1-\epsilon)z^{-1}, 1]$ for an $\epsilon < 1$ (by (b)). The above inequality implies that the right-hand side of (5.22) is less than 1. That is, $\delta(n)$ has an order not greater than z^n, and the limiting coefficient of z^n is less than 1. Therefore, we have the following result.

Proposition 5.4.2 For the single note network with batch movements, $\overline{\Pi}^*$ gives the best geometric bound, under assumptions (a) and (b), and the lower bound of the limiting coefficient is given by

$$\liminf_{n\to\infty}\frac{\overline{\Pi}(n)}{\overline{\Pi}^*(n)} \ge \frac{\frac{p_0z}{1-z}\left(\tilde{A}(z^{-1})-1\right) - p_1z^2\tilde{B}'(z)}{p_0\tilde{A}'(z^{-1})-p_1z^2\tilde{B}'(z)} > 0. \quad (5.23)$$

Remark 5.4.3 By using the lower bound (5.18), we can improve (5.23) by adding the following term to the numerator.

$$p_1 \sum_{m=1}^{\infty} \sum_{n=1}^{m} \sum_{k=n}^{\infty} (z^{m-k} - z^{m-n}) a(k) b(m)$$

If $a(k)$ and $b(m)$ are positive only for small k, m, we may be able to calculate this, but it is generally difficult to do so.

The above argument breaks down even for the single node network if there is a feedback loop. There is another way to verify the tightness of the bound $\overline{\Pi}^*$. From (5.17), we have, for $n > 0$

$$\begin{aligned} \delta(n) &= p_1 \sum_{k=1}^{\infty} \sum_{m=1}^{n} \delta(n+k-m) c_{1,1}(k,m) + p_1 \sum_{k=1}^{\infty} \delta(n+k) c_{1,0}(k) \\ &\quad + p_0 \sum_{m=1}^{n-1} \delta(n-m) c_{0,1}(m) + (1-z_1) p_1^+ \sum_{m=n}^{\infty} a_1^+(m). \end{aligned} \tag{5.24}$$

Define $\delta^{(l)}$ $(l = 1, 2, \cdots)$ recursively by

$$\begin{aligned} \delta^{(l+1)}(n) &= p_1 \sum_{k=1}^{\infty} \sum_{m=1}^{n} \delta^{(l)}(n+k-m) c_{1,1}(k,m) \\ &\quad + p_1 \sum_{k=1}^{\infty} \delta^{(l)}(n+k) c_{1,0}(k) + p_0 \sum_{m=1}^{n-1} \delta^{(l)}(n-m) c_{0,1}(m) \\ &\quad + (1-z_1) p_1^+ \sum_{m=n}^{\infty} a_1^+(m). \end{aligned}$$

We put $\delta^{(l)}(0) = \delta(0) = 0$. Then, it is easily seen that $\delta^{(l)}$ monotonically converges to δ as l tends to infinity. Clearly we can take $\delta^{(1)}(n) = 0$ for all $n \geq 0$, which implies that $\delta^{(l)}$ is non-decreasing. We apply the arguments due to Kingman [9] for the inequality concerning the $GI/G/1$ type queues. That is, suppose that there exists a function h such that

$$\begin{aligned} h(n) &\geq p_1 \sum_{k=1}^{\infty} \sum_{m=1}^{n} h(n+k-m) c_{1,1}(k,m) + p_1 \sum_{k=1}^{\infty} h(n+k) c_{1,0}(k) \\ &\quad + p_0 \sum_{m=1}^{n-1} h(n-m) c_{0,1}(m) + (1-z_1) p_1^+ \sum_{m=n}^{\infty} a_1^+(m). \end{aligned} \tag{5.25}$$

Then, we can see that $\delta^{(l)}(n) \leq h(n)$, and hence $\delta(n) \leq h(n)$ for all $l \geq 1, n \geq 0$. Let $h(n) = C_1 z_1^n$ for a positive constant C_1, and substitute it to

(5.25). By reminding that $\overline{\Pi}^*(n)(= z_1^n)$ satisfies (5.16), we have

$$C_1 z_1^n \geq C_1 \left(z_1^n - p_1 \sum_{k=1}^{\infty} \sum_{m=n+k}^{\infty} c_{1,1}(k,m) - p_0 \sum_{m=n}^{\infty} c_{0,1}(m) - (1-z_1)p_1^+ \sum_{m=n}^{\infty} a_1^+(m) \right) + (1-z_1)p_1^+ \sum_{m=n}^{\infty} a_1^+(m).$$

This yields:

$$C_1 \geq \frac{(1-z_1)p_1^+ \sum_{m=n}^{\infty} a_1^+(m)}{p_1 \sum_{k=1}^{\infty} \sum_{m=n+k}^{\infty} c_{1,1}(k,m) + p_0 \sum_{m=n}^{\infty} c_{0,1}(m) + (1-z_1)p_1^+ \sum_{m=n}^{\infty} a_1^+(m)}.$$

Note that, if the numerator of the above equation is zero, we can choose an arbitrary C_1 for (5.25). Hence, if there are only finite numbers of m such that $c_{1,1}(k,m)$ or $c_{0,1}(m)$ are positive, we can find a lower bound for C_1 such that $0 < C_1 < 1$. That is, (5.25) is satisfied by $h(n) = C_1 z_1^n$ for this C_1. This concludes the following result.

Proposition 5.4.4 For the single node network with a feedback loop, if the arrival and feedback batch sizes are bounded, $\overline{\Pi}^*$ gives the best geometric bound.

Although we have no results for the multiple node network, Propositions 4.3 and 4.4 suggests that $\overline{\Pi}^*$ gives the best bound in general.

5.5 Algorithm for Calculating Geometric Rates

Lemma 3.1 tells us that the decay rates of the product form geometric distribution can be determined by solving (5.8) and (5.9) with respect to z_i, which simultaneously gives sufficient conditions for the stability. However, those equations are highly non-linear in general. Usually, one applies the fixed point theorem for such non-linear equations (cf. Gelenbe and Schassberger [5]). In [11], such an approach was taken, but it only succeeded under certain restrictions. We here take another look at the problem. That is, we consider an algorithm to get z_i numerically. This might be more realistic in actual applications. Because this algorithm is applicable for non-linear traffic equations in general, we discuss it under a general setting.

Let $\alpha_i(k)$ and $\beta_i(k)$ be the arrival and departure rates (or probabilities for the discrete-time case) of type k customers at node i in a queueing network, which have routing probabilities $\{r((i,k),(j,l))\}$ and arrival rates (or probabilities) $\lambda_i(k)$ from the outside. For notational convenience, we introduce vector descriptions, $\boldsymbol{\alpha} = \{\alpha_i(k)\}$, $\boldsymbol{\beta} = \{\beta_i(k)\}$, $\boldsymbol{\lambda} = \{\lambda_i(k)\}$, and matrix $R = \{r((i,k),(j,l))\}$, where R presents the routing inside of the network. Then, the traffic equations become:

$$\boldsymbol{\alpha} = \boldsymbol{\lambda} + \boldsymbol{\beta} R \,. \tag{5.26}$$

On the other hand, we suppose that $\beta_i(k)$ is a function of $\boldsymbol{\alpha}$, i.e. $\phi_{i,k}(\boldsymbol{\alpha})$, and denote its vector version by $\boldsymbol{\phi}(\boldsymbol{\alpha})$. That is, $\boldsymbol{\beta}$ is determined by

$$\boldsymbol{\beta} = \boldsymbol{\phi}(\boldsymbol{\alpha}) \,. \tag{5.27}$$

(5.26) and (5.27) yields the following equation for $\boldsymbol{\alpha}$:

$$\boldsymbol{\alpha} = \boldsymbol{\lambda} + \boldsymbol{\phi}(\boldsymbol{\alpha}) R \,. \tag{5.28}$$

The problem is to find a solution for (5.28) (then $\boldsymbol{\beta}$ is obtained from (5.27)). Our key assumption is:

(c-1) $\boldsymbol{\phi}(\boldsymbol{\alpha})R$ is non-decreasing in $\boldsymbol{\alpha}$,

where vectors are ordered in component-wise. (c-1) is plausible in the sense that more input yields more output. In many cases, the stronger condition that $\boldsymbol{\phi}(\boldsymbol{\alpha})$ is non-decreasing holds. We will see this for the model in Section 4 later. We define $\boldsymbol{\alpha}^{(n)}$ inductively by

$$\begin{aligned} \boldsymbol{\alpha}^{(0)} &= \mathbf{o} \,, \\ \boldsymbol{\alpha}^{(n)} &= \boldsymbol{\lambda} + \boldsymbol{\phi}(\boldsymbol{\alpha}^{(n-1)})R \qquad (n = 1, 2, \cdots) \,. \end{aligned}$$

Since

$$\boldsymbol{\alpha}^{(n+1)} - \boldsymbol{\alpha}^{(n)} = \left(\boldsymbol{\phi}(\boldsymbol{\alpha}^{(n)}) - \boldsymbol{\phi}(\boldsymbol{\alpha}^{(n-1)})\right) R \,,$$

$\boldsymbol{\alpha}^{(n)}$ is non-decreasing in n, so its limit exists, which may be infinite. We denote this limit by $\boldsymbol{\alpha}^{(\infty)}$.

Let us define the mean value vector $\overline{\boldsymbol{\alpha}} = \{\sum_k k\alpha_i(k)\}$. Similarly, $\overline{\boldsymbol{\alpha}^{(\infty)}}$ and $\overline{\boldsymbol{\mu}_b}$ are defined, where $(\mu_b)_i(k) = \mu_i b_i(k)$, which is the departure rate of type k customers at node i, when sufficiently many type k customers are in node i. We suppose

(c-2) The system is stable, i.e., the queue length vector has the stationary distribution if and only if there exists a solution $\boldsymbol{\alpha}$ of (5.28) such that $\overline{\boldsymbol{\alpha}} < \overline{\boldsymbol{\mu}_b}$.

For examining the above stability condition, we make one technical assumption:

(c-3) $\lim_{\boldsymbol{h} \downarrow \mathbf{o}} \boldsymbol{\phi}(\boldsymbol{\alpha} - \boldsymbol{h}) = \boldsymbol{\phi}(\boldsymbol{\alpha})$ if $\overline{\boldsymbol{\alpha}} < \overline{\boldsymbol{\mu}_b}$ and $\boldsymbol{\alpha} > \mathbf{o}$.

This upper continuity is easily verified in many cases. Then the question is answered in the following way. First note that, if (5.28) has a solution $\boldsymbol{\alpha}$, $\boldsymbol{\alpha} \geq \boldsymbol{\alpha}^{(\infty)}$ due to the definition of $\boldsymbol{\alpha}^{(\infty)}$ and the monotonicity (c-1). Then either one of the following cases holds.

(Case 1) $\overline{\boldsymbol{\alpha}^{(\infty)}} \not< \overline{\boldsymbol{\mu}_b}$: Since $\boldsymbol{\alpha} \geq \boldsymbol{\alpha}^{(\infty)}$ if $\boldsymbol{\alpha}$ exists, there is no solution such that $\overline{\boldsymbol{\alpha}} < \overline{\boldsymbol{\mu}_b}$.

(Case 2) $\overline{\boldsymbol{\alpha}^{(\infty)}} < \overline{\boldsymbol{\mu}_b}$: From the assumption (c-2) and the monotone convergence theorem, we have

$$\boldsymbol{\alpha}^{(\infty)} = \boldsymbol{\lambda} + \boldsymbol{\phi}(\boldsymbol{\alpha}^{(\infty)})R\,. \tag{5.29}$$

Hence, $\boldsymbol{\alpha} = \boldsymbol{\alpha}^{(\infty)}$ is the desired solution.

This method is easily implemented as computation software. All the work we need is to verify conditions (c-1), (c-2) and (c-3). Let us consider them for the modified model of Section 3 for the discrete-time Jackson network. We set $\beta_i(k) = p_i z_i^k b_i(k)$, $\lambda_i(k) = p_0 c_{0,j}(k)$ and $\alpha_i(k) = \gamma_i(k)$. Then, (5.26) exactly corresponds with (5.8). On the other hand, (5.27) corresponds with (5.9). Clearly z_i's are continuous with respect to $\gamma_i(k)$, so condition (c-3) is satisfied, while Remark 3.2 leads to (c-2). It is not hard to see (c-1) through (5.9). Thus all the conditions hold for the modified network of Section 3.

5.6 Discussions

There are two theoretical issues on the stochastic bound obtained in Section 3. First one is the reason that the bound is of geometric product form. As noted in [11], this comes from the quasi-reversibility of the nodes under separation. To make the nodes to be quasi-reversible (in the sense of discrete-time), we set two conditions. One is that only batches which attain the requested numbers of customers for departure are routed. Another is the extra arrivals when the nodes are empty, which is only used in the modified model so to get the bound. The former condition might be too strong, but we conjecture that the geometric decay rates are unchanged even if those incomplete batches are routed as well.

To get the quasi-reversibility, we regard regular batch arrivals (excluding extra batches) as arriving customers, and routed batch departures (excluding incomplete batches) as departing customers, where batch sizes are considered as types of customers (see [11] for their details). However, one should be careful to apply the conventional quasi-reversibility, since the nodes are not internally balanced, i.e. their arrival rates do not equal their departure rates, and state transitions due to regular arrivals are overlapped with those due to extra arrivals, which are counted as internal transitions. These issues on the quasi-reversibility are recently discussed in [4].

Another issue is an interpretation of the bound from the view of large deviation. For the geometric product form distribution Π^*, let $\theta_j \equiv -\log(z_j)$, which is called the exponential rate of the j-th marginal. Chang [3] recently proposed a three step algorithm to determine the exponential rates of nodes in intree networks, in which arrival batches are not necessarily independent. Since we assume batch sizes to be independent, the first step of the algorithm is obvious. It is not hard to see that the second and third steps of the algorithm correspond with (5.8) and (5.9), respectively. We here identify the total input stream of node j to be the batch Bernoulli process with batch size distribution γ_j/q_j and the mean rate $q_j \equiv \tilde{\Gamma}_j(1)$. Note that (5.8) is not exactly the same as (4) of [3], since all nodes simultaneously release output flows with constant rates in [3]. Those observations suggest that θ_j is the exponential rate, so the bound obtained in Section 3 is indeed tight.

Acknowledgments: The author is grateful to Peter G. Taylor for cooperation on queueing network problems, and to Karl Sigman for his hospitality during the workshop. This paper would not be completed without their help.

5.7 References

[1] Boucherie, R.J. and van Dijk N.M., Product forms for queueing networks with state-dependent multiple job transitions, *Adv. Appl. Prob.* 23 (1991), 152-187.

[2] Chang, C.S., Stability, queue length and delay of deterministic and stochastic queueing networks, *IEEE Trans. Autom. Cont.* 39 (1994), 913-931.

[3] Chang, C.S., Sample path large deviations and intree networks, *Queueing Systems* 20 (1995), 7-36.

[4] Chao, X. and Miyazawa, M., A Probabilistic Decomposition Approach to Quasi-Reversibility and Its Applications in Coupling of Queues, preprint, 1996.

[5] Gelenbe, E. and Schassberger, R., Stability of product form G-network, *Probability in the Engineering and Informational Sciences* 6 (1992), 271-276.

[6] Henderson, W. and Taylor, P.G., Product form in networks of queues with batch arrivals and batch services, *Queueing Systems* 6 (1990), 71-88.

[7] Kelly, F.P., *Reversibility and Stochastic Networks,* John Wiley & Sons, New York, 1979.

[8] KELLY, F.P., Effective bandwidths at multi-class queues *Queueing Systems* 9 (1991), 5-16.

[9] KINGMAN, J.F.C. Inequalities in the theory of queues, *J. Roy. Statist. Soc.* B. 32 (1970), 102-110.

[10] MIYAZAWA, M., On the characterization of departure rules for discrete-time queueing networks with batch movements and its applications, *Queueing Systems* 18 (1994), 149-166.

[11] MIYAZAWA, M. AND TAYLOR, P.G., A geometric product-form distribution for a queueing network with nonstandard batch arrivals and batch transfers, preprint, 1995.

[12] SERFOZO, R.F., Queueing networks with dependent nodes and concurrent movements, *Queueing Systems* 13 (1993), 143-182.

6

Stability for Queues with Time Varying Rates

William A. Massey

ABSTRACT An $M(t)/M(t)/1$ queue or $M/M/1$ queue with time varying rates, may alternate through periods of underloading, overloading, and critical loading. We analyze this model by using a general asymptotic method called *uniform acceleration*, which we will show is the appropriate time-varying analogue to steady state analysis. Applying this method to the transition probabilities of the queue length process, we obtain necessary and sufficient conditions for underloading which we will show is the time-varying analogue to steady state stability.

Using the theory of strong approximations. we can also apply a similar asymptotic analysis directly to the random sample paths of the queueing process. In obtaining a functional strong law of large numbers and central limit theorem for the $M(t)/M(t)/1$ queue, we obtain a rigorous basis for the fluid and diffusion approximations that are used to analyze this system. Moreover, the will be many candidates for the time-varying analogue to heavy traffic limit processes. The results are presented to suggest new methods for the asymptotic analysis of nonstationary, continuous time Markov chains.

6.1 The $M(t)/M(t)/1$ Queue and Poisson Processes

The initial motivation for developing limit theorems for queues with time varying rates is a simple one, queueing models with time-varying arrival and service rates are more realistic. Unfortunately much of the supporting theory for continuous time Markov chains with constant rates (or time homogeneous) does not work for the time-varying case or is no longer appropriate. Moreover the requisite analysis will be more difficult. A recurring theme in the study of queues with time-varying rates is that we must constantly reevaluate and reinterpret what should be analyzed.

Since the $M/M/1$ queue is one of the fundamental models of queueing theory we want to discuss its time-varying rate analogue, the $M(t)/M(t)/1$ queue as a pretext for developing a general analysis for queues with time-varying rates. Although the $M(t)/G/\infty$ queue is one of the few time-varying queueing systems that can be solved explicitly (see Prèkopa [17] as well as [13] with Whitt, [2] and [3] with Eick and Whitt for references to other papers in this area), we want to focus on a queueing model that can be made tractable only through the use of limit theorems that we will review in this paper.

Consider a queueing system where customers arrive to a single server but only one of them can receive service at any given time. For convenience, we can assume that the order in which the customers are serviced is according to the first-in first-out (FIFO) service discipline, but this will not affect the number of customers in the queueing system at any given time. Finally, besides the one customer being serviced, all the remaining customers wait their turn to be served in a waiting buffer that has infinite capacity. A telecommunications motivation for the $M(t)/M(t)/1$ queue is to model a switching node for transmitting messages. We will let $Q \equiv \{ Q(t) \mid t \geq 0 \}$ be the queueing process for this system where $Q(t)$ is the number of customers waiting in the buffer plus the one customer in service.

Now define $\Lambda \equiv \{ \Lambda(t) \mid t \geq 0 \}$ to be arrival process for this queue. This means that $\Lambda(t)$ equals the number of customers who arrived to the queue before and including time t. We will define $\lambda \equiv \{ \lambda(t) \mid t \geq 0 \}$ to be the mean arrival rate function for Λ, where assume that for all $t \geq 0$,

$$\mathsf{E}[\Lambda(t)] = \int_0^t \lambda(s)ds. \tag{6.1}$$

We will assume that Λ is a non-homogeneous Poisson process. This is simply a stochastic process with independent Poisson increments which means that for all $s < t$, we have

$$\mathsf{P}(\Lambda(t) - \Lambda(s) = n) = \frac{(\int_s^t \lambda(r)dr)^n}{n!} \exp\left(- \int_s^t \lambda(r)dr \right). \tag{6.2}$$

The appropriateness of modeling an arrival process as non-homogeneous Poisson can be justified by the following theorem.

Theorem 6.1.1 (Prèkopa, 1957) Let $\{ \Lambda(t) \mid t \geq 0 \}$ be a simple counting process (increasing, unit jumps) with a locally integrable rate function λ such that for all t, $\mathsf{E}[\Lambda(t)] = \int_0^t \lambda(s)ds$. It follows that this process is non-homogeneous Poisson if and only if it has independent increments.

This suggests that if we assume that our arriving (one at a time) customers originate from an infinite pool of independent agents, then such a process can only be modelled by a nonhomogeneous Poisson process and nothing else. For simplicity, we will model the service times to be interarrival times of a nonhomogeneous Poisson process with average rate function $\mu \equiv \{ \mu(t) \mid t \geq 0 \}$. This will be the time-varying analogue to the exponential distribution. The true departure process for Q will behave like this Poisson process only when the queue length is non-zero. By imposing this distributional structure we have turned Q into a Markov process. This system is referred to as the $M(t)/M(t)/1$ queue.

The basic paradigm of queueing theory is to find the distribution of Q given λ and μ or at least $\mathsf{E}[Q(t)]$ for all $t \geq 0$. This can be done when λ and μ are constants but there are no closed form solutions when λ and μ are time-varying (and their ratio is *not* constant). Over the years, many ad-hoc approximations have been formulated to estimate the average behavior of the $M(t)/M(t)/1$ queue (see Hall [5], Newell [14] and [15], Rothkopf and Oren [19]). Two of the standard methods are the following:

1. If the mean queue size is "small" and $\rho(t) = \lambda(t)/\mu(t) < 1$, we can use the *quasi-equilibrium* approximation which is
$$\mathrm{E}[Q(t)] \approx \frac{\rho(t)}{1-\rho(t)}. \tag{6.3}$$
2. If the mean queue size is "large", then we have a *fluid* approximation
$$\mathrm{E}[Q(t)] \approx \sup_{0\leq s\leq t} \int_s^t \lambda(r) - \mu(r)dr \tag{6.4}$$
assuming $Q(0) = 0$.

Our goal is to develop an asymptotic analysis to justify and refine these approximations. In so doing we hope to lay the groundwork for developing a general set of asymptotic methods for nonstationary Markov chains. In particular, we want to define nonstationary analogs steady state analysis, traffic intensity parameters and heavy traffic limits. We will first do this in Section 2 by defining a special asymptotic analysis for the transition probabilities of the $M(t)/M(t)/1$ queueing process. In Section 3, we applying these asymptotic methods directly to the random sample paths of the process. Finally, in Section 4 we explore the different types of heavy traffic behavior that arise for a queue with time-varying rates.

6.2 Uniform Acceleration for the Transition Probabilities

We can construct the transition probabilities for the $M(t)/M(t)/1$ queue length process $\{ Q(t) \mid t \geq 0 \}$ by specifying the forward equations for the probability distribution of $Q(t)$, which we will represent as a *probability vector*
$$\mathbf{p}(t) \equiv \sum_{n=0}^{\infty} \mathrm{P}(Q(t) = n)\mathbf{e}_n.$$
Since the probabilities sum to one, it is natural to view $\mathbf{p}(t)$ as an infinite-dimensional row vector whose components are absolutely summable. Such vectors comprise the Banach space of ℓ_1.

The forward equations for the $M(t)/M(t)/1$ can be written as

$$\frac{d}{dt}\mathbf{p}(t) = \mathbf{p}(t)\mathbf{A}(t),$$

where $\mathbf{A}(t)$ is the tridiagonal bounded operator

$$\mathbf{A}(t) \equiv \begin{bmatrix} -\lambda(t) & \lambda(t) & & \\ \mu & -(\lambda(t)+\mu) & \lambda(t) & \\ & \mu & -(\lambda(t)+\mu) & \ddots \\ & & \ddots & \ddots \end{bmatrix}.$$

For the special case of constant rates, we can formally solve (6.2) as

$$\mathbf{p}(t) = \mathbf{p}(0)\exp(\mathbf{A}t). \tag{6.5}$$

However, for the general time varying case, the best we can say is

$$\mathbf{p}(t) = \mathbf{p}(0)\mathbf{E}_{\mathbf{A}}(t), \tag{6.6}$$

where $\mathbf{E}_{\mathbf{A}}(t)$ is the unique operation such that:

1. For all $t \geq 0$,

$$\frac{d}{dt}\mathbf{E}_{\mathbf{A}}(t) = \mathbf{E}_{\mathbf{A}}(t)\mathbf{A}(t). \tag{6.7}$$

2. If $\mathbf{I}$ is the identity operator, then

$$\mathbf{E}_{\mathbf{A}}(0) = \mathbf{I}. \tag{6.8}$$

The operator $\mathbf{E}_{\mathbf{A}}(0)$ is referred to as the *time-ordered exponential.* In general, it is the case that

$$\mathbf{E}_{\mathbf{A}}(t) \neq \exp\left(\int_0^t \mathbf{A}(s)ds\right). \tag{6.9}$$

Despite the componentwise solution for the vector process (6.5) in terms of modified Bessel functions (see Prabhu [16]), it is more desirable to develop an asymptotic approach. In general, we seek to find the *steady state distribution* (if it exists) which is the limiting distribution obtained when $t \to \infty$. If we define

$$\boldsymbol{\pi} \equiv \lim_{t\to\infty} \mathbf{p}(t) \quad \text{where} \quad \boldsymbol{\pi} \equiv \sum_{n=0}^{\infty} \pi_n \mathbf{e}_n, \tag{6.10}$$

then defining $\rho \equiv \lambda/\mu$, we have

$$\pi_n = \begin{cases} (1-\rho)\rho^n & \text{if } \rho < 1, \\ 0 & \text{if } \rho \geq 1. \end{cases} \tag{6.11}$$

This result is equivalent to stating that $\boldsymbol{\pi}$ is the unique absolutely summable vector that solves

$$\boldsymbol{\pi}\mathbf{A} = \mathbf{0}. \tag{6.12}$$

What also follows from (6.5) is that the queue length process in the constant rate case will be *stationary* if its initial distribution equals $\boldsymbol{\pi}$, since we will have $\boldsymbol{\pi} = \boldsymbol{\pi}\exp(t\mathbf{A})$.

For queueing systems with constant rates, the fact that steady state behavior yields stationary behavior justifies the application of the steady state distribution to approximate the current behavior of the system. For time-varying rates however, large time limiting behavior (assuming it exists) cannot approximate what happens at some time t since any large time analysis by definition must be the result of arrival and service rates yet to come. Therefore, such a definition of the steady state behavior for time-varying queues will not give a meaningful method for analyzing queues with time-varying rates (an exception can be made for periodic rates, see Asmussen and Thorisson [1], Harrison and Lemoine [6], Lemoine [9], and Rolski [18]). A more appropriate asymptotic analysis can be constructed as follows:

Define $\{\, Q^\epsilon(t) \mid t \geq 0 \,\}$ to be an $M(t)/M(t)/1$ queueing process with a probability vector that solves

$$\frac{d}{dt}\mathbf{p}^\epsilon(t) = \mathbf{p}^\epsilon(t)\mathbf{A}(\epsilon t).$$

This is equivalent to saying that the arrival and service rate functions vary slowly in time since they are now of the form $\lambda(\epsilon t)$ and $\mu(\epsilon t)$ respectively. If we assume that λ and *mu* are differentiable functions of time, then we can show that as $\epsilon \downarrow 0$

$$\mathbf{p}^\epsilon(t) = \mathbf{p}^{(0)}(t) + O(\epsilon),$$

with respect to the ℓ_1 norm where

$$\frac{d}{dt}\mathbf{p}^{(0)}(t) = \mathbf{p}^{(0)}(t)\mathbf{A}(0).$$

Using the terminology of asymptotic expansions (see [8]), this is the *inner expansion* which approximates the evolution of the probability distribution for $t \approx 0$.

We can also find the asymptotics for the outer expansion by defining a new time scale $\tau = \epsilon t$. Holding τ fixed gives us a controlled way of having $t \to \infty$ and $\epsilon \downarrow 0$. Since τ is the time scale for the arrival and service rates, we now have a means of doing large time asymptotics in t without doing large time asymptotics for λ and μ in τ. If we set $\mathbf{q}^\epsilon(\tau) \equiv \mathbf{p}^\epsilon(\tau/\epsilon)$, then the vector process $\mathbf{q}^\epsilon$ solves the ℓ_1 differential equation

$$\epsilon\frac{d}{d\tau}\mathbf{q}^\epsilon(\tau) = \mathbf{q}^\epsilon(\tau)\mathbf{A}(\tau).$$

If we assume that the initial distribution is independent of ϵ, then for the constant rate case we obtain

$$\mathbf{q}^{\epsilon}(\tau) = \mathbf{q}(0)\exp(\mathbf{A}\tau/\epsilon).$$

Notice that the asymptotic behavior here as $\tau \to \infty$ will be identical to what is obtained by letting $\epsilon \downarrow 0$. In general, these two limits will *not* be the same for time-varying λ and μ. We will make the case that the desired limit is $\epsilon \downarrow 0$ and we will call this type of analysis *uniform acceleration*.

The formal asymptotic expansion as $\epsilon \downarrow 0$ for a given uniform acceleration at time τ will be

$$\mathbf{q}^{\epsilon}(\tau) \cong \sum_{n=0}^{\infty} \epsilon^{n} \boldsymbol{\pi}^{(n)}(\tau)$$

where

$$\boldsymbol{\pi}^{(0)}(\tau)\mathbf{A}(\tau) = \mathbf{0} \text{ and } \boldsymbol{\pi}^{(0)}(\tau) \cdot \mathbf{1}^{\mathrm{T}} = 1$$

and

$$\boldsymbol{\pi}^{(n+1)}(\tau)\mathbf{A}(\tau) = \frac{d}{d\tau}\boldsymbol{\pi}^{(n)}(\tau). \text{ and } \boldsymbol{\pi}^{(n+1)}(\tau) \cdot \mathbf{1}^{\mathrm{T}} = 0$$

where $\mathbf{1}$ is the vector of all ones. Observe that $\lim_{t \to \infty} \mathbf{p}^{(0)}(t) = \lim_{\tau \downarrow 0} \boldsymbol{\pi}^{(0)}(\tau)$ which show that both asymptotic expansions match at their common boundary layer.

The outer expansion will be of more interest so we will henceforth let t be the time scale for λ and μ and apply uniform acceleration to the $\mathbf{q}^{\epsilon}$ process. The M/M/1 queue is stable (as $t \to \infty$) whenever its *traffic intensity parameter* $\rho \equiv \lambda/\mu$ is less than one. The asymptotic analysis of uniform acceleration reveals an analogous traffic intensity parameter for the $M(t)/M(t)/1$ queue, namely

$$\rho^{*}(t) = \sup_{0 \le s < t} \frac{\int_s^t \lambda(r)\,dr}{\int_s^t \mu(r)\,dr}.$$

Moreover, it will be the case that the formal asymptotic expansion is valid and non-zero at time t if and only if $\rho^{*}(t) < 1$. The resulting expansion gives us formulas like

$$\mathsf{P}(Q^{\epsilon}(t) > 0) = \rho(t) - \epsilon \frac{\rho'(t)}{\mu(t)(1-\rho(t))^2} + O(\epsilon^2) \tag{6.13}$$

and

$$\mathsf{E}[Q^{\epsilon}(t)] = \frac{\rho(t)}{1-\rho(t)} - \epsilon \frac{\rho'(t)(1+\rho(t))}{\mu(t)(1-\rho(t))^4} + O(\epsilon^2). \tag{6.14}$$

Hence we obtain the quasi-equilibrium approximation as the leading term in the uniform acceleration expansion.

Now we present what we will call the *first limit theorem* for the $M(t)/M(t)/1$ queue. The proof can be found in [12] or [11].

Theorem 6.2.1 (Massey, 1981) If λ and μ are infinitely differentiable on the positive reals, then for all $t > 0$, the following three statements are equivalent:

1. The random variable $Q^\epsilon(t)$ converges in distribution as $\epsilon \downarrow 0$, to a probability measure on the non-negative integers (i.e. stable).

2. The probability vector for $Q^\epsilon(t)$ has a non-zero asymptotic expansion as $\epsilon \downarrow 0$.

3. Our traffic intensity parameter satisfies the condition $\rho^*(t) < 1$.

6.3 Uniform Acceleration for the Sample Paths

In this section we will generalize the notion of uniform acceleration so that we can apply these types of asymptotics directly to the random sample path behavior of the queueing process. If λ and μ are defined to be locally integrable on the positive reals and $Q(0) = 0$, then we can construct the random sample paths for the $M(t)/M(t)/1$ queue as

$$Q(t) = X(t) - \inf_{0 \leq s \leq t} X(s)$$

where

$$X(t) = N_1\left(\int_0^t \lambda(r)\, dr\right) - N_2\left(\int_0^t \mu(r)\, dr\right)$$

and where N_1 and N_2 are two independent, standard (rate 1) Poisson processes. So the $M(t)/M(t)/1$ queueing process is a functional applied to the difference of two independent Poisson processes.

We define Q^ϵ, the uniformly accelerated queueing process to be

$$Q^\epsilon(t) = X^\epsilon(t) - \inf_{0 \leq s \leq t} X^\epsilon(s),$$

where

$$X^\epsilon(t) = N_1\left(\frac{1}{\epsilon}\int_0^t \lambda(r)\, dr\right) - N_2\left(\frac{1}{\epsilon}\int_0^t \mu(r)\, dr\right).$$

The reflection mapping representation of $Q(t)$ asserts that its random sample paths can be built up out of the random sample paths of Poisson processes. By appealing to the theory of strong approximations for Levy processes, we obtain a sample path asymptotic analysis of Poisson processes. This in turn leads to an asymptotic analysis of the uniformly accelerated $M(t)/M(t)/1$ queue.

Now we present the *second limit theorem* for the $M(t)/M(t)/1$ queue, which is a functional strong law of large numbers result. The proof can be found in [10].

Theorem 6.3.1 (Mandelbaum and Massey, 1995) For all $t \geq 0$, we have

$$\lim_{\epsilon \downarrow 0} \epsilon\, Q^{\epsilon}(t) = Q^{(0)}(t) \quad \text{a.s.}$$

where

$$Q^{(0)}(t) \equiv \sup_{0 \leq s \leq t} \int_s^t [\lambda - \mu](r)\, dr,$$

and the convergence is uniform on compact subsets of $t \geq 0$. Note that

$$Q^{(0)}(t) = \int_0^t [\lambda - \mu](r)\, dr - \inf_{0 \leq s \leq t} \int_0^s [\lambda - \mu](r)\, dr.$$

By this second fundamental limit theorem, we obtain the fluid approximation as the leading asymptotic term of the uniformly accelerated random sample paths of the $M(t)/M(t)/1$ queueing process.

Finally, the strong approximation results that give us the strong law limits enable us to prove a *third limit theorem* for the $M(t)/M(t)/1$ queue, which is a functional central limit theorem (proof is also in [10]).

Theorem 6.3.2 (Mandelbaum and Massey, 1995) For all $t \geq 0$, we have

$$\lim_{\epsilon \downarrow 0} \sqrt{\epsilon} \left(Q^{\epsilon}(t) - \frac{1}{\epsilon} Q^{(0)}(t) \right) \stackrel{d}{=} Q^{(1)}(t)$$

where

$$Q^{(1)}(t) \equiv \hat{W}(t) - \inf_{s \in \Phi_t} \hat{W}(s),$$

and

$$\Phi_t \equiv \left\{ 0 \leq s \leq t \;\middle|\; \int_s^t [\lambda - \mu](r)\, dr = Q^{(0)}(t) \right\}$$

with

$$\hat{W}(t) \equiv W \left(\int_0^t [\lambda + \mu](r)\, dr \right),$$

$W = \{ W(t) \mid t \geq 0 \}$ is standard Brownian motion, and the convergence is weak with respect to the Skorohod topology on $D[0, \infty)$.

While it would be tempting to refer to $Q^{(1)}(t)$ as a diffusion limit, it would be misleading since we can show (see [10]) that it is at best an upper-semicontinuous function. Moreover, we can show the times where overloading or critical loading end are precisely the times (and no other) that a sample path of the process will be discontinuous with a non-zero probability.

These results will following strong approximation theorem for Lévy processes discussed in Ethier and Kurtz [4] based on work by Kolmòs, Major and Tusnàdy [7].

Theorem 6.3.3 Any Lévy process M for which $E[\exp M(1)] < \infty$ holds, has a realization such that

$$\sup_{t \geq 0} \frac{|M(t) - \mu t - W(\sigma^2 t)|}{\log(2 \vee t)} < \infty \text{ a.s.},$$

where $\mu = E[M(1)]$, $\sigma^2 = Var[M(1)]$, and $W(t)$ is standard Brownian motion. From this it follows that

$$M(t) = \mu t + W(\sigma^2 t) + O(\log t) \text{ a.s.}$$

for large t.

Observing that integrals of integrable functions are continuous and sample paths for Brownian motion can be constructed to be continuous also then the asymptotic analysis of the random sample paths for $Q^\epsilon(t)$ reduces to analyzing the asymptotic behavior of the following functional of continuous functions:

Lemma 6.3.4 Let a and b be continuous functions on the interval $[0, t]$. We then have as $\epsilon \downarrow 0$,

$$\sup_{0 \leq s \leq t} \left(\frac{1}{\epsilon} a(s) + \frac{1}{\sqrt{\epsilon}} b(s) \right) = \frac{1}{\epsilon} \bar{a}(t) + \frac{1}{\sqrt{\epsilon}} \bar{b}_\Phi(t) + o\left(\frac{1}{\sqrt{\epsilon}}\right), \tag{6.15}$$

where

$$\bar{a}(t) \equiv \sup_{0 \leq s \leq t} a(s), \; \bar{b}_\Phi(t) \equiv \sup_{s \in \Phi_t} b(s),$$

and

$$\Phi_t \equiv \left\{ 0 \leq s \leq t \;\middle|\; a(s) = \sup_{0 \leq u \leq t} a(u) \right\}.$$

In effect the contribution of the second term will only be from the values of b that are "sitting" on top of the maximal values of a. In [10] page 54, we give an example to show that for the remainder term $o(1/\sqrt{\epsilon})$ is the tightest estimate that can be made of its asymptotic behavior.

6.4 Asymptotic Regimes

Having established the three major limit theorems, we can use them to do a more refined analysis of $Q^\epsilon(t)$. For the constant rate case there are only three different types of asymptotic behavior, where we have stability and null-recurrent instability is distinguished from transient instability. The null-recurrent situation is where we obtain the various heavy-traffic limits for single station queues and queueing networks.

For the time varying case, we will have six regions of asymptotic behavior:

1. Underloaded: $\rho^*(t) < 1$.
2. Overloaded: $\rho^*(t) > 1$.
3. Critically Loaded: $\rho^*(t) = 1$.
   (a) Onset of Critical Loading: $\ell_n \uparrow t$ with $\rho^*(\ell_n) < 1$.
   (b) Middle of Critical Loading: $\rho^* \geq 1$ on $(t-\delta, t+\delta)$, and $\ell_n \uparrow t$ with $\rho^*(\ell_n) = 1$.
   (c) End of Critical Loading: $\rho^* \geq 1$ on $(t-\delta, t]$, $\ell_n \uparrow t$ with $\rho^*(\ell_n) = 1$, and $r_n \downarrow t$ such that $\rho^*(r_n) < 1$.
   (d) End of Overloading: $\rho^* > 1$ on $(t-\delta, t)$.

Unlike the constant rate case, a single queueing system can experience all six types of asymptotic behavior as it evolves through time. The heavy traffic limiting behavior, coming in four varieties, is responsible for the six types of asymptotic results. As we have discussed in the previous sections, the quasi-stationary approximation is the leading asymptotic terms for the underloaded case. Similarly, the fluid approximation is the leading asymptotic term for the overloaded case. So the only this left to refine is a more detailed analysis of the four types of critical loading. The following theorems quantitatively describe the different types of asymptotic behavior. The proofs to these theorems can be found in [10].

First consider a stable queue that is about to experience a significant surge in the arrival rate. Near the time that the process changes from underloaded to overloaded, we can show that it will resemble reflected Brownian motion with drift.

Theorem 6.4.1 (Onset of Critical Loading) If the onset of critical loading is at time t, λ and μ are differentiable in a neighborhood about t, such that

$$\lambda(t+s) = \mu(t) + \frac{\lambda^{(k)}(t)}{k!} s^k + o(s^k),$$

for some $k \geq 1$, then

$$\lim_{\epsilon \downarrow 0} \epsilon^{\frac{k}{2k+1}} Q^\epsilon(t + \epsilon^{\frac{1}{2k+1}} T_0, t + \epsilon^{\frac{1}{2k+1}} T) \stackrel{\mathrm{d}}{=} \tilde{W}(T_0, T) - \inf_{T_0 \leq \sigma \leq T} \tilde{W}(T_0, \sigma), \tag{6.16}$$

where

$$\tilde{W}(T_0, T) \equiv \frac{\lambda^{(k)}(t)}{(k+1)!}[T^{k+1} - T_0^{k+1}] + W(2\mu(t)(T - T_0)). \tag{6.17}$$

For the case of onset of critical loading, our asymptotic results resemble the usual type of reflecting diffusions encountered in heavy traffic analysis. This will not be the case for the other regions.

Queues with time-varying rates can initially experience asymptotic stability, evolve to instability but later return to stability. Our analysis confirms that the fluid approximation tells us that these periods of stability correspond to the times that the fluid approximation is zero. This means that even at a time t such that $\lambda(t) < \mu(t)$ we could still have $\rho^*(t) > 1$. Despite the fact that formally the first condition $\lambda(t) < \mu(t)$ allows us to compute the terms for a non-zero uniform acceleration expansion, we can show that when $\rho^*(t) > 1$ that expansion is *not* valid. It is in the region of overloading that the fluid approximation is most appropriate. We then encounter our next transition into underloading or stability when the fluid level is zero which we call the end of overloading.

Theorem 6.4.2 (End of Overloading) If overloading ends at time t, and λ and μ are continuous in a neighborhood of t, then for all T,

$$Q^\epsilon(t + \sqrt{\epsilon}T) \stackrel{d}{=} \frac{1}{\sqrt{\epsilon}} \left[\sup_{s \in \Phi_t \setminus \{t\}} \hat{W}(s,t) + (\lambda(t) - \mu(t))T \right]^+ + o\left(\frac{1}{\sqrt{\epsilon}}\right). \tag{6.18}$$

Here the process that bridges the transition from overloading to underloading is a deterministic function (of T) that is transforming a fixed random variable.

The remain two cases have the same form asymptotically, but with slightly different consequences.

Theorem 6.4.3 (Middle and End of Critical Loading) If the system is in the middle or end of critical loading at time t, then

$$Q^\epsilon(t) \stackrel{d}{=} \frac{1}{\sqrt{\epsilon}} \sup_{s \in \Phi_t} \hat{W}(s,t) + o\left(\frac{1}{\sqrt{\epsilon}}\right).$$

where $t = \sup \Phi_t$. Moreover, if t is in the middle of critical loading, then $\sup_{s \in \Phi_t} \hat{W}(s,t)$ is a continuous function. But, if t is the end of critical loading, then $\sup_{s \in \Phi_t} \hat{W}(s,t)$ is left continuous, but for a non-zero probability not right continuous at t.

Thus it follows that leading asymptotic term is (for some non-zero probability) not a continuous function when t is a time for the end of critical loading.

Acknowledgments: Thanks to Joseph B. Keller who first suggested to me the subject of queues with time-varying rates at a Ph.D. thesis topic. Thanks also to Avi Mandelbaum and Ward Whitt for their collaborations and insights into this topic.

6.5 References

[1] Asmussen, S. and Thorisson, H., A Markov Chain Approach to Periodic Queues, *J. Appl. Probab.*, **24**, 215–225, 1987.

[2] Eick, S., Massey, W.A. and Whitt, W., $M_t/G/\infty$ Queues with Sinusoidal Rates, *Management Science*, **39**, 241–252, 1993.

[3] Eick, S., Massey, W.A. and Whitt, W., The Physics of the $M_t/G/\infty$ Queue, *Operations Research*, **41** (1993 to appear).

[4] Ethier, S.N. and Kurtz, T.G., *Markov Processes, Characterization and Convergence*, John Wiley & Sons, 1986.

[5] Hall, Randolph W., *Queueing Methods for Services and Manufacturing*, Prentice Hall Publishers, 1991.

[6] Harrison, J.M. and Lemoine, A.J., Limit Theorems for Periodic Queues, *J. Appl. Probab.*, **14**, 566–576, 1977.

[7] Kolmòs, Major and Tusnàdy, An Approximation of Partial Sums of Independent RV's and the Sample DF, I; II , *ZW*, **32**, 111–131, 1975; **34**, 33–58, 1976.

[8] Lagerstrom, P.A., *Matched Asymptotic Expansions*, Springer Verlag, 1988.

[9] Lemoine, A.J., Waiting Time and Workload in Queues with Periodic Poisson Input, *J. Appl. Probab.*, **26**, 390–397, 1989.

[10] Mandelbaum, A. and Massey, W.A., Strong Approximations for Time-Dependent Queues,. *Math. of Op. Res.*, **20**, No. 1, 33–64, 1995.

[11] Massey, W.A., Asymptotic Analysis of the Time Dependent M/M/1 Queue, *Math. of Op. Res.*, **10**, 305–327, 1985.

[12] Massey, W.A., Nonstationary Queues, Thesis, Stanford University, 1981.

[13] Massey, W.A. and Whitt, W., Networks of Infinite-Server Queues with Nonstationary Poisson Input, *Queueing Systems*, **13**, No. 1–3, 183–250, 1993.

[14] Newell, G.F., Queues with Time-Dependent Arrival Rates, I, II, III, *J. Appl. Probab.*, **5**, 436–451 (I); 570–590 (II); 591–606 (III), 1968.

[15] Newell, G.F., *Applications of Queueing Theory*, Chapman and Hall (Second Edition), 1982.

[16] Prabhu, N.U., *Stochastic Storage Processes: Queues, Insurance Risk, and Dams*, Springer Verlag, 1980.

[17] A. PRÉKOPA, On Poisson and Composed Poisson Stochastic Set Functions, *Stud. Math.* **16** (1957) 142–155.

[18] ROLSKI, T., Queues with Non-Stationary Input Stream: Ross's Conjecture, *Adv. Appl. Prob.*, **13**, 603–618, 1981.

[19] ROTHKOPF, M.H. AND OREN, S.S., A Closure Approximation for the Nonstationary M/M/s Queue, *Mgmt. Sci.*, **25**, 522–534, 1979.

7
Nonparametric Estimation of Tail Probabilities for the Single-Server Queue

Peter W. Glynn and Marcelo Torres

ABSTRACT We consider the estimation of tail probabilities in queues via the nonparametric estimator constructed by simply computing the observed fraction of time that the queue is out in the tail. We show that for reflected Brownian motion, the M/M/1 queue-length process, and the GI/G/1 waiting time sequence that the amount of time over which one must observe the queue grows exponentially in the tail parameter when such a nonparametric estimator is used.

7.1 Introduction

This paper is concerned with the question of how long one must observe a queue in order to accurately estimate a tail probability corresponding to a performance measure like queue-length or waiting time. The motivation for the problem stems from the current interest in developing robust admission control schemes for high-speed packet-based telecommunications networks. Assuming that a principle objective of such a scheme would likely be to minimize packet loss at the buffers associated with the switches in such a network, the value of estimating the loss probabilities from observed traffic becomes clear. As indicated above, we choose here to focus on tail probabilities for queues having an infinite buffer, in the belief that such a theory is likely to also describe the asymptotic behavior of estimators for loss networks of the type arising in the above telecommunications context. We note, in passing, that replacing a finite buffer system by an infinite buffer analog is a commonly used approximation in the queueing community.

There are many different types of estimation methodologies that can be used in this context. Our focus, in this paper, will be on a nonparametric formulation. Assuming that essentially nothing is known about the traffic source to the queue, one may be tempted to try to estimate the tail probability by the observed fraction of time that the queue is out in the tail. (Any other estimator will typically take advantage of additional structure that we may be unwilling to assume.) This observed fraction of time is precisely the nonparametric tail estimator considered in this paper. A companion paper (Torres and Glynn [12]) describes a competing estimator, based on parametric statistical modeling of the input source, that assumes more about the structure of the queue.

Because so little is assumed about the queue, one might expect such nonparametric estimators to be very inefficient. In particular, one expects that the amount of time over which the system needs to be observed in order to accurately estimate a tail probability increases as a function of the tail parameter. In this paper, we establish this result rigorously. In fact, we compute the critical rate at which the observed time horizon needs to grow as a function of the tail parameter; the growth is typically exponential. This suggests that for the finite buffer queues arising in packet networks that the amount of traffic observed needs to grow exponentially in the size of the buffers. From a pragmatic standpoint, this likely renders the nonparametric methodology considered here impractical. As a consequence, the analysis in this paper makes a strong case for the need to impose more model structure on the input sources in attempting to estimate such loss probabilities.

In addition to the above conclusion, we view the following as the main results in the paper:

1. We develop general theory (Proposition 7.2.1), relevant to non-queues as well as queues, on how much time a stochastic process needs to be observed in order to accurately estimate the probability of a rare event. In addition to the statistical implications, this result is also relevant to simulation, in which the nonparametric estimators considered in this paper are especially widely used.

2. We develop central limit theorems, with explicitly computed time-average variance constants, describing the rate of convergence of non-parametric tail estimators for both reflected Brownian motion and the M/M/1 queue. These results illustrate the use of stochastic calculus, martingale methods, regeneration, and Poisson's equation in obtaining central limit theorems for stochastic processes.

3. In the setting of the GI/G/1 waiting time sequence, we obtain a quite precise solution to the question of how long one must observe the queue in order to accurately estimate a tail probability. The critical rate is exponential and depends on the Cramér-Lundberg parameter; see Theorems 7.5.1 and 7.5.2.

This paper is organized as follows. General theory is developed in Section 7.2, whereas Sections 7.3 and 7.4 are devoted to the analysis of reflected Brownian motion and the M/M/1 queue, respectively. Section 7.5 undertakes the analysis of the GI/G/1 queue, while Section 7.6 offers conclusions, including some discussion of the competing parametric methodology.

7.2 A Bit of General Theory

Our concern, in this paper, is with the nonparametric estimation of (extreme) tail probabilities associated with the single-server queue. This can be cast in more general terms, as follows.

Let $X = (X_n : n \geq 0)$ be a (strictly) stationary stochastic process taking values in a state space S. For a given sequence of subsets $(A_m : m \geq 1)$ contained in S, consider the problem of estimating the quantity $p_m = P(X_0 \in A_m)$ for m large. We assume throughout that $p_m \downarrow 0$ as $m \to \infty$. In the single-server queue context that interests us, X_n might, for example, be the (stationary) waiting time of the n'th customer to arrive to the queue, with $A_m = [m, \infty)$, in which case p_m is an (extreme) tail probability for the steady-state waiting time distribution.

The obvious non-parametric estimator for p_m is clearly

$$\alpha(n;m) \stackrel{\Delta}{=} \frac{1}{n} \sum_{j=0}^{n-1} I(X_j \in A_m).$$

The issue, then, is how large one must choose n, as a function of m, in order that $\alpha(n;m)$ be an accurate estimator of p_m. This can be re-formulated, more precisely, in terms of the relative error given by $|\alpha(n;m)/p_m - 1|$. In particular, how fast must $n_m \to \infty$ in order that

$$\frac{1}{p_m}\alpha(n_m;n) \Rightarrow 1 \tag{7.1}$$

as $m \to \infty$? The answer is easy if the X_n's are i.i.d.

In this case, note that (7.1) holds, provided that

$$p_m^{-2} \text{var } \alpha(n_m;m) \to 0 \tag{7.2}$$

as $m \to \infty$. But var $\alpha(n;m) = n^{-1}p_m(1-p_m)$. Hence, (7.2) follows if we require that $p_m n_m \to \infty$ as $m \to \infty$. On the other hand, if $p_m n_m \to c > 0$ as $m \to \infty$, then

$$n_m \alpha(n_m;m) \stackrel{\mathcal{D}}{=} \text{Binomial}(n_m, p_m) \tag{7.3}$$

$$\Rightarrow \text{Poisson}(c) \tag{7.4}$$

as $m \to \infty$, from which it is evident that

$$\frac{1}{p_m}\alpha(n_m;m) \Rightarrow \frac{1}{c}\text{Poisson}(c)$$

as $m \to \infty$, violating (7.1). Consequently, $n_m \gg 1/p_m$ is a necessary and sufficient condition for the (relatively) accurate determination of p_m.

Our goal is now to see how far this analysis can be extended when X is a dependent sequence, as is typical of queueing applications. First, note that simple algebra verifies that

$$n \cdot \text{var } \alpha(n;m) = \text{var } I_0(m) + 2\sum_{j=1}^{n-1}(1-j/n)\text{cov } (I_0(m), I_j(m)) \quad (7.5)$$

where $I_j(m) \stackrel{\Delta}{=} I(X_j \in A_m)$. Hence, if the stationary sequence $(I_j(m) : j \geq 0)$ has non-negative autocorrelations (i.e. cov $(I_0(m), I_j(m)) \geq 0$ for $j \geq 1$), it follows that

$$\begin{aligned} \text{var } \alpha(n;m) &\geq \frac{1}{n}\text{var } I_0(m) \\ &= \frac{1}{n}p_m(1-p_m). \end{aligned}$$

Thus, assuming $\limsup_{m\to\infty} n_m p_m < +\infty$, it is evident that

$$\liminf_{m\to\infty} \text{var } [p_m^{-1}\alpha(n_m;m)] > 0.$$

So, in the presence of non-negative autocorrelations, one requires that $n_m \gg 1/p_m$ in order that the variance of $p_m^{-1}\alpha(n_m;m)$ converge to zero.

This has important implications for estimation of tail probabilities in the single-server queue. Consider, for example, the stationary waiting time sequence $W = (W_n : n \geq 0)$ associated with the GI/G/1 queue with traffic intensity lass than unity. Suppose that our interest is in estimating the tail probability $p_m = P(W_0 \geq m)$. It is well known that W is a stochastically monotone Markov chain, so that the stationary waiting times $(W_n : n \geq 0)$ form an associated sequence; see, for example, Stoyan [11]. Note that for each $m \geq 1$, $I_j(m) = I(W_j \geq m)$ is a non-decreasing function of W_j, and consequently $(I_j(m) : j \geq 0)$ is itself an associated sequence. Thus, it is evident that cov $(I_0(m), I_j(m)) \geq 0$ for $j \geq 1$, establishing the fact that our non-negative autocorrelation condition holds in this setting. Hence, in order that the variance of $p_m^{-1}\alpha(n_m;m)$ go to zero as $n \to \infty$, it is necessary that $n_m \gg p_m^{-1}$. Of course, the Cramér-Lundberg approximation states that, under general conditions, $P(W_0 \geq x) \sim a\exp(-\theta^* x)$ as $x \to \infty$, for suitable positive constants a and θ^*; see, for example, Asmussen [1]. As a result, we may conclude that we need to at least observe W over a time horizon of order $\exp(\theta^* m)$ in order to estimate p_m to a reasonable degree of accuracy. However, given the complex dependency structure of W, it is conceivable that one might need to observe W over a time scale much longer than p_m^{-1}. We address this issue next.

A standard concept used in the stochastic process setting to describe dependence is that of mixing. Without loss of generality, we may extend $(X_j : j \geq 0)$ to a "two-sided" (strictly) stationary sequence $(X_j : -\infty <$

$j < \infty$). Let $\mathcal{F}_j = \sigma(X_k : k \leq j)$ and $\mathcal{F}^j = \sigma(X_k : k \geq j)$. The j'th ($j \geq 1$) uniform mixing coefficient of X is defined by

$$\varphi_\infty(j) \triangleq \sup_{A \in \mathcal{F}^j} \sup_{\substack{B \in \mathcal{F}_0 \\ P(B)>0}} |P(A|B) - P(A)|.$$

Assume that X is uniformly mixing in the sense that

$$\sum_{j=1}^{\infty} \varphi_\infty(j) < \infty. \tag{7.6}$$

Then, a standard inequality (see, for example, Ethier and Kurtz [5]) states that

$$|\text{cov}\ (I_0(m), I_j(m))| \leq 2\varphi_\infty(j) E I_0(m). \tag{7.7}$$

It follows from (7.5) - (7.7) that

$$n \cdot \text{var}\ \alpha(n;m) \leq \text{var}\ I_0(m) + 4\sum_{j=1}^{\infty} \varphi(j) E I_0(m) \tag{7.8}$$

$$\leq p_m(1 + 4\sum_{j=1}^{\infty} \varphi(j)) \tag{7.9}$$

and hence

$$\text{var}\ [p_m^{-1}\alpha(n_m;m)] \leq \frac{1}{n_m p_m}(1 + 4\sum_{j=1}^{\infty} \varphi(j)) \to 0$$

provided that $n_m p_m \to +\infty$. We summarize the above discussion with the following result.

Proposition 7.2.1 Suppose that X satisfies the uniform mixing condition (7.6). Then, if $p_m n_m \to +\infty$,

$$p_m^{-1}\alpha(n_m;m) \Rightarrow 1$$

as $m \to \infty$. Conversely, if $(I_j(m) : j \geq 0)$ has non-negative autocorrelations for each $m \geq 1$, then it is necessary that $p_m n_m \to \infty$ in order that

$$\text{var}\ [p_m^{-1}\alpha(n_m;m)] \to 0$$

as $m \to \infty$.

Thus Proposition 7.2.1 gives fairly general conditions under which observing X for a time period $n_m \gg 1/p_m$ is necessary and sufficient for (relatively) accurate estimation of p_m. It effectively shows that, under the conditions stated, the dependence is such that the relative magnitude of n_m is the same as in the i.i.d. case.

It turns out that the uniform mixing condition is fundamental to the sufficiency half of Proposition 7.2.1. To see this, consider the stationary Markov chain X on the non-negative integers $\{0, 1, ...\}$ with transition probabilities $P(0, y) = a_y > 0$ for $y \geq 1$, $P(y, 0) = b_y > 0$ for $y \geq 1$, $P(y, y) = 1 - b_y$ for $y \geq 1$, and $P(x, y) = 0$ for all other pairs $(x, y) \in S \times S$. Assume $\sum_{k=1}^{\infty} a_k/b_k < \infty$. Then, X is irreducible and positive recurrent with $\pi_0 = (1 + \sum_{j=1}^{\infty} a_j/b_j)^{-1}$ and

$$\pi_j = \pi_0 a_j/b_j$$

for $j \geq 1$. Let $A_m = \{m\}$, so that $p_m = \pi_m$. Then,

$$\begin{aligned} P(\frac{1}{n_m} \sum_{j=0}^{n_m - 1} I(X_j = m) = 0) & \\ \geq \; & \pi_0 P(\frac{1}{n_m} \sum_{j=0}^{n_m - 1} I(X_j = m) = 0 | X_0 = 0) \\ = \; & \pi_0 P(T_m \geq n_m | X_0 = 0) \end{aligned}$$

where $T_m = \inf\{j \geq 0 : X_j = m\}$. Observe that

$$T_m \geq \sum_{j=0}^{T_m - 1} I(X_j = 0) \stackrel{\Delta}{=} \beta_m,$$

where β_m is geometric with parameter a_m. So,

$$P(\frac{1}{n_m} \sum_{j=0}^{n_m - 1} I(X_j = m) = 0) \geq \pi_0 (1 - a_m)^{n_m}.$$

Choose $a_j = 2^{-j}$, $b_j = 1/j$, and $n_j = 2^j$. Observe that $p_m n_m \to \infty$ as $m \to \infty$. On the other hand,

$$\liminf_{m \to \infty} P(\frac{1}{n_m} \sum_{j=0}^{n_m - 1} I(X_j = m) = 0) \geq \pi_0/e > 0,$$

violating (7.1). The difficulty here is that this chain, although strong mixing (see Ethier and Kurtz (1986), p. 345), is not uniformly mixing.

Unfortunately, it turns out that the standard processes that arise in conjunction with the single-server queue are not uniformly mixing. Take, for example, the waiting time sequence W. Because W is Markov, it is evident that

$$\varphi_\infty(j) = \sup_A \sup_x |P(W_j \in A | W_0 = x) - P(W_j \in A)|.$$

But for each $j \geq 1$ and $b > 0$, it is clear that

$$P(W_j \in [0,b] | W_0 = x) \to 0$$

as $x \to \infty$, and hence $\varphi_\infty(j) = 1$ for $j \geq 1$, violating (7.6). Consequently, we can not invoke the sufficiency half of Proposition 7.2.1, and must consider other methods of attack for studying this problem in the context of queues.

7.3 Analysis of Reflected Brownian Motion

In this section, we consider the problem of estimating tail probabilities for one-dimensional reflected Brownian motion (RBM). As is well known, this stochastic process is the limit process that arises when studying the single-server queue in heavy traffic; see, for example, Glynn [8]. Thus, we would minimally expect the theory developed for RBM to be representative of the qualitative behavior of queues in heavy traffic. In addition, RBM is a particularly tractable stochastic process from a mathematical viewpoint, making the required analysis especially straightforward.

Let $X = (X(t) : t \geq 0)$ be a stationary version of a one-dimensional RBM living on $[0, \infty]$, having drift $-\mu < 0$ and infinitesimal variance $\sigma^2 > 0$. Then, in order that X be stationary, it follows that for each $t \geq 0$, $X(t)$ has an exponential distribution with parameter $\theta^* \stackrel{\Delta}{=} 2\mu/\sigma^2$; see Asmussen [1]. Our goal here is to estimate the tail probability

$$\alpha(y) \stackrel{\Delta}{=} P(X(0) \geq y) = \exp(-\theta^* y).$$

Our focus will be on estimating $\alpha(y)$ via the nonparametric time-average

$$\alpha(t; y) \stackrel{\Delta}{=} \frac{1}{t} \int_0^t I(X(s) \geq y) ds.$$

Clearly, in practice, if we know a priori that the observed process X is such an RBM, we would choose to estimate $\alpha(y)$ via a parametric estimator of the form $\exp(-2\hat{\mu}y/\sigma)$, where $\hat{\mu}$ is (for example) the maximum likelihood estimator of μ. (Note that σ^2 can be estimated without error by computing the quadratic variation of X.) However, our interest is in considering the estimation of $\alpha(y)$ in settings in which we are unwilling to make any a priori parametric assumptions about the observed process. In such a context, $\alpha(t; y)$ is perhaps the most natural estimator for $\alpha(y)$.

To proceed with our analysis, let $f(x; y) = I(x \geq y)$ and $f_c(x; y) = f(x; y) - \alpha(y)$. Then

$$\alpha(t; y) - \alpha(y) = \frac{1}{t} \int_0^t f_c(X(s); y) ds.$$

A key idea in studying $\int_0^t f_c(X(s);y)ds$ is to represent this additive functional of X as a semimartingale. We can construct such a representation via judicious use of the Itô calculus.

Let $u(\cdot;y)$ satisfy the conditions of Itô's lemma so that, in particular, it has a continuous first derivative and second derivative existing a.e. Then, recall that

$$dX(t) = -\mu dt + \sigma dB(t) + dL(t),$$

where $B = (B(t) : t \geq 0)$ is a one-dimensional standard Brownian motion and $L = (L(t) : t \geq 0)$ is a non-decreasing process that increases only when X is zero; see Harrison [14] for details. Itô's formula states that for $u(\cdot;y)$ chosen as above,

$$du(X(t);y) = u'(X(t);y) + \frac{1}{2}u''(X(t);y)\sigma^2 dt,$$

where the derivatives are computed with respect to the first component of u. Hence,

$$\begin{aligned} u(X(t);y) - u(X(0);y) &= \int_0^t \{-\mu u'(X(s);y) + \frac{\sigma^2}{2}u''(X(s);y)\}ds \\ &\quad + \int_0^t u'(X(s);y)\sigma dB(s) \qquad (7.10) \\ &\quad + \int_0^t u'(X(s))dL(s). \end{aligned}$$

Suppose that we choose $u(\cdot;y)$ to satisfy "Poisson's equation" for RBM:

$$-\mu u'(\cdot;y) + \tfrac{\sigma^2}{2}u''(\cdot;y) = -f_c(\cdot;y) \qquad (7.11)$$

such that

$$u'(0;y) = 0.$$

Because $L(\cdot)$ increases only when $X(\cdot)$ is zero, evidently, the last term on the right-hand side of (7.10) vanishes. By our choice of $u(\cdot;y)$, it then follows that

$$\int_0^t f_c(X(s);y)ds + u(X(t);y) - u(X(0);y) = \int_0^t u'(X(s);y)\sigma dB(s). \qquad (7.12)$$

But the stochastic integral $\int_0^t u'(X(s);y)\sigma dB(s)$ is a martingale, provided that $u'(\cdot;y)$ is appropriately integrable. Thus, (7.12) provides us with a martingale representation for our additive functional $\int_0^t f_c(X(s);y)ds$.

Note that (7.11) only determines $u(\cdot;y)$ up to an additive constant. Arbitrarily setting $u(0;y) = 0$, it is straightforward to verify that the unique solution to (7.11) is then given by

$$u(x;y) = \begin{cases} \frac{e^{-\theta^* y}}{\mu\theta^*}(e^{\theta^* x} - 1) - \frac{\alpha(y)}{\mu}x & ; 0 \leq x \leq y, \\ \frac{(1-e^{-\theta^* y} - y\theta^* e^{-\theta^* y})}{\mu\theta^*} + (\frac{1-\alpha(y)}{\mu})(x - y) & ; x \geq y. \end{cases}$$

This solution has a continuous first derivative and satisfies the necessary regularity hypothesis required for the application of Itô's lemma; see Chung and Williams [4]. In addition, since $u'(\cdot;y)$ is bounded, it follows that the stochastic integral $\int_0^t u'(X(s);y)\sigma dB(s)$ is in fact a martingale.

We are interested in knowing over how long a time interval t_y one must observe X in order to obtain a (relatively) accurate estimate of $\alpha(y)$. This amounts to asking how fast t_y must grow in order that

$$\alpha(t_y;y)/\alpha(y) \Rightarrow 1 \tag{7.13}$$

as $y \to \infty$. Relation (7.13) is equivalent to showing that

$$\frac{\alpha(y)^{-1}}{t_y}\int_0^{t_y} f_c(X(s);y)ds \Rightarrow 0$$

as $y \to \infty$. Hence by (7.12), we need to show that

$$\frac{\alpha(y)^{-1}}{t_y}\int_0^{t_y} u'(X(s);y)\sigma dB(s)+\frac{\alpha(y)^{-1}}{t_y}u(X(0);y)-\frac{\alpha(y)^{-1}}{t_y}u(X(t);y) \Rightarrow 0 \tag{7.14}$$

as $y \to \infty$. Now,

$$\begin{aligned}
\alpha(y)^{-1}E|u(X(t_y);y)| &= \alpha(y)^{-1}E|u(X(0);y)| && (7.15)\\
&= \alpha(y)^{-1}|\int_0^\infty |u(x;y)|\theta^* e^{-\theta^* x}dx && (7.16)\\
&= O(y) && (7.17)
\end{aligned}$$

and consequently we require that $t_y/y \to +\infty$ in order that the latter two terms in (7.14) converge to zero in expectation, implying convergence in probability to zero.

As for the first term on the left-hand side of (7.14), the L^2-isometry property of stochastic integrals implies that

$$\begin{aligned}
&E[\frac{\alpha(y)^{-2}}{t_y^2}(\int_0^{t_y} u'(X(s);y)\sigma dB(s))^2]\\
&= \frac{\alpha(y)^{-2}}{t_y^2}E\int_0^{t_y} u'(X(s);y)^2\sigma^2 ds\\
&= \frac{\alpha(y)^{-2}}{t_y}\sigma^2 Eu'(X(0);y)^2\\
&= \frac{\alpha(y)^{-2}}{t_y}\sigma^2\int_0^\infty u'(x;y)^2\theta^* e^{-\theta^* x}dx
\end{aligned}$$

But

$$\alpha(y)^{-2}\int_0^\infty u'(x;y)^2\theta^* e^{-\theta^* x}dx = O(e^{\theta^* y})$$

as $y \to \infty$, so it follows that if $e^{\theta^* y}/t_y \to 0$, then the L^2-norm of the first term in (7.14) converges to zero, proving convergence in probability. We have therefore proved the following result.

Theorem 7.3.1 If $e^{\theta^* y}/t_y \to 0$ as $y \to \infty$, then

$$\alpha(t_y; y)/\alpha(y) \Rightarrow 1$$

as $y \to \infty$.

Consequently, $t_y \gg \alpha(y)^{-1}$ is sufficient for (relatively) accurate determination of $\alpha(y)$ via the nonparametric estimator $\alpha(t_y; y)$. Thus, it is enough to observe X over an amount of time that increases exponentially (at rate θ^*) in the level y.

In developing estimators for $\alpha(y)$, it is often of some interest to obtain a rate of convergence, as expressed via a central limit theorem (CLT). Our first CLT characterizes the rate of convergence, for fixed y, as $t \to \infty$.

Theorem 7.3.2 Fix $y > 0$. Then,

$$t^{1/2}(\alpha(t; y) - \alpha(y)) \Rightarrow \sigma(y) N(0, 1)$$

as $t \to \infty$, where $\sigma^2(y) = 2(\frac{\sigma}{\mu})^2 \alpha(y)^2 \{\alpha(y)^{-1} - y - 1\}$.

Proof. Because of (7.12), Theorem 7.3.2 follows from an invocation of the martingale central limit theorem (see p. 339 of Ethier and Kurtz (1986)). To prove Theorem 7.3.2, the martingale CLT requires that we show

$$\frac{1}{t}[M_y](t) \Rightarrow \sigma^2(y) \tag{7.18}$$

as $t \to \infty$, where $[M_y](t)$ is the quadratic variation over $[0, t]$ of the martingale $M_y(t) \triangleq \int_0^t u'(X(s); y) \sigma dB(s)$. But

$$[M_y](t) = \sigma^2 \int_0^t u'(X(s); y)^2 ds.$$

By the ergodic theorem, it follows that

$$\frac{1}{t}[M_y](t) \to \sigma^2 E u'(X(0); y)^2$$

a.s. as $t \to \infty$. Noting that $u'(x; y) = \mu^{-1}\alpha(y)(\exp(\theta^*(x \wedge y)) - 1)$, it is easily verified that $\sigma^2 E u'(X(0); y)^2$ equals the given expression for $\sigma^2(y)$; this establishes (7.18) and proves that

$$t^{-1/2} M_y(t) \Rightarrow \sigma(y) N(0, 1)$$

as $t \to \infty$. The theorem is then an immediate consequence of the representation (7.12) and the fact that $u(X(t); y)/t^{1/2}$ and $u(X(0); y)/t^{1/2}$ both have the same distributions and converge in probability to zero. □

Of course, given our interest in the behavior of $\alpha(t;y)$ for large y, the more interesting question is perhaps the CLT behavior of $\alpha(t_y;y)$ as $y \to \infty$. Theorem 7.3.2 suggests the approximation

$$\alpha(t_y;y) \overset{\mathcal{D}}{\approx} \alpha(y) + \sqrt{\frac{\sigma^2(y)}{t_y}} N(0,1), \tag{7.19}$$

where $\overset{\mathcal{D}}{\approx}$ denotes "approximately equal in distribution to." Noting that $\sigma^2(y) \sim 2\sigma^2\alpha(y)/\mu^2$ as $y \to \infty$, (7.19) can be re-cast in the form

$$\frac{\alpha(t_y;y)}{\alpha(y)} \overset{\mathcal{D}}{\approx} 1 + \frac{\sigma}{\mu}\sqrt{\frac{2}{\alpha(y)t_y}} N(0,1). \tag{7.20}$$

As a consequence, the relative error is of the order of $(\alpha(y)t_y)^{-1/2}$, showing clearly the benefits of observing X over a time horizon $[0,t_y]$, with $t_y \gg \alpha(y)^{-1}$. Our final result of this section is a CLT that makes (7.20) rigorous.

Theorem 7.3.3 Suppose $t_y\alpha(y) \to +\infty$ as $y \to \infty$. Then,

$$\sqrt{\alpha(y)t_y}\left(\frac{\alpha(t_y;y)}{\alpha(y)} - 1\right) \Rightarrow \sqrt{2}(\sigma/\mu)N(0,1)$$

as $y \to \infty$.

Proof. As in the proof of Theorem 7.3.2, the key step is showing that

$$\frac{1}{t_y\alpha(y)}[M_y](t_y) \Rightarrow \frac{2\sigma^2}{\mu^2}$$

as $y \to \infty$. Equivalently, we need to prove that

$$\frac{1}{t_y\alpha(y)}\int_0^{t_y} u'(X(s);y)^2\,ds \Rightarrow \frac{2\sigma^2}{\mu^2} \tag{7.21}$$

as $y \to \infty$. But $\alpha(y)^{-1}u'(x;y)^2 = \mu^{-2}\alpha(y)\sigma^2\exp(2\theta^*(x \wedge y)) + O(1)$, where $O(1)$ is uniform in x and y. Since $t_y \to \infty$, (7.21) holds, provided that

$$\frac{\alpha(y)}{t_y}\int_0^{t_y} h(X(s);y)ds \Rightarrow 2 \tag{7.22}$$

as $y \to \infty$, where $h(x;y) \triangleq \exp(2\theta^*(x \wedge y))$. Let $h_c(x;y) = h(x;y) - 2\alpha(y)^{-1}$ and set

$$w(x;y) = \begin{cases} \frac{(2+4\alpha(y)^{-1})}{\sigma^2\theta^*}e^{\theta^* x} - \frac{4\alpha(y)}{\sigma^2\theta^*}x - \frac{1}{\sigma^2(\theta^*)^2}e^{2\theta^* x} & ; 0 \le x \le y \\ w(y;y) + 2(\alpha(y)^{-2} - \alpha(y)^{-1})(x-y)/\mu & ; x > y. \end{cases}$$

Then, it is straightforward to verify that $w(\cdot;y)$ has a continuous first derivative and satisfies at $x \neq y$ the differential equation

$$\frac{\sigma^2}{2}w''(\cdot;y) - \mu w'(\cdot;y) = -h_c(\cdot;y)$$

subject to $w'(0;y) = 0$. Proceeding as in the development of (7.12), we can then show that

$$\int_0^{t_y} h_c(X(s);y)ds + w(X(t_y);y) - w(X(0);y) = \int_0^{t_y} w'(X(s);y)\sigma dB(s). \tag{7.23}$$

Now, $|w(x;y)| \leq a_1 + a_2\alpha(y)^{-2}x$ uniformly in x and y (for suitably chosen a_1, a_2) so that $\alpha(y)|w(X(t_y);y)|/t_y \leq |X(t_y)|/t_y\alpha(y)$. Since $t_y\alpha(y) \to \infty$ and $X(t_y)$ has a distribution independent of t_y, it follows that

$$w(X(t_y);y)\alpha(y)/t_y \Rightarrow 0$$

as $y \to \infty$. A similar, but simpler, argument shows that

$$w(X(0);y)\alpha(y)/t_y \Rightarrow 0$$

as $y \to \infty$. As for the stochastic integral,

$$E[(\int_0^{t_y} w'(X(s);y)dB(s))^2\alpha(y)^2/t_y^2] = Ew'(X(0);y)^2\alpha(y)^2/t_y \to 0$$

as $y \to \infty$, proving that $\int_0^{t_y} w'(X(s);y)dB(s)\alpha(y)/t_y \to 0$ in mean square and hence in probability. The representation (7.23) therefore proves that $\int_0^{t_y} h_c(X(s);y)ds \cdot \alpha(y)/t_y \Rightarrow 0$ as $y \to \infty$, yielding (7.22).

The martingale CLT then implies that

$$(t_y\alpha(y))^{-1/2}M_y(t_y) \Rightarrow \frac{\sigma}{\mu}\sqrt{2}N(0,1) \tag{7.24}$$

as $y \to \infty$. But since $u(x;y) \leq b_1 + b_2x$ uniformly in x and y,

$$(t_y\alpha(y))^{-1/2}u(X(t);y) \Rightarrow 0$$

and

$$(t_y\alpha(y))^{-1/2}u(X(0);y) \Rightarrow 0$$

as $y \to \infty$. Relations (7.24) and (7.12) then finish the proof of the theorem. □

7.4 Analysis of the M/M/1 Queue

In this section, we turn to the analysis of the M/M/1 queue. As perhaps the simplest model that exhibits true queueing behavior (as opposed to the "approximate" queueing behavior of RBM), we would hope to see the same qualitative behavior as that obtained in Section 7.3 for RBM.

We start with an analysis of the "number in system" process for the M/M/1 queue. To be precise, assume that $\lambda < \mu$ and let $Q = (Q(t) : t \geq 0)$ be a stationary version of the birth-death process on the non-negative integers $\{0, 1, 2, ...\}$ with birth rates $\lambda_n = \lambda (n \geq 0)$ and death rates $\mu_n = \mu (n \geq 1)$. Then, λ can, of course, be interpreted as the arrival rate to the queue, μ can be viewed as the service intensity, and $Q(t)$ represents the number of customers in the system as time t.

Suppose that we are interested in estimating the tail probability $\alpha(m) = P(Q(0) \geq m)$ via the nonparametric estimator

$$\alpha(t; m) \stackrel{\Delta}{=} \frac{1}{t} \int_0^t I(Q(s) \geq m) ds.$$

The issue here is how fast t must grow as a function of m, in order that $\alpha(t; m)$ be a (relatively) accurate estimator of $\alpha(m)$. To answer this question, we adopt the same basic approach as used in our analysis of RBM in Section 7.3.

Proceeding as in Section 7.3, let $f(x; m) = I(x \geq m)$ and set $f_c(x; m) = f(x; m) - P(Q(0) \geq m) = f(x; m) - \rho^m$, where $\rho \stackrel{\Delta}{=} \lambda/\mu$. We wish to express $\int_0^t f_c(Q(s); m) ds$ in terms of an appropriately defined martingale. To do so, we solve "Poisson's equation" for $u(\cdot; m)$:

$$\sum_{j=0}^{\infty} A_{ij} u(j; m) = -f_c(i; m), (i \geq 0)$$

where $A = (A_{ij} : i, j \geq 0)$ is the infinitesimal generator of $(Q(t) : t \geq 0)$. In other words, we must compute the solution $u(\cdot; m)$ to

$$\lambda u(i+1; m) - (\lambda + \mu) u(i; m) + \mu u(i-1; m) = -f_c(i; m),$$

for values $i \geq 1$, and

$$\lambda u(1; m) - \lambda u(0; m) = -f_c.$$

A solution to the above linear system is easily verified to be

$$u(i; m) = \begin{cases} \frac{\rho^m}{\mu(1-\rho)^2}(\rho^i - 1) - \frac{\rho^m i}{\mu(1-\rho)} & ; 0 \leq i < m \\ \frac{1-\rho^m}{\mu(1-\rho)^2} - \frac{\rho^m m}{\mu(1-\rho)} + \frac{(1-\rho^m)}{\mu(1-\rho)}(i - m) & ; i \geq m. \end{cases}$$

(Note the similarity to the solution $u(\cdot; y)$ obtained in the RBM context.) Noting that $u(Q(t); m)$ and $f_c(Q(t); m)$ are both integrable r.v.'s for each

$t \geq 0$, it is a well known fact (see, for example, p. 298 of Karlin and Taylor [10]) that for each $m \geq 1$,

$$M_m(t) \stackrel{\Delta}{=} u(Q(t); m) - u(Q(0); m) + \int_0^t f_c(Q(s); m) ds$$

is a martingale. To study the behavior of the relative error

$$\alpha(t; m)/\alpha(m)) - 1,$$

observe that

$$\frac{\alpha(t; m)}{\alpha(m)} - 1 = \frac{\alpha(m)^{-1}}{t}[M_m(t) - u(Q(t); m) + u(Q(0); m)]. \qquad (7.25)$$

Now, $|u(Q(t); m))\alpha(m)^{-1}/t| \leq (c_1 + c_2|Q(t)|)/t$ independent of m (for suitable constants c_1, c_2). Since $E|Q(t)| = \rho(1-\rho)^{-1} < \infty$, it is evident that $\alpha(m)^{-1}u(Q(t_m); m)/t_m \Rightarrow 0$ as $m \to \infty$, provided that $t_m \to \infty$ (as does $\alpha(m)^{-1}u(Q(0); m)/t_m$). As for the first term on the right-hand side of (7.25), note that

$$[M_m](t) = \sum_{i=1}^{J(t)} [u(Y_j; m) - u(Y_{j-1}; m)]^2,$$

where $Y = (Y_j : j \geq 0)$ is the embedded discrete-time Markov chain associated with $(Q(t) : t \geq 0)$, and $J(t)$ is the number of jumps of Q over $[0, t]$.

This quadratic variation process is a bit easier to handle if we uniformize. (Note, for example, that $Y = (Y_j : j \geq 0)$ is non-stationary, despite the fact that we have chosen Q to be stationary.) Let $\Gamma = (\Gamma(t) : t \geq 0)$ be a Poisson process having rate $\lambda + \mu$ and let $Z = (Z_n : n \geq 0)$ be a stationary version of the discrete-time Markov chain living on the non-negative integers having transition matrix $P = (P_{ij} : i, j \geq 0)$ with $P_{i,i+1} = p \stackrel{\Delta}{=} \lambda/(\lambda + \mu)(i \geq 0)$ and $P_{i,i-1} = P_{0,0} = 1 - p$ for $i \geq 1$. Then, provided that Γ and Z are independent of one another,

$$[M_m](t) \stackrel{\mathcal{D}}{=} \sum_{i=1}^{\Gamma(t)} [u(Z_i; m) - u(Z_{i-1}; m)]^2,$$

where $\stackrel{\mathcal{D}}{=}$ denotes "equality in distribution". Then,

$$\begin{aligned}
EM_m^2(t) &= E[M_m](t) \\
&= E\Gamma(t) \cdot E[u(Z_1; m) - u(Z_0; m)]^2 \\
&= 2(\lambda + \mu)t \cdot E[u(Z_0; m)^2 - u(Z_0; m)u(Z_1; m)] \\
&= -2(\lambda + \mu)t \cdot E[u(Z_0; m) \cdot [\sum_{j=0}^{\infty} P_{Z_0, j} u(j; m) - u(Z_0; m)]]
\end{aligned}$$

By virtue of the fact that we constructed our solution $u(\cdot; m)$ so that $u(0; m) = 0$, it is straightforward to verify that

$$(\lambda + \mu)(\sum_{j=0}^{\infty} P_{ij} u(j; m) - u(i; m)) = \sum_{j=0}^{\infty} A_{ij} u(j; m) = -f_c(i; m)$$

and thus

$$EM_m^2(t) = 2tEu(Z_0; m) f_c(Z_0; m). \tag{7.26}$$

Since $Z_0 \stackrel{\mathcal{D}}{=} Q(0)$ (i.e. geometric with parameter $1 - \rho$),

$$Eu(Z_0; m) f_c(Z_0; m) = \frac{(1 + \rho)(1 - \rho^m)\rho^m}{\mu(1 - \rho)^2} - \frac{2\rho^{2m} m}{\mu(1 - \rho)}. \tag{7.27}$$

Consequently, in view of (7.26), it is evident that

$$\frac{\alpha(m)^{-2}}{t_m^2} EM_m^2(t_m) \to 0,$$

provided that $t_m \rho^m \to \infty$. Thus, $\alpha(m)^{-1} M_m(t_m)/t_m \to 0$ in mean square, and therefore in probability, under the condition $t_m \gg \rho^{-m}$. Based on (7.25), we have proved the following result, which is the M/M/1 queue-length analog to Theorem 7.3.1.

Theorem 7.4.1 If $t_m \rho^m \to \infty$ as $m \to \infty$, then

$$\alpha(t_m; m)/\alpha(m) \Rightarrow 1$$

as $m \to \infty$.

We have also done essentially all the necessary work required to establish a CLT for $\alpha(t; m)$ for fixed m.

Theorem 7.4.2 Fix $m > 0$. Then,

$$t^{1/2}(\alpha(t; m) - \alpha(m)) \Rightarrow \sigma_1(y) N(0, 1)$$

as $m \to \infty$, where

$$\sigma_1^2(y) = \frac{2(1 + \rho)(1 - \rho^m)\rho^m}{\mu(1 - \rho)^2} - \frac{4\rho^{2m} m}{\mu(1 - \rho)}.$$

Proof. The strong law for the chain $(Z_n : n \geq 0)$ guarantees that

$$\frac{1}{t}[M_m](t) \to 2Eu(Z_0; m) f(Z_0; m) \quad a.s.$$

as $t \to \infty$. Since $M_m(\cdot)$ is a martingale with discontinuous sample paths, the martingale CLT requires that we verify the additional condition

$$t^{-1/2} E[\sup_{0 \leq s \leq t} |M_m(s) - M_m(s-)|] \to 0$$

as $t \to \infty$. But

$$\sup_{0 \leq s \leq t} |M_m(s) - M_m(s-)| = \max_{1 \leq i \leq \Gamma(t)} |u(Z_i; m) - u(Z_{i-1}; m)| \quad (7.28)$$

$$\leq d_1 + d_2 \max_{1 \leq i \leq \Gamma(t)} |Z_i| \quad (7.29)$$

for suitable constructs d_1, d_2. Since $n^{-1} \sum_{i=1}^n Z_i^2 \to EZ^2 < \infty$ a.s. as $n \to \infty$, it is evident that

$$\max_{1 \leq i \leq n} Z_i^2/n \to 0 \text{ a.s.},$$

and thus

$$t^{-1/2} \max_{1 \leq i \leq \Gamma(t)} |Z_i| \to 0 \text{ a.s.}$$

as $t \to \infty$. It remains to verify that this latter limit also holds in expectation.

Now,

$$t^{-1/2} \max_{1 \leq i \leq \Gamma(t)} |Z_i| \leq 1 + t^{-1} \max_{1 \leq i \leq \Gamma(t)} Z_i^2 \quad (7.30)$$

$$\leq 1 + t^{-1} \sum_{i=1}^{\Gamma(t)} Z_i^2. \quad (7.31)$$

Since $t^{-1} \sum_{i=1}^{\Gamma(t)} Z_i^2 \to (\lambda + \mu) EZ_1^2 < \infty$ a.s. and

$$E \sum_{i=1}^{\Gamma(t)} Z_i^2 = EE[\sum_{i=1}^{\Gamma(t)} Z_i^2 | \Gamma(t)] = E\Gamma(t) \cdot EZ_1^2 = (\lambda + \mu) \cdot EZ_1^2 \cdot t,$$

it follows that $(t^{-1} \sum_{i=1}^{\Gamma(t)} Z_i^2 : t > 0)$ is a uniformly integrable family of r.v.'s, and consequently the same must be true of $(t^{-1/2} \max_{1 \leq i \leq \Gamma(t)} |Z_i| : t > 0)$.□

The above CLT suggests the approximation

$$(\rho^m t_m)^{-1/2} (\frac{\alpha(t_m; m)}{\alpha(m)} - 1) \overset{\mathcal{D}}{\approx} (\frac{2(1+\rho)}{\mu(1-\rho)^2})^{1/2} N(0,1)$$

for m large, provided that we choose the time horizon t_m so that $t_m \gg \rho^{-m}$. (A rigorous proof of this result would follow an argument similar to that used in establishing Theorem 7.3.3.)

We conclude this section with a brief discussion of the corresponding computation for the workload process $(W(t) : t \geq 0)$ associated with the M/M/1 queue. Because $(W(t) : t \geq 0)$ is neither a continuous-time Markov chain nor a diffusion, developing a martingale representation like (7.12) (or that involving $M_m(\cdot)$) for the tail probability estimator is not straightforward. Instead, we illustrate here a different approach, involving the use of regeneration.

To precisely define $W(\cdot)$, let $N = (N(t) : t \geq 0)$ be a Poisson process running at rate $\lambda > 0$ and let $V = (V_n : n \geq 1)$ be an independent sequence of i.i.d. exponential(μ) r.v.'s, with $\mu > \lambda$. Set

$$S(t) = W(0) + \sum_{i=1}^{N(t)} V_i - t.$$

Then, assuming $W(0) = 0$, $W(\cdot)$ may be represented in terms of $S(\cdot)$ as

$$W(t) = S(t) - \min_{0 \leq u \leq t} S(u).$$

It is well known that $W(t) \Rightarrow W(\infty)$ as $t \to \infty$, where $P(W(\infty) \geq y) = \rho \exp(-(\mu - \lambda)y) \triangleq \alpha(y)$ for $y > 0$. The obvious nonparametric estimator for the tail probability $\alpha(y)$ is

$$\alpha(t; y) = \frac{1}{t} \int_0^t I(W(s) \geq y) ds.$$

The key computation in developing an understanding of the asymptotics of $\alpha(t; y)$ is the calculation of the time-average variance constant $\sigma_2^2(y)$ appearing in the CLT

$$t^{1/2}(\alpha(t; y) - \alpha(y)) \Rightarrow \sigma_2(y) N(0, 1)$$

as $t \to \infty$. Noting that the origin is a regeneration state for $(W(t) : t \geq 0)$, the regenerative CLT expresses $\sigma_2^2(y)$ as

$$\sigma_2^2(y) = E_0 R^2 / E_0 T_0$$

where $E_x(\cdot) = E(\cdot | W(0) = x), T_x = \inf\{t \geq 0 : W(t) = x, W(t-) \neq x\}$, and $R = \int_0^{T_0} [I(W(t) \geq y) - \alpha(y)] dt$. Given that $(1 - \rho) = P(W(\infty) = 0) = \lambda^{-1} / E_0 T_0$, it is evident that $\sigma_2^2(y)$ can be computed via successive differentiation of the moment generating function of R. Set $a = -\alpha(y)$, $b = 1 - \alpha(y)$, and observe that conditioning on the time of the first jump yields the identity

$$E_0 \exp(\theta R) = \frac{\lambda}{\lambda - \theta a} \int_0^\infty E_x \exp(\theta R) \mu e^{-\mu x} dx.$$

The key to computing $\sigma_2^2(y)$ is therefore the calculation of $E_x \exp(\theta R)$ for $x > 0$. Note, first of all, that for all $x \geq y$, the absence of downward jumps in $W(\cdot)$ guarantees that y will be hit prior to the origin, and consequently, the strong Markov property guarantees that

$$\begin{aligned} E_x \exp(\theta R) &= E_x \exp(\theta b T_y) E_y \exp(\theta R) && (7.32) \\ &= E_{x-y} \exp(\theta b T_0) E_y \exp(\theta R). && (7.33) \end{aligned}$$

Now, since $S(\cdot)$ is a compound Poisson process with exponential jumps, it is easily verified that

$$M(t) = \exp(\kappa(S(t) - S(0)) - \kappa t(\lambda - \mu + \kappa)/(\mu - \kappa))$$

is a P_x-martingale for $\kappa < \mu$. Set $\theta = -\kappa(\lambda - \mu + \kappa)/(\mu - \kappa)$ and note that $\theta \leq \mu + \lambda - 2\sqrt{\lambda\mu} \triangleq \Lambda$. So, for $\theta \leq \Lambda$,

$$\kappa_1(\theta) = [(\theta + \mu - \lambda) - ((\lambda - \mu - \theta)^2 - 4\theta\mu)^{1/2}]/2$$

and

$$\kappa_2(\theta) = [(\theta + \mu - \lambda) + ((\lambda - \mu - \theta)^2 - 4\theta\mu)^{1/2}]/2$$

both satisfy $\theta = -\kappa(\lambda - \mu + \kappa)/(\mu - \kappa)$ and hence each of the processes

$$M_i(t) = \exp(\kappa_i(\theta)(S(t) - S(0)) + \theta t)$$

$(i = 1, 2)$ are P_x-martingales for $\theta \leq \Lambda$. For $0 < x < z$, let $\tilde{T}_z = T_z \wedge T_0$ and note that $W(t) = S(t)$ for $t \leq \tilde{T}_z$. The optional sampling theorem implies that

$$E_x M_i(\tilde{T}_z \wedge t) = 1$$

for $i = 1, 2$ and $t \geq 0$. But the memoryless structure of the exponential "overshoot" above level z allows us to re-write the above identity as

$$\begin{aligned} 1 \;=\; & \exp(-\kappa_i(\theta)x)E_x[\exp(\theta T_0); T_0 < T_z \wedge t] \\ & + \exp(\kappa_i(\theta)(z - x))(\mu/(\mu - x_i(\theta))E_x[\exp(\theta T_z); T_z < T_0 \wedge t] \\ & + E_x[\exp(\kappa_i(\theta)(S(t) - S(0)) + \theta t); T_0 \wedge T_z > t] \end{aligned} \tag{7.34}$$

For $\theta \leq 0$, we can use the dominated convergence theorem and the fact that $|S(t) - S(0)| \leq z$ on $\{\tilde{T}_z > t\}$ to obtain the identity

$$\begin{aligned} \exp(\kappa_i(\theta)x) \;=\; & E_x[\exp(\theta T_0); T_0 < T_z] \\ & + \mu E_x[\exp(\theta T_z); T_z < T_0] \exp(\kappa_i(\theta)z)/(\mu - \kappa_i(\theta)) \end{aligned} \tag{7.35}$$

for $i = 1, 2$. For $\theta > 0$, Stein's lemma (see Feller [5], p. 601) guarantees that $E_x \exp(\theta\tilde{T}_z) < \infty$ for θ in a neighborhood of the origin. For such θ's, the dominated convergence theorem ensures that the third term on the right-hand side of (7.34) converges to zero as $t \to \infty$. The monotone convergence then applies to the first two terms, thereby verifying (7.35) for positive θ's in a neighborhood of the origin.

Since (7.35) holds for both $i = 1$ and $i = 2$, we have two linear equations in the two unknowns $E_x[\exp(\theta T_0); T_0 < T_z]$ and $E_x[\exp(\theta T_z); T_z < T_0]$, from which the two unknowns may be computed. Now, for $0 < x \leq y$,

$$\begin{aligned} E_x \exp(\theta R) \;=\; & E_x[\exp(\theta a T_0); T_0 < T_y] \\ & + E_x[\exp(\theta a T_y); T_y < T_0] \cdot \mu \int_0^\infty e^{-\mu r} E_{y+r} \exp(\theta R) dr. \end{aligned} \tag{7.36}$$

But $E_{y+r}\exp(\theta R) = E_r\exp(\theta b T_0)E_y\exp(\theta R)$ for $r \geq 0$. The quantity $E_r\exp(\theta b T_0)$ is, by monotone convergence, the limit of $E_r[\exp(\theta b T_0); T_0 < T_z]$ as $z \to \infty$, and is therefore calculable. Setting $x = y$ in (7.36) then leaves (7.36) as an equation in one unknown, namely $E_y\exp(\theta R)$, from which $E_x\exp(\theta R)$ can be obtained for all $x \geq 0$.

Thus, the regenerative approach outlined here provides, in principle, a closed form for the Laplace transform of R, and the potential to analytically compute its distribution.

7.5 General Theory for the GI/G/1 Queue

In Sections 7.3 and 7.4, we computed very explicit asymptotics for RBM and the M/M/1 queue. These asymptotics provided insight into the asymptotic efficiency of our proposed nonparametric estimator for the (extreme) tail probabilities associated with the single server queue. In this section, we develop some general theory for the GI/G/1 queue. As one might expect, however, our resulting asymptotic theory is somewhat less explicit than that presented in Sections 7.3 and 7.4.

We shall focus here on the waiting time sequence $W = (W_n : n \geq 0)$ of the GI/G/1 queue. It is to be expected that similar theory may be developed for the workload process and the queue-length process of the GI/G/1 queue, using methods similar to those we will exploit in the waiting time context. Assuming that $W_0 = 0$, W may be constructed from a random walk $(S_n : n \geq 0)$ via the identity

$$W_n = S_n - \min_{0\leq k\leq n} S_k,$$

for $n \geq 0$. Letting $X_n = S_n - S_{n-1}$ be the n'th increment of the random walk, we assume that $(X_n : n \geq 1)$ is i.i.d. with $EX_1 < 0$; this "negative drift" condition on $(S_n : n \geq 0)$ is, of course, merely an assertion that the traffic intensity of the queue is strictly less than one.

Under the above "negative drift" condition, it is well known that

$$W_n \Rightarrow W_\infty$$

as $n \to \infty$. Furthermore, under certain conditions on the X_i's, it is possible to give a fairly precise characterization of the tail behavior of W_∞. Specifically, suppose that the X_i's are non-lattice r.v.'s and let $\varphi(\theta) = E\exp(\theta X_i)$ be the moment generating function of X_i. Assume that there exists a positive root θ^* to the equation $\varphi(\theta) = 1$, and assume further that $\varphi(\cdot)$ converges in a neighborhood of θ^*. Then, the Cramér-Lundberg approximation (see, for example, p. 269 of Asmussen [1]) states that

$$P(W_\infty \geq x) \sim a\exp(-\theta^* x) \tag{7.37}$$

as $x \to \infty$, for some positive constant a. The natural nonparametric estimator for $\alpha(y) \triangleq P(W(\infty) \geq y)$ is

$$\alpha(n;y) = \frac{1}{n} \sum_{j=0}^{n-1} I(W_j \geq y).$$

In view of (7.37), the theory developed thus far in this paper suggests that in order to obtain (relatively) accurate estimates of $\alpha(y)$, one must take the time horizon n large enough so that $n \gg \exp(\theta^* y)$. The remainder of this section is devoted to verifying this result rigorously.

Our argument takes advantage of the regenerative structure of W. In particular, the origin acts as a regenerative state for W. Let $T(0) = 0$, $T(n) = \inf\{m > T(n-1) : W_m = 0\}$ for $n \geq 1$, and set $\tau_i = T(i) - T(i-1)$. Put $\ell(n) = \max\{k \geq 0 : T(k) \leq n\}$, so that $\ell(n)$ counts the number of completed regenerative cycles to occur in $[0, n]$. If we let

$$\chi_j(y) = \sum_{k=T(j-1)}^{T(j)-1} (I(W_k \geq y) - \alpha(y)),$$

then $(\chi_j(y) : j \geq 1)$ is a sequence of i.i.d. random variables; furthermore, it is a standard fact of regenerative process theory that $E\chi_j(y) = 0$. (Unless otherwise stated, all expectations and probabilities in this section are computed conditionally on $W_0 = S_0 = 0$.) In addition, we may write the relative error for our nonparametric estimator in the form

$$\alpha(n;y)/\alpha(y) - 1 = n^{-1} \sum_{j=1}^{\ell(n)} \chi_j(y)/\alpha(y) + n^{-1} R(n;y) \tag{7.38}$$

where $R(n;y)$ is the "remainder term" given by

$$R(n;y) = \sum_{j=T(\ell(n))}^{n-1} [I(W_j \geq y) - \alpha(y)]/\alpha(y).$$

In view of the representation (7.38), it seems clear that the variance of $\chi_j(y)$ as $y \to \infty$ will play an important role in our subsequent analysis.

The key to this asymptotic analysis is a "change-of-measure" argument similar to that used to obtain the Cramér-Lundberg approximation itself. We start by observing that W is identical to the random walk up to time $\tau \triangleq \tau_1$, so that

$$\chi_1(y) = \sum_{j=0}^{\tau-1} (I(S_j \geq y) - \alpha(y))$$

where $\tau = \inf\{n \geq 1 : S_n \leq 0\}$. We now define a change-of-measure on the paths of the random walk. To be precise, for $y > 0$, let $T_y = \inf\{n \geq 0 : S_n \geq y\}$, and define the measure

$$\begin{aligned} &P^*(S_1 \in dx_1, \ldots, S_n \in dx_n; T_y = j) \\ &\stackrel{\Delta}{=} \quad e^{\theta^* x_j} P(S_1 \in dx_1, \ldots S_n \in dx_n; T_y = j), \end{aligned}$$

with j, n arbitrary integers satisfying $j \leq n$. If $E^*(\cdot)$ denotes the expectation corresponding to P^*, the above identity implies that

$$E^*[\xi; T_y < \infty] = E[\xi \exp(\theta^* S_{T_y}); T_y < \infty]$$

for all non-negative r.v.'s ξ. It immediately follows that

$$E[\xi; T_y < \infty] = E^*[\xi \exp(-\theta^* S_{T_y}); T_y < \infty].$$

Consequently,

$$\begin{aligned} E[(\sum_{j=0}^{\tau-1} I(S_j \geq y))^2] &= E^*[(\sum_{j=0}^{\tau-1} I(S_j \geq y))^2 \exp(-\theta^* S_{T_y}); T_y < \infty] \\ &= E^*[(\sum_{j=T_y}^{\tau-1} I(S_j \geq y))^2 \exp(-\theta^* S_{T_y}); T_y < \tau < \infty] \\ &= E^*[g(S_{T_y} - y; y) \exp(-\theta^* S_{T_y}); T_y < \tau < \infty], \end{aligned}$$

where $g(x; y) = E[(\sum_{j=0}^{T_{-y}} I(S_j \geq 0))^2 | S_0 = x]$ and $T_{-y} = \inf\{n \geq 0 : S_n \leq -y\}$ for $-y \leq 0$. It is a standard fact about such changes-of-measure that, conditional on $\{T_y \geq j\}$, P^* makes $(X_i : 1 \leq i \leq j)$ independent, with common distribution $\exp(\theta^* x) P(X_1 \in dx)$, whereas conditional on $\{T_y = j\}$, the X_i's are independent for $i > j$, with common distribution $P(X_1 \in dx)$. The "twisted" distribution $\exp(\theta^* x) P(X_1 \in dx)$ gives X_1 positive mean (equal to $\varphi'(\theta^*)$) so that the random walk $(S_n : n \geq 0)$ initially has positive drift under P^*. Consequently, $T_y < \infty$ P^* a.s. Subsequent to T_y, the random walk evolves according to its original dynamics (having negative drift) so that $\tau < \infty$ P^* a.s. Hence,

$$E[(\sum_{j=0}^{\tau-1} I(S_j \geq y))^2] = \exp(-\theta^* y) E^*[g(S_{T_y} - y; y) \exp(-\theta^*(S_{T_y} - y)); T_y < \tau]. \tag{7.39}$$

Now the monotone convergence theorem guarantees that

$$g(x; y) \nearrow g(x) \stackrel{\Delta}{=} E[(\sum_{j=0}^{\infty} I(S_j \geq 0))^2 | S_0 = x]$$

as $y \to \infty$. In addition, $I(T_y < \tau) \downarrow I(\tau = +\infty)$ and, under our non-lattice hypothesis on the X_i's, the "overshoot" $\Psi(y) \stackrel{\Delta}{=} S_{T_y} - y$ converges weakly to a limit r.v. $\Psi(\infty)$ as $y \to \infty$ (see p.168 of Asmussen [1]). These observations suggest the approximation

$$E[(\sum_{j=0}^{\tau-1} I(S_j \geq y))^2] \approx \exp(-\theta^* y) E^*[g(\Psi(\infty)) \exp(-\theta^* \Psi(\infty))] P^*[\tau = +\infty]. \tag{7.40}$$

Because it seems reasonable to expect

$$E\chi_1(y)^2 (= \text{var } \chi_1(y)) \approx E[(\sum_{j=0}^{\tau-1} I(S_j \geq y))^2]$$

for y large, this analysis suggests the following theorem.

Theorem 7.5.1 Suppose that $(X_i : i \geq 1)$ is a sequence of bounded i.i.d. non-lattice r.v.'s. If $P(X_1 > 0) > 0$ and $EX_1 < 0$, then

$$E\chi_1^2(y) \sim \exp(-\theta^* y) E^*[g(\Psi(\infty)) \exp(-\theta^* \Psi(\infty))] \cdot P^*[\tau = +\infty] \tag{7.41}$$

as $y \to \infty$.

Proof. We start by making rigorous our approximation (7.40). Since $P(X_1 > 0) > 0$, it follows that $E \exp(\theta X_1) = \varphi(\theta) \to +\infty$ as $y \to \infty$. Since $\varphi'(0) = EX_1 < 0$, the convexity of $\varphi(\cdot)$ guarantees existence of a unique $\theta^* > 0$ such that $\varphi(\theta^*) = \varphi(0) = 1$. Furthermore, the boundedness of the X_i's implies that $\varphi(\cdot)$ is everywhere finite-valued. Hence, (7.39) is valid and $\Psi(y) \Rightarrow \Psi(\infty)$ as $y \to \infty$.

We turn now to the analysis of $g(x; y)$. We first wish to prove that $g(\Psi(y); y)$ is a uniformly bounded sequence of r.v.'s under P^* so that, in particular, $g(\Psi(\infty))$ is bounded; a sketch of the proof follows. Observe that

$$\sum_{j=0}^{\infty} I(S_j \geq 0) \leq T_{-1} + \sum_{i=1}^{\beta} T_{-1,i}, \tag{7.42}$$

where β is the number of times that the random walk proceeds from below level -1 to above the origin, and $T_{-1,i}$ is the time required, on the i'th excursion above the origin, to go below level -1 again. Each time the random walk goes below level -1, there is a probability at most equal to $P_0(T_1 < \infty) < 1$ of an additional excursion. Consequently, $P(\beta > k) \leq P_0(T_1 < \infty)^k$ so β has moments of all orders. In addition, because the X_i's are bounded and living in $[-K, K]$, say,

$$\begin{aligned} P(T_{-1,i} > t) &\leq P(T_{-1} \geq t | S_0 = K) \\ &= P(\tau \geq t | S_0 = K + 1). \end{aligned}$$

Set $\psi(\theta) = \log \varphi(\theta)$ and observe that $\exp(\theta S_n - \psi(\theta)n)$ is a martingale. Hence, the optional sampling theorem implies that

$$E[\exp(\theta S_{\tau\wedge n} - \psi(\theta)(\tau \wedge n))|S_0 = x] = \exp(\theta x).$$

Consequently,

$$E[\exp(\theta S_\tau - \psi(\theta)\tau)I(\tau \leq n)|S_0 = x] \leq \exp(\theta x),$$

and the monotone convergence theorem then implies that for $\theta > 0$,

$$\exp(-\theta K)E[\exp(-\psi(\theta)\tau)|S_0 = x] \leq E[\exp(\theta S_\tau - \psi(\theta)\tau)|S_0 = x] \leq \exp(\theta x).$$

Thus, if we choose $\theta > 0$ so that $\psi(\theta) < 0$, we may conclude that the $T_{-1,i}$'s have uniformly bounded exponential moments, as does T_1 (provided that S_0 lies in a compact set). We conclude from (7.42) that $g(\cdot)$ is bounded on compact sets. But $g(\Psi(y); y) \leq g(\Psi(y)) \leq \sup\{g(x) : 0 \leq x \leq K\}$, and hence the $g(\Psi(y); y)$'s are a family of uniformly bounded r.v.'s.

Our next task is to prove that we can replace $g(\Psi(y); y)$ in (7.41) by $g(\Psi(y))$. This can be done provided that we show that

$$g(\Psi(y); y) - g(\Psi(y)) \to 0$$

P^* a.s. as $y \to \infty$. Because of the boundedness of $\Psi(y)$ (by K), it suffices to prove that

$$\sup_{0 \leq x \leq K} |g(x; y) - g(x)| \to 0 \tag{7.43}$$

as $y \to \infty$. Recall that

$$\begin{aligned} g(x) &= E[(\sum_{k=0}^{\infty} I(S_k \geq 0))^2|S_0 = x] \\ &= g(x; y) + 2E[(\sum_{j=0}^{T_{-y}} I(S_j \geq 0)) \cdot (\sum_{j=T_{-y}+1}^{\infty} I(S_j \geq 0))|S_0 = x] \\ &\quad + E[(\sum_{j=T_{-y}+1}^{\infty} I(S_j \geq 0))^2|S_0 = x]. \end{aligned}$$

Relation (7.43) will therefore follow from the Cauchy-Schwarz inequality if we show that

$$\sup_{0 \leq x \leq K} E[(\sum_{j=T_{-y}+1}^{\infty} I(S_j \geq 0))^2|S_0 = x] \to 0$$

as $y \to \infty$. Note that the above r.v. vanishes unless the random walk exceeds level 0 subsequent to time T_{-y}. Applying the strong Markov property at time T_{-y} and taking advantage of the stochastic monotonicity and spatial homogeneity of random walk, we can bound the above by

$$E[(\sum_{j=0}^{\infty} I(S_j \geq y))^2; T_y < \infty].$$

Applying the strong Markov property, at time T_y, to the above expectation, we obtain

$$E[g(\Psi(y)); T_y < \infty].$$

But $g(\Psi(y))$ is bounded in y and $P(T_y < \infty) \to 0$ as $y \to \infty$, proving (7.43).

The proof of (7.40) is therefore complete if we can show that

$$\begin{aligned} & E^*[g(\Psi(y))\exp(-\theta^*\Psi(y))I(T_y < \tau)] \\ \to \quad & E^*[g(\Psi(\infty))\exp(-\theta^*\Psi(\infty))]P^*[\tau = +\infty]. \end{aligned} \tag{7.44}$$

To prove the above limit relation, we first show that

$$\exp(-\theta^*\Psi(y))g(\Psi(y)) \Rightarrow \exp(-\Psi(\infty))g(\Psi(\infty))$$

as $y \to \infty$; this will follow if we establish that $g(\cdot)$ is continuous a.s. at $\Psi(\infty)$. Since X_1 is non-lattice, it is evident that $\Psi(\infty)$ is a continuous r.v. (see p. 168 of Asmussen [1]), and therefore it suffices to prove that $g(\cdot)$ has at most countably many discontinuities. But the stochastic monotonicity of random walk implies that $g(\cdot)$ is non-decreasing. Such a function can have at most countably many discontinuities, and thus it follows that $\exp(-\theta^*\Psi(y))g(\Psi(y)) \Rightarrow \exp(-\Psi(\infty))g(\Psi(\infty))$ as $y \to \infty$.

Now the left-hand side of (7.44) can be written as

$$E^*[g(\Psi(y))\exp(-\theta^*\Psi(y))] - E^*[g(\Psi(y))\exp(-\theta^*\Psi(y))I(\tau < T_y)]. \tag{7.45}$$

Denote the first expectation in (7.45) as $\tilde{g}(y)$. Since $(g(\Psi(y))\exp(-\theta^*\Psi(y)) : y > 0)$ is bounded in y and converges weakly, it is evident that $\tilde{g}(y)$ converges to $E^*g(\Psi(\infty))\exp(-\theta^*\Psi(\infty))$ as $y \to \infty$. As for the second expectation in (7.45), applying the strong Markov property at time τ allows us to re-write it as

$$E^*[\tilde{g}(y - S_\tau)I(\tau < T_y)].$$

Since $\tilde{g}(y - S_\tau) \to E^*g(\Psi(\infty))\exp(-\theta^*\Psi(\infty))$ as $y \to \infty$ and $I(\tau < T_y) \to I(\tau < \infty)$ P^* a.s. as $y \to \infty$, (7.44) is therefore proved. Our proof thus far yields the conclusion that

$$\begin{aligned} & E[(\sum_{j=0}^{\tau-1} I(S_j \geq y))^2] \\ \sim \quad & \exp(-\theta^* y)E^*[g(\Psi(\infty))\exp(-\theta^*\Psi(\infty))] \cdot P^*[\tau = +\infty] \end{aligned} \tag{7.46}$$

as $y \to \infty$. But

$$E\chi_1^2(y) = E[(\sum_{j=0}^{\tau-1} I(S_j \geq y))^2] - 2\alpha(y)E[\tau \sum_{j=0}^{\tau-1} I(S_j \geq y)] + \alpha^2(y)E\tau^2. \tag{7.47}$$

We earlier showed that τ has moments of all orders. Since

$$\tau \sum_{j=0}^{\tau-1} I(S_j \geq y) \leq \tau^2,$$

it is evident, via an application of the dominated convergence theorem, that

$$E[\tau \cdot \sum_{j=0}^{\tau-1} I(S_j \geq y)] \to 0$$

as $y \to \infty$, as does $\alpha(y)E\tau^2$. Relations (7.46) and (7.47), together with the Cramér-Lundberg approximation for $\alpha(y)$, then prove the theorem. □

Theorem 7.5.1 provides a precise asymptotic for the behavior of $E\chi_1^2(y)$. This, in turn, sheds light on the asymptotic behavior of the relative error $\alpha(n; y)/\alpha(y) - 1$. The CLT for regenerative processes guarantees that for y fixed,

$$n^{1/2}(\frac{\alpha(n;y)}{\alpha(y)} - 1) \Rightarrow (\frac{E\chi_1^2(y)}{\alpha^2(y)E\tau})^{1/2} N(0,1)$$

as $n \to \infty$. Relation (7.48) suggests that

$$\frac{\alpha(n;y)}{\alpha(y)} - 1 \overset{\mathcal{D}}{\approx} (\frac{E\chi_1^2(y)}{n\alpha^2(y)E\tau})^{1/2} N(0,1) \tag{7.48}$$

for n large. But, according to (7.37) and Theorem 7.5.1,

$$\frac{E\chi_1^2(y)}{\alpha^2(y)E\tau} \sim \exp(\theta^* y) r \tag{7.49}$$

as $y \to \infty$, where

$$r \triangleq \frac{E^*[g(\Psi(\infty)) \exp(-\theta^* \Psi(\infty))] P^*[\tau = +\infty]}{(E^*[\exp(-\theta^* \Psi(\infty))])^2 E\tau}.$$

(See Asmussen [1], p. 269, for the exact form of the constant a appearing in (7.37).) Thus (7.48) and (7.49) together suggest the approximation

$$\frac{\alpha(n_y;y)}{\alpha(y)} - 1 \overset{\mathcal{D}}{\approx} (\frac{\exp(\theta^* y)}{n_y} r)^{1/2} N(0,1),$$

provided $n_y \gg \exp(\theta^* y)$. Our final theorem of this section makes this approximation rigorous.

Theorem 7.5.2 Assume the conditions of Theorem 7.5.1 and suppose that $n_y \exp(-\theta^* y) \to +\infty$ as $y \to \infty$. Then,

$$(n_y \exp(-\theta^* y))^{1/2}(\frac{\alpha(n_y; y)}{\alpha(y)} - 1) \Rightarrow r^{1/2} N(0,1)$$

as $y \to \infty$.

Proof. We return to the representation (7.38) for the relative error. Our first task is to deal with the remainder term. We wish to show that

$$(n_y \exp(-\theta^* y))^{1/2} R(n_y; y)/n_y \Rightarrow 0 \tag{7.50}$$

as $y \to \infty$. Let $V_j(y) = \sum_{k=T(j-1)}^{T(j)-1} I(W_k \geq y)$ and observe that

$$|R(n_y; y)/n_y| \leq \max_{1 \leq j \leq n_y} V_j(y)/n_y \alpha(y) + \max_{1 \leq j \leq n_y} \tau_j / n_y.$$

Because $E\tau_j^2 < \infty$ (see the proof of Theorem 7.5.1), it follows that

$$\max_{1 \leq j \leq n_y} \tau_j / n_y^{1/2} \to 0 \text{ a.s.}$$

as $y \to \infty$. To finish the proof of (7.50), we therefore need to show that

$$\max_{1 \leq j \leq n_y} V_j(y)/(n_y \alpha(y))^{1/2} \Rightarrow 0.$$

We accomplish this task by showing that the sequence converges to zero in mean square error. Note that

$$\frac{1}{n_y \alpha(y)} E \max_{1 \leq j \leq n_y} V_j(y)^2 = \frac{1}{n_y \alpha(y)} \int_0^\infty P(\max_{1 \leq j \leq n_y} V_j(y)^2 > t) dt. \tag{7.51}$$

For each fixed t, $P(\max_{1 \leq j \leq n_y} V_j(y)^2 > t)/n_y \alpha(y) \to 0$ as $y \to \infty$. On the other hand,

$$\begin{aligned} P(\max_{1 \leq j \leq n_y} V_j(y)^2 > t)/n_y \alpha(y) &\leq P(V_1(y)^2 > t)/\alpha(y) \\ &\leq EV_1(y)^4 / t^2 \alpha(y). \end{aligned}$$

The proof of Theorem 7.5.1 establishes that $EV_1(y)^2 = O(\alpha(y))$ as $y \to \infty$. A similar change-of-measure argument proves that $EV_1(y)^p = O(\alpha(y))$ as $y \to \infty$, for $p > 0$. Consequently, we can apply the dominated convergence theorem to (7.51), thereby obtaining the asymptotic negligibility of the remainder term.

The proof is therefore complete if we can show that

$$(n_y \alpha(y))^{-1/2} \sum_{j=1}^{\ell(n_y)} \chi_j(y) \Rightarrow (ar)^{1/2} N(0,1)$$

as $y \to \infty$, where a is the constant appearing in (7.37). This, in turn, follows if we can prove that

$$(n_y\alpha(y))^{-1/2}\Big(\sum_{j=1}^{\ell(n_y)} \chi_j(y) - \sum_{j=1}^{\lfloor \ell n_y \rfloor} \chi_j(y)\Big) \Rightarrow 0 \tag{7.52}$$

and

$$(n_y\alpha(y))^{-1/2} \sum_{j=1}^{\lfloor \ell n_y \rfloor} \chi_j(y) \Rightarrow (ar)^{1/2} N(0,1) \tag{7.53}$$

as $y \to \infty$, where $\ell \stackrel{\Delta}{=} 1/E\tau$. We note that $\ell(n)/n \to \ell$ a.s. as $n \to \infty$, so that $I(|\ell(n)/n - \ell| < \epsilon) \to 1$ a.s. as $n \to \infty$ for any $\epsilon > 0$. To obtain (7.42), fix $\epsilon > 0$ and observe that on $\{|\ell(n)/n - \ell| < \epsilon\}$,

$$\Big|\sum_{j=1}^{\ell(n_y)} \chi_j(y) - \sum_{j=1}^{\lfloor \ell n_y \rfloor} \chi_j(y)\Big| \le \max_{|k| \le \lceil n_y \epsilon \rceil} \Big|\sum_{j=\lceil \ell n_y \rceil}^{\ell n_y + k} \chi_j(y)\Big|.$$

Kolmogorov's maximal inequality implies, however, that

$$P\Big(\max_{k \le \lceil n_y \epsilon \rceil} \Big|\sum_{j=\lfloor \ell n_y \rfloor}^{\lfloor \ell n_y \rfloor + k} \chi_j(y)\Big| > x(n_y\alpha(y))^{1/2}\Big) \le 2\lceil n_y \epsilon \rceil E\chi_1^2(y)/(x^2 n_y \alpha(y)).$$

Theorem 7.5.1 and the fact that ϵ may be made arbitrarily small then yield (7.52).

To prove (7.53), we use the Lindeberg-Feller theorem for triangular arrays of i.i.d. r.v.'s; see, for example, p. 205 of Chung [3]. The triangular array that arises here is trivially holospoudic. As for the Lindeberg condition, this requires showing that

$$E[\chi_1^2(y)/\alpha(y); \chi_1^2(y) > x\alpha(y)n_y] \to 0 \tag{7.54}$$

as $y \to \infty$. But the above expectation is bounded by $E\chi_1^4(y)/(x\alpha^2(y)n_y)$. It is easily seen that

$$E\chi_1^4(y) = EV_1^4(y) + o(\alpha(y)) = O(\alpha(y)),$$

from which (7.54) follows. Application of the Lindeberg-Feller CLT gives (7.53), proving the theorem. □

Theorem 7.5.2 provides a precise rate of convergence for the relative error of our nonparametric tail probability estimator. Thus, we have established, in significant generality, the relative magnitude of the time horizon over which a queue needs to be observed, in order that one obtain (relatively) accurate estimates of the tail probability.

7.6 Some Concluding Remarks

In the preceding sections of this paper, we analyzed a number of different queueing models, with the aim of producing as strong an argument as possible for the following claim:

> If a tail probability for a queue is to be estimated via the observed proportion of time that the corresponding performance measure is out in the tail, then, in order that the probability be (relatively) accurately estimated, it is necessary and sufficient that the time horizon over which the queue is observed be large relative to the reciprocal of the tail probability.

Given that queues typically exhibit non-negative autocorrelations, the necessity was covered by Proposition 7.2.1. As for sufficiency, an asymptotic theory in which the various limiting constants were explicitly identified was developed for both the M/M/1 queue and RBM. In addition, we verified the sufficiency of such a time horizon for uniformly mixing processes and for the waiting time sequence of the single-server queue. Because tail probabilities for queues typically decay exponentially fast, the above claim establishes that the time interval over which a queue needs to be observed in order to (relatively) accurately estimate a tail probability grows exponentially rapidly in the tail parameter. This implies that, if such a nonparametric estimation approach is adapted, then the amount of data which one must collect in order to accurately estimate the loss probability for a buffered queue is potentially enormous.

This suggests that perhaps one should explore alternative means, for estimating such tail probabilities. In particular, suppose that the queue can be well described by a (finite-dimensional) parametric model. This would typically occur when, for example, enough is known about the behavior of the input sources to the queue that a parameterized stochastic model describing the input can be developed. In such a setting, the parametric estimator of the tail probability associated with level y typically takes the form $\gamma(\hat{\theta}; y)$, where $\hat{\theta}$ is an estimator of the parametric vector θ describing (for example) the input model, and $\gamma(\theta; y)$ is the tail probability associated with level y under parameter θ. Torres and Glynn [12] show that, in contrast to the estimators described in the current paper, the tail probability associated with tail parameter y can be (relatively) accurately estimated provided that the time interval over which the queue is observed is large relative to y^2. Thus, although tail estimation continues to get harder in the parametric context, the rate at which it gets harder is (much) better behaved than in the nonparametric setting. Of course, the disadvantage of the parametric approach is the need to fit an appropriate parametric family to the (for example) input sources to the system.

When this parametric methodology is applied, it is clearly necessary to evaluate $\gamma(\theta; y)$ at $\theta = \hat{\theta}$. Furthermore, if confidence intervals for the tail probability are required, the derivative $\gamma'(\hat{\theta}; y)$ (with respect to θ) needs to be computed; while such computations are trivial for simple parametric models like the M/M/1 queue, they are decidedly less so for more complex models, such as queues in which the arrival process is described by a Markov modulated Poisson process. For more complex parametric models, simulation may be needed to compute $\gamma(\hat{\theta}; y)$ and/or $\gamma'(\hat{\theta}; y)$.

The results developed in this paper have important implications for such simulations. The conventional simulation-based estimator for a tail probability is precisely the non-parametric tail estimator considered in this paper, namely the proportion of time that the simulated queue lies out in the tail. The theory developed in this paper makes clear the enormous computation time demanded by such estimators. This suggests that one consider alternative means of computing such tail probabilities. Fortunately, this is a problem to which the simulation-based efficiency improvement technique known as importance sampling is ideally suited. Chang et al. [2] and Glynn [7] show that when this method is applied to the tail probability estimation problem, the computational difficulty increases only linearly in the tail parameter y. (The variance per run is roughly constant in y; however, the computation time per simulated run increases linearly in y because this growth describes the time required for the queue, under the change-of-measure, to hit level y.) This is to be compared with the exponential growth in difficulty associated with the conventional simulation-based estimator. Thus, enormous improvements in efficiency are to be afforded by using importance sampling in this context.

Of course, if parametric tail probability estimators are to be computed in "real time", one would ideally pre-compute $\gamma(\cdot; y)$ and $\gamma'(\cdot; y)$ off-line and store the corresponding function values in memory. This would avoid the need to (possibly) simulate the queue in real time in order to evaluate $\gamma(\hat{\theta}; y)$ and $\gamma'(\hat{\theta}; y)$.

Acknowledgments: This research was supported by the Army Research Office under contract no. DAAL03-91-G-0319.

7.7 References

[1] Asmussen, S., *Applied Probability and Queues,* John Wiley & Sons, New York, 1987.

[2] Chang, C.S., Heidelberger, P., Juneja, S., and Shahabuddin, P., Effective bandwidth and fast simulation of ATM intree networks, *Performance Evaluation,* **20**, 45–65, 1994.

[3] Chung, K.L., *A Course in Probability Theory,* Academic Press, New York, 1974.

[4] CHUNG, K.L. AND WILLIAMS R.G., *Stochastic Integration,* Birkhäuser, Boston, 1983.

[5] ETHIER, S.N. AND KURTZ, T.G., *Markov Processes: Characterization and Convergence,* John Wiley & Sons, New York, 1986.

[6] FELLER, W., *An Introduction to Probability Theory and its Applications, Volume 2,* John Wiley & Sons, New York, 1971.

[7] GLYNN, P.W., Efficiency Improvement Techniques, *Annals of Operations Research,* **53**, 175–197, 1994.

[8] GLYNN, P.W., Diffusion Approximations in *Stochastic Models, Handbook of OR and MS Volume 2,* (D. Heyman and M. Sobel, eds.,) Elsevier Science Publishers, 1990.

[9] HARRISON, J.M., *Brownian Motion and Stochastic Flow Systems,* John Wiley & Sons, New York, 1985.

[10] KARLIN, S. AND TAYLOR, H.M., *A Second Course in Stochastic Processes,* Academic Press, New York, 1981.

[11] STOYAN, D., *Comparison Methods for Queues and Other Stochastic Models,* John Wiley & Sons, New York, 1983.

[12] TORRES, M. AND GLYNN, P.W., Parametric estimation of tail probabilties for the single-server queue, Technical Report, Department of Operations Research, Stanford University, Stanford, CA, 1996.

8
Rational Interpolation for Rare Event Probabilities

Wei-Bo Gong and Soracha Nananukul

ABSTRACT We propose to use rational interpolants to tackle some computationally complex performance analysis problems such as rare-event probabilities in stochastic networks. Our main example is the computation of the cell loss probabilities in ATM multiplexers. The basic idea is to use the values of the performance function when the system size is small, together with the asymptotic behaviour when the size is very large, to obtain a rational interpolant which can be used for medium or large systems. This approach involves the asymptotic analysis of the rare-event probability as a function of the system size, the convergence analysis of rational interpolants on the positive real line, and the quasi-Monte Carlo analysis of discrete event simulation.

8.1 Motivation: Padé Approximation for the GI/GI/1 Queue

The introduction of rational interpolation for evaluating rare event probabilities in [22] was motivated by the earlier work on the application of Padé approximants to single-server queues with renewal arrival processes [7]. To obtain the Padé approximants of a performance function we first need to obtain its MacLaurin series. This has been done for several systems [6, 7, 10, 23]. The basic idea is to expand the innermost part of the performance function (for example the k-fold convolution of the interarrival-time density functions) into a MacLaurin series. Then interchange the operations (integrations, expectations, and summations) so that the summation operation is in the outermost position. The interchanges of operations are usually justifiable using the dominated convergence theorem.

We first derive the MacLaurin series of the k-fold convolution of a function. Let $f(t)$ and $g(t)$ be defined on the real line and are 0 when $t < 0$. The convolution of $f(t)$ and $g(t)$ is denoted by $(f * g)(t)$, and the k-fold convolution of $f(t)$ is denoted by $f_{*k}(t)$, where $f_{*1}(t) \triangleq f(t)$. Assume that $\sum_{i=0}^{\infty} \frac{f^{(i)}(0)}{i!} t^i$ and $\sum_{i=0}^{\infty} \frac{g^{(i)}(0)}{j!} t^i$ converge at any finite $t \geq 0$. The convolution of $f(t)$ and $g(t)$ is

$$(f * g)(t) = \int_0^t f(t-\tau)g(\tau)d\tau.$$

Expand $f(t-\tau)$ and $g(t)$ around the origin, multiply the two series, and do the integrations term by term, we have

$$\begin{aligned}
(f * g)(t) &= \int_0^t \sum_{i=0}^{\infty} \frac{f^{(i)}(0)}{i!}(t-\tau)^i \sum_{j=0}^{\infty} \frac{g^{(j)}(0)}{j!}\tau^j \, d\tau \\
&= \int_0^t \sum_{n=0}^{\infty}\sum_{k=0}^{n} \frac{f^{(k)}(0)}{k!}(t-\tau)^k \frac{g^{(n-k)}(0)}{(n-k)!}\tau^{n-k} d\tau \\
&= \sum_{n=0}^{\infty}\sum_{k=0}^{n} \frac{f^{(k)}(0)}{k!}\frac{g^{(n-k)}(0)}{(n-k)!}\int_0^t (t-\tau)^k \tau^{n-k} d\tau \\
&= \sum_{n=0}^{\infty}\sum_{k=0}^{n} \frac{f^{(k)}(0)}{k!}\frac{g^{(n-k)}(0)}{(n-k)!}\frac{k!(n-k)!}{(n+1)!}t^{n+1} \\
&= \sum_{n=0}^{\infty}\sum_{k=0}^{n} \frac{f^{(k)}(0)g^{(n-k)}(0)}{(n+1)!}t^{n+1}. \qquad (8.1)
\end{aligned}$$

Therefore $(f * g)^{(n)}(0) = \sum_{i=0}^{n-1} f^{(i)}(0)g^{(n-1-i)}(0)$, for $n = 1, 2, ...,$ and $(f * g)(0) = 0$. Substitute g with $f_{*(k-1)}$, we have

$$f_{*k}^{(n)}(0) = \sum_{i=0}^{n-1} f^{(i)}(0) f_{*(k-1)}^{(n-1-i)}(0), \qquad (8.2)$$

for $n = 1, 2, ...,$ and $f_{*k}(0) = 0$. Note that $f_{*k}^{(n)}(0) = 0$ when $n < k-1$. Therefore, $f_{*k}(t) = \sum_{j=k-1}^{\infty} f_{*k}^{(j)}(0)\frac{t^j}{j!}$.

Next we give a brief description of Padé approximants. For details see [2]. Consider a formal power series $J(z) = \sum_{j=0}^{\infty} c_j z^j$. The $[L/M]$ Padé approximant to $J(z)$ is a rational function $[L/M] = \frac{P_L(z)}{Q_M(z)}$, where $P_L(z) = a_0 + a_1 z + \ldots + a_L z^L$ and $Q_M(z) = b_0 + b_1 z + \ldots + b_M z^M$, which satisfies

$$J(z) - \frac{P_L(z)}{Q_M(z)} = O(z^{L+M+1}).$$

Coefficients of $P_L(z)$ and $Q_M(z)$ can be obtained by multiplying both sides by $Q_M(z)$ and equating the coefficients of $z^0, z^1, \ldots, z^{L+M}$ and let $b_0 = 1$ (by convention). We have

$$\begin{pmatrix} c_{L-M+1} & c_{L-M+2} & c_{L-M+3} & \cdots & c_L \\ c_{L-M+2} & c_{L-M+3} & c_{L-M+4} & \cdots & c_{L+1} \\ \vdots & \vdots & \vdots & \vdots & \vdots \\ c_L & c_{L+1} & c_{L+2} & \cdots & c_{L+M-1} \end{pmatrix} \begin{pmatrix} b_M \\ b_{M-1} \\ \vdots \\ b_1 \end{pmatrix} = - \begin{pmatrix} c_{L+1} \\ c_{L+2} \\ \vdots \\ c_{L+M} \end{pmatrix}$$

and

$$\begin{aligned} a_0 &= c_0, \\ a_1 &= c_1 + b_1 c_0, \\ &\vdots \\ a_L &= c_L + \sum_{i=1}^{\min(L,M)} b_i c_{L-i}. \end{aligned}$$

We now turn to the $GI/G/1$ queue with an i.i.d. interarrival-time process $\{A_n\}$ and an i.i.d. service-time process $\{B_n\}$. They are independent. Assume that B_n can be expressed as θX_n, where $\theta \geq 0$ is a scaling parameter, and X_n's are independent samples of a random variable X with unity mean and is independent of θ. The order of service is FIFO. Let W_n be the waiting time of the nth customer. At steady state, we will denote it by W.

For $n = 1, 2, 3, \ldots$, define

$$\begin{aligned} U_n &\triangleq B_n - A_n, \\ S_0 &\triangleq 0, \\ S_n &\triangleq U_1 + \ldots + U_n, \\ M_n &\triangleq max(S_0, \ldots, S_n). \end{aligned}$$

Each sample path of $\{M_n\}$ is a nondecreasing sequence, and M_n converges w.p.1 to a random variable M. It can be shown that [21]

$$E[M_n](\theta) = \sum_{k=1}^{n} \frac{1}{k} E[S_k^+](\theta), \quad n = 1, 2, \ldots, \tag{8.3}$$

when the queue is initially empty, and

$$E[W](\theta) = E[M](\theta) = \sum_{k=1}^{\infty} \frac{1}{k} E[S_k^+](\theta),$$

where $x^+ = \max(x, 0)$. A condition for $E[W](\theta)$ to be finite is $\theta < E[A_n]$ and $E[B_n^2] < \infty$.

Assume that

1. The probability density function of A_n, denoted by f, has a strictly proper rational Laplace transform.

2. $E[e^{\theta_1 X}] < \infty$ for some $\theta_1 > 0$.

Let $Y_k \triangleq \sum_{i=1}^{k} X_i$. Then we have

$$E[S_k^+ | X_1, \ldots, X_k](\theta) = E[(U_1 + \ldots + U_k)^+ | X_1, \ldots, X_k]$$

$$= \int_0^{\theta Y_k} (\theta Y_k - y) f_{*(k)}(y) dy$$

$$= \int_0^{\theta Y_k} (\theta Y_k - y)(\sum_{i=k-1}^{N} f_{*(k)}^{(i)}(0) \frac{y^i}{i!}) dy + R_2(N, \theta, k),$$

where $f_{*k}(t)$ is the k-fold convolution of $f(t)$ and $f_{*1}(t) \triangleq f(t)$, and

$$R_2(N, \theta, k) = \int_0^{\theta Y_k} (\theta Y_k - y) \sum_{j=N+1}^{\infty} f_{*(k)}^{(j)}(0) \frac{y^j}{j!} dy$$

and $N \geq k-1$. Do the integrations term by term.

$$\begin{aligned} E[S_k^+ | X_1, ..., X_k](\theta) &= \sum_{i=k-1}^{N} \int_0^{\theta Y_k} (\theta Y_k - y) f_{*(k)}^{(i)}(0) \frac{y^i}{i!} dy + R_2(N, \theta, k) \\ &= \sum_{i=k-1}^{N} \frac{f_{*(k)}^{(i)}(0)}{(i+2)!} (\theta Y_k)^{i+2} + R_2(N, \theta, k). \end{aligned}$$

Using the fact that $E\{E[S_k^+ | X_1, ..., X_k]\}(\theta) = E[S_k^+](\theta)$, we have

$$E[S_k^+](\theta) = \sum_{i=k-1}^{N} \frac{f_{*(k)}^{(i)}(0)}{(i+2)!} E[(\theta Y_k)^{i+2}] + E[R_2(N, \theta, k)],$$

So from (8.3), we have

$$\begin{aligned} E[M_n](\theta) &= \sum_{k=1}^{n} \sum_{i=k-1}^{N} \frac{1}{k} \frac{f_{*(k)}^{(i)}(0)}{(i+2)!} E[(\theta Y_k)^{i+2}] + \sum_{k=1}^{n} \frac{1}{k} E[R_2(N, \theta, k)] \\ &= \sum_{i=0}^{N} \sum_{k=1}^{\min(i+1,n)} \frac{1}{k} \frac{f_{*(k)}^{(i)}(0) E[(\theta Y_k)^{i+2}]}{(i+2)!} + \sum_{k=1}^{n} \frac{1}{k} E[R_2(N, \theta, k)]. \end{aligned}$$

Set n to $N+1$.

$$E[M_{N+1}](\theta) = \sum_{i=0}^{N} \sum_{k=1}^{i+1} \frac{1}{k} \frac{f_{*(k)}^{(i)}(0) E[(\theta Y_k)^{i+2}]}{(i+2)!} + \sum_{k=1}^{N+1} \frac{1}{k} E[R_2(N, \theta, k)].$$

Let $N \to \infty$. When $\theta < E[A_n]$, $\lim_{N\to\infty} E[M_{N+1}](\theta) = E[M](\theta)$ is finite. It can be shown [7] that

$$\lim_{N\to\infty} \sum_{k=1}^{n} \frac{1}{k} E[R_2(N, \theta, k)] = \lim_{N\to\infty} \sum_{k=1}^{N+1} \frac{1}{k} E[R_2(N, \theta, k)] = 0,$$

	$E[W]$		$E[W_5]$	
θ	$[2/2]\frac{\theta^3}{(1-\theta)}$	Simulation	$[8/8]\theta^3$	Simulation
0.1	0.0017	0.0017	0.0017	0.0017
0.2	0.0121	0.0122	0.0121	0.0120
0.3	0.0393	0.0394	0.0387	0.0390
0.4	0.0944	0.0943	0.0892	0.0891
0.5	0.1961	0.1949	0.1697	0.1692
0.6	0.3812	0.3780	0.2840	0.2836
0.7	0.7352	0.7305	0.4333	0.4336
0.8	1.5140	1.4723	0.6162	0.6165
0.9	3.9969	4.0099	0.8306	0.8293

TABLE 8.1. $E[W](\theta)$ and $E[W_5](\theta)$ of $E_2/E_2/1$

when $\theta \in [0, c)$ for some $c > 0$. Therefore, we have that, when $\theta \in [0, \min(c, E[A_n]))$,

$$E[M_n](\theta) = \sum_{i=0}^{\infty} \sum_{k=1}^{\min(i+1,n)} \frac{1}{k} \frac{f_{*(k)}^{(i)}(0) E[(Y_k)^{i+2}]}{(i+2)!} \theta^{i+2},$$

and

$$E[M](\theta) = \sum_{i=0}^{\infty} \sum_{k=1}^{i+1} \frac{1}{k} \frac{f_{*(k)}^{(i)}(0)}{(i+2)!} E[(Y_k)^{i+2}] \theta^{i+2}.$$

As an example, we consider the $E_2/E_2/1$ queue. Without loss of generality let the mean interarrival time be 1. Since $E[W](\theta) \to \infty$ as $\theta \to 1$, and $\lim_{\theta \to 0} \frac{E[W](\theta)}{\theta^3}$ is finite, we write

$$E[W](\theta) = \frac{\theta^3}{(1-\theta)} \left[\frac{(1-\theta)}{\theta^3} E[W](\theta) \right],$$

and approximate $\frac{(1-\theta)}{\theta^3} E[W](\theta)$ by a [2/2]. We have

$$[2/2] = \frac{2 + 7.6667\theta + 0.7619\theta^2}{1 + 7.3333\theta + 12.0476\theta^2}.$$

In Table 8.1, we compare the approximation of $E[W](\theta)$ with simulation results using 1,000,000 departures. A plot of the results is shown in Figure 1.

Next consider $E[W_5] = E[M_4]$ in the $E_2/E_2/1$ queue, initially empty. Since $\lim_{\theta \to 0} \frac{E[W_5](\theta)}{\theta^3}$ is finite, we write $E[W_5](\theta) = \theta^3 \left[\frac{E[W_5](\theta)}{\theta^3} \right]$, and

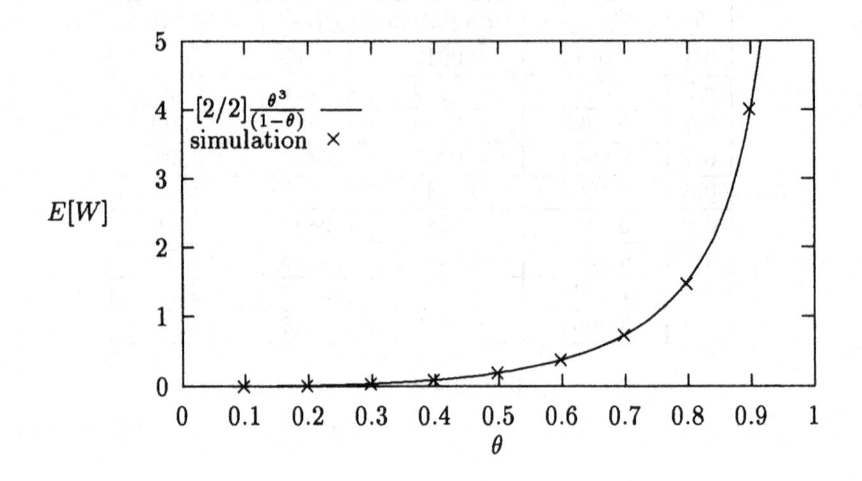

FIGURE 1. $E[W](\theta)$ of $E_2/E_2/1$

approximate $\frac{E[W_5](\theta)}{\theta^3}$ by [8/8]. We have [8/8] =

$$\frac{2+20.65\theta+108.1\theta^2+342\theta^3+800.2\theta^4+1189\theta^5+2261\theta^6-1062\theta^7+339.7\theta^8}{1+12.83\theta+74.64\theta^2+256.1\theta^3+592\theta^4+922.9\theta^5+976.3\theta^6+659.2\theta^7+228.3\theta^8}.$$

In Table 8.1, we compare the approximation of $E[W_5](\theta)$ with simulation results using 1,000,000 departures. A plot of the results is shown in Figure 2.

As we can see, the numerical results are good in spite of the fact that we used only low-order Padé approximants. This hints that sequences of Padé approximants $[n/n]$ may generally converge quite fast to the intended functions. A general convergence theorem claiming geometric convergence will be given in the following section. It covers not only the convergence of sequences of Padé approximants, but also the convergence of sequences of the more general rational interpolants. Before we get to that, we first introduce the concept of rational interpolants.

8.2 Rational Interpolation

Padé approximation method only works when the MacLaurin series of the performance function is known. This greatly limits its application. However, in many cases, although it is very costly to calculate the performance

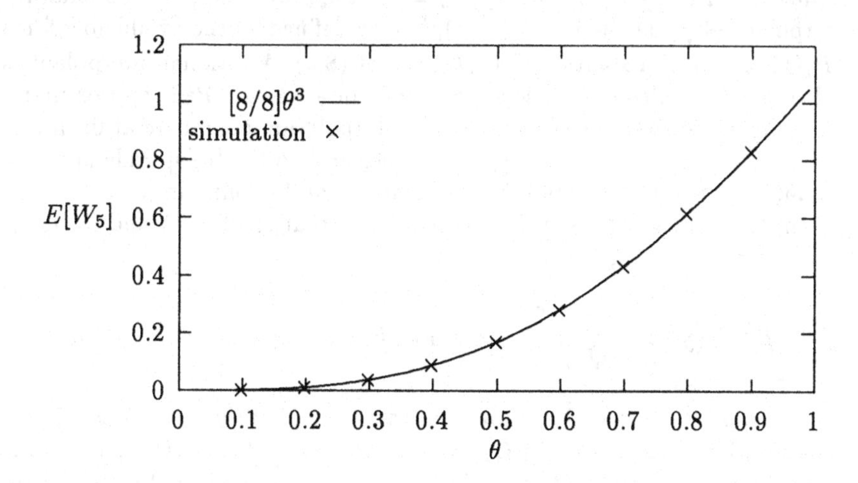

FIGURE 2. $E[W_5](\theta)$ of $E_2/E_2/1$

function f at a particular point $\hat{\theta}$ directly, it is much easier to calculate it at some other points $\theta_1, \theta_2, ..., \theta_n$. [8] suggest the idea of calculating $f(\hat{\theta})$ by first calculate values of f at $\theta_1, \theta_2, ..., \theta_n$, and then use a rational interpolant of f, with $\theta_1, \theta_2, ..., \theta_n$ as interpolation points, to extrapolate $f(\hat{\theta})$.

First we look at the definition of a rational interpolant. Let A be a set of interpolation points $\{b_1, ..., b_{n+v+1}\} \subset C$, where n and v are nonnegative integers. They are not necessarily distinct. Let f be a function which is analytic at least at the interpolation points. The rational Hermite interpolation problem is to find $P_n(z) = \sum_{i=0}^{n} a_i z^i$ and $Q_v(z) = \sum_{i=0}^{v} b_i z^i$ $(Q_v(z) \neq 0)$ such that

$$\frac{f(z)Q_v(z) - P_n(z)}{\omega(z)} \text{ is analytic at } b_1, ..., b_{n+v+1}, \tag{8.4}$$

where $\omega(z) = \prod_{i=1}^{n+v+1}(z - b_i)$. The problem can also be stated in the following equivalent way. Let $\{y_1, ..., y_m\}$ be a set of distinct points in A, and, for $i = 1, 2, ..., m$, y_i appears r_i times in A. So, $\sum_{i=1}^{m} r_i = n + v + 1$. We want to find $P_n(z) = \sum_{i=0}^{n} a_i z^i$ and $Q_v(z) = \sum_{i=0}^{v} b_i z^i$ $(Q_v(z) \neq 0)$ such that

$$(f \cdot Q_v - P_n)^{(k-1)}(y_i) = 0 \text{ for } k = 1, ..., r_i \text{ and } i = 1, ..., m. \tag{8.5}$$

(8.5) is a system of $n + v + 1$ linear equations in $n + v + 2$ unknowns $a_0, a_1, ..., a_n$ and $b_0, b_1, ..., b_v$, so it has at least one nontrivial solution. It can be shown (see, e.g., [5]) that if $P_n(z)/Q_v(z)$ and $P_n^*(z)/Q_v^*(z)$ are both solutions of (8.5), then $P_n(z)Q_v^*(z) = P_n^*(z)Q_v(z)$. The (n, v) rational interpolant of f, denoted by $R_{n,v}(z)$, is now defined as the irreducible form $\hat{P}_n(z)/\hat{Q}_v(z)$ of a solution $P_n(z)/Q_v(z)$ of (8.4). A rational interpolant is also called a multipoint Padé approximant or a Newton-Padé approximant. An important special case is when all interpolation points are at 0. In this case, the (n, v) rational interpolant is identical to the $[n/v]$ Padé approximant. Algorithms for rational interpolation can be found in [3].

In [17], it is shown that if $P_n(z)/Q_v(z)$ is a solution of (8.5), and $Q_v(y_i) \neq 0$, for $i = 1, 2, ..., m$, then

$$f^{(k-1)}(y_i) = \left(\frac{P_n}{Q_v}\right)^{(k-1)}(y_i) \quad \text{for} \quad k = 1, ..., r_i \quad \text{and} \quad i = 1, ..., m.$$

In general, $\hat{P}_n(z)$ and $\hat{Q}_v(z)$ may not satisfy (8.5). When $\hat{P}_n(z)$ and $\hat{Q}_v(z)$ satisfy (8.5), it implies that $\hat{Q}_v(y_i) \neq 0$, for $i = 1, 2, ..., m$ (because of the irreducibility of $\hat{P}_n(z)/\hat{Q}_v(z)$). Therefore, if $\hat{P}_n(z)$ and $\hat{Q}_v(z)$ satisfy (8.5), then $R_{n,v}(z)$ satisfies the following interpolation property.

$$f^{(k-1)}(y_i) = R_{n,v}^{(k-1)}(y_i) \quad \text{for} \quad k = 1, ..., r_i \quad \text{and} \quad i = 1, ..., m.$$

For our applications, we want to be able to choose a low-order rational interpolant (i.e. using a small number of interpolation points) to approximate the performance function accurately. Therefore, we need to know how R_{nv} converges to the performance function as n, v grows. It is natural to hope for the appealing concept of uniform convergence of rational interpolants. However, Wallin [19] shows that there exists an entire function such that 1) each point in C is a limit point of poles of $\{[n/n]\}_{n=1}^{\infty}$ Padé approximants, and 2) $\{[n/n]\}_{n=1}^{\infty}$ diverges everywhere except 0. Because, as we can see, poles of rational interpolants are unpredictable and can be just about anywhere, and, obviously, rational interpolants cannot converge at a pole; therefore, uniform convergence is not an appropriate concept. In [15], John Nuttall introduced the concept of convergence in measure. This concept is appropriate because it takes into account exceptional sets around the poles of Padé approximants where the error of approximation is large. In [16], Pommerenke introduced the concept of convergence in capacity, which uses the capacity for measuring the size of exceptional sets instead of the planar Lebesgue measure. We now discuss briefly the capacity concept.

First for a signed meaure μ we define the "Energy"

$$I(\mu) = \int_E \int_E \ln\left(\frac{1}{|z_1 - z_2|}\right) d\mu(z_1) d\mu(z_2)$$

and the "Potential"

$$V = \inf_{\mu} I(\mu).$$

The capacity of a set E is then defined as

$$\text{cap}(E) = e^{-V}.$$

Capacity is dominated by the measure μ. Specifically, for a compact set E, let λ be the planar Lebesgue measure. Then

$$\lambda(E) \leq \pi[\text{cap}(E)]^2.$$

This means that capacity is a finer measurement of size than the planar Lebesgue measure. Any countable set has capacity zero; however, although an interval $[a, b)$ on a plane has planar Lebesgue measure 0, we have

$$\text{cap}([a, b)) = (b - a)/4.$$

This is important in rational interpolation when we want to know how good the approximation is on the real line. If we use the concept of convergence in measure, we cannot say anything about the approximation on the real line. This is because no matter how small the measure of the exceptional set is, the whole real line may still be in the exceptional set. Therefore, the convergence in capacity is essential.

The following theorem gives the conditions under which a sequence of rational interpolants converges in capacity.

Theorem 8.2.1 (Wallin [20]) Let E and F be closed sets in the extended complex plane $\bar{C}$ such that E and F are disjoint and $\text{cap}(F) = 0$. Let f be analytic in $\bar{C} \setminus F$. Let R_{nv} be the rational interpolant of type (n, v) to f using the interpolation points $b_j \in E \cap C$, $1 \leq j \leq n + v + 1$. Then, for $\epsilon, \delta, R > 0$ and $\lambda \geq 1$, there exists n_0 such that

$$|f(z) - R_{nv}(z)| < \epsilon^n$$

if $n \geq n_0$ and $\frac{1}{\lambda} \leq \frac{n}{v} \leq \lambda$, for $|z| \leq R$ except on a set A_{nv} which has $\text{cap}(A_{nv}) \leq \delta$.

Note that the convergence here is geometric. This implies that we may only need a small number of interpolation points to obtain a satisfactory accuracy. This can be seen from the numerical examples in the previous section and in the following sections. A similar theorem also exists for functions which have branch points. See [18].

We should emphasize that in order to approximate a function with singularities, rational functions are superior to polynomials because poles of rational functions can create an effect similar to singularities of the function. Polynomials, on the contrary, is analytic everywhere and cannot "imitate" singularities. Most performance functions do have singularities. For example, in the $E_2/M/1$ queue, we have

$$E[W] = \theta \frac{(1 + 2\lambda\theta - \sqrt{1 + 4\lambda\theta})}{(1 - 2\lambda\theta + \sqrt{1 + 4\lambda\theta})}.$$

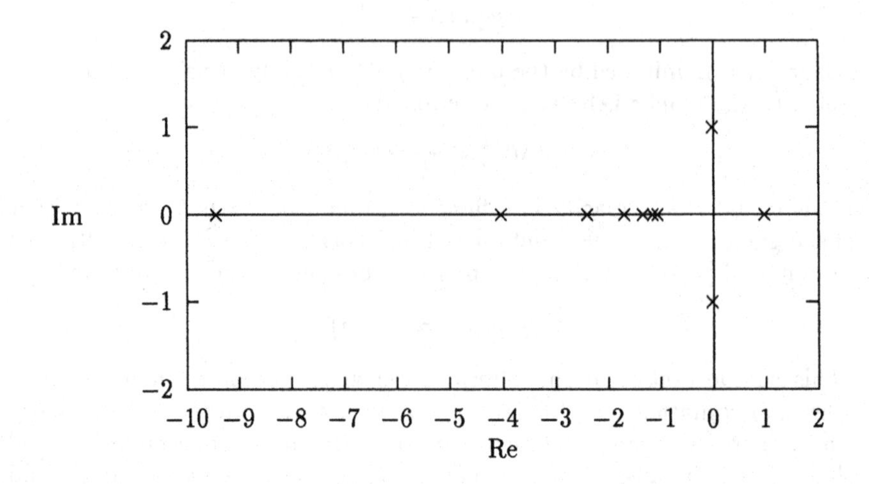

FIGURE 3. Poles of [10/10] of $J(x)$

Figure 3 shows how poles of [10/10] Padé approximant of a function $J(x) = \frac{\sqrt{x+1}}{(x-1)(x^2+1)}$ behave to create an effect of singuralities. Note that $J(x)$ has a branch cut $(-\infty, -1]$ and poles at 1, i, and $-i$.

8.3 Rational Interpolation for Integer-parameterized function $f(n)$

Integer-parametrized function is often encountered in the performance analysis of computer systems and communication networks. Our main example is the cell loss probability in ATM multiplexer as the function of the buffer size or the number of sources. Such function often can be evaluated when the parameter n is small. However, the desired parameter value is usually too big for conventional evaluation, but not big enough to use asymptotic analysis. The algorithm for computing rational interpolants is very similiar to the one for computing the Padé approximant. The procedure is as follows:

- Analyze asymptotic behavior when $n \to \infty$;
- Find appropriate transformation $T(\cdot)$ based on the asymptotic behavior of $f(n)$;

- Evaluate $T[f(n)]$ when n is small;
- Calculate rational interpolants for $T[f(n)]$;
- Extrapolate $f(n)$ for big n's.

Like in the case of Padé approximants, convergence analysis of rational interpolants needs to be done in the complex plane. We therefore need to extend the definition of $f(n)$ to the complex plane. This can be done via the Gregory-Newton series, which is defined by

$$F(x) \triangleq \sum_{k=0}^{\infty} a_k \frac{(x-1)(x-2)\dots(x-k)}{k!},$$

where a_k is k-th order difference for $f(1), f(2), \cdots$.

Theorem 8.3.1 (1)Gregory-Newton series converges in a half plane $\{z : R(z) > \lambda\}$ to a holomorphic function. The abscissa of convergence

$$\lambda = \limsup_{n \to \infty} \frac{\log |a_n|}{\log n}.$$

(2) If

$$\limsup_{n \to \infty} |a_n|^{1/n} = \alpha < 1$$

then $\lambda = -\infty$.

Theorem 8.2.1 can be modified for the rational interpolation of $f(n)$. In this case, we want to find an error bound for

$$|f(m) - R_{nv}(m)|,$$

where m is a positive integer, and R_{nv} is the rational interpolant of type (n, v) to $f(z)$ using the interpolation points $b_j = j$, $1 \le j \le n + v + 1$. However, it is more convenient to consider instead the equivalent case where we find an error bound for

$$\left| f(1/m) - \hat{R}_{nv}(1/m) \right|,$$

where $\hat{R}_{nv}$ is the rational interpolant of type (n, v) to $f(1/z)$ using the interpolation points $b_j = 1/j$, $1 \le j \le n + v + 1$.

Theorem 8.3.2 Let f be analytic everywhere (including the infinity) except at the origin. Let $\frac{P_n}{Q_n}$ be the rational interpolant of type (n, n) to f with the interpolation points $b_j = 1/j$, $1 \le j \le 2n + 1$. Then, for any number $0 < \eta < \frac{1}{2n+1}$,

$$\left| f(z) - \frac{P_n(z)}{Q_n(z)} \right| \le c(\eta, f, z) \eta^n \frac{\prod_{j=1}^{2n+1} |z - \frac{1}{j}|}{\prod_{j=1}^{2n+1} |\eta - \frac{1}{j}|} \left(\frac{4e}{\delta} \right)^{2n},$$

where

$$c(\eta, f, z) = \frac{1}{2\pi} \int_\Gamma \frac{|f(\xi)|}{|(\xi - z)|} d\xi,$$

$\eta < |z| \leq 1$ and the minimum distance from z to the zeros of Q_n and the origin is $> \delta$.

Remark: For our applications, the choice of z is quite flexible. Therefore, if some zeros of Q_n are very close to z, we can rechoose z such that it is not too close to any zeros of Q_n.

Proof. Let $h(z) = z$. Let $D = \{z : |z| < \eta\}$. Since $h^n(fQ_n - P_n)/w_{nn}$ is analytic in $\bar{C} \setminus D$ and zero at infinity, by using Cauchy's integral formula for an unbounded domain (see, e.g., [12] p. 319), we have that, for $|z| > \eta$,

$$\frac{h^n(z)(fQ_n - P_n)(z)}{w_{nn}(z)} = \frac{1}{2\pi i} \int_\Gamma \frac{h^n(\xi)(fQ_n - P_n)(\xi)}{w_{nn}(\xi)(\xi - z)} d\xi, \tag{8.6}$$

where Γ is the boundary of D with clockwise direction. $w_{nn}(z)$ can be expressed as

$$w_{nn}(z) = \prod_{j=1}^{2n+1} (z - \frac{1}{j}).$$

So we have

$$|w_{nn}(\xi)| \geq \prod_{j=1}^{2n+1} |\eta - \frac{1}{j}|,$$

for any $\xi \in \Gamma$. Also, the term

$$\frac{1}{2\pi i} \int_\Gamma \frac{h^n(\xi) P_n(\xi)}{w_{nn}(\xi)(\xi - z)} d\xi$$

is zero because the integrant is analytic in D. Therefore, from (8.6), and the fact that $|h(\xi)| = \eta, \xi \in \Gamma$, we have that, for $|z| > \eta$,

$$|f(z) - \frac{P_n(z)}{Q_n(z)}| \leq c(\eta, f, z) \frac{\eta^n}{|h(z)|^n} \frac{\prod_{j=1}^{2n+1} |z - \frac{1}{j}|}{\prod_{j=1}^{2n+1} |\eta - \frac{1}{j}|} \max_{\xi \in \Gamma} \left| \frac{Q_n(\xi)}{Q_n(z)} \right|,$$

where

$$c(\eta, f, z) = \frac{1}{2\pi i} \int_\Gamma \frac{|f(\xi)|}{(\xi - z)} d\xi.$$

The term $\left| \frac{Q_n(\xi)}{Q_n(z)} \right|$ can be estimated by using the following lemma.

Lemma 1 (Wallin [20] p. 439) Let $Q(z) = \prod_{i=1}^m (z - w_i)$ and $R \geq 1$. Assume that $|w_j| \leq 2R$ for $1 \leq j \leq m^*, m^* \leq m$. Let $Q^*(z) = \prod_{i=1}^{m^*} (z - w_i)$. Then

$$\max_{\{\xi : |\xi| \leq R\}} \left| \frac{Q_n(\xi)}{Q_n(z)} \right| \leq \frac{(3R)^m}{|Q^*(z)|}$$

for $|z| \leq R$.

In our case, let $R = 1$, and P_n and Q_n are normalized so that Q_n has leading coefficient 1. We have that, for $\eta < |z| \leq 1$,

$$|f(z) - \frac{P_n(z)}{Q_n(z)}| \leq c(\eta, f, z)\eta^n \frac{\prod_{j=1}^{2n+1} |z - \frac{1}{j}|}{\prod_{j=1}^{2n+1} |\eta - \frac{1}{j}|} \frac{1}{|h^n(z)Q_n^*(z)|}. \tag{8.7}$$

We can estimate $|h^n(z)Q_n^*(z)|$ by using the following lemma.

Lemma 2 (Wallin [19] p. 241) Let $g(z) = \prod_{i=1}^{n}(z - z_i)$ be an arbitrary polynomial, and H be a positive number. Then

$$|g(z)| > H^n$$

for all z except on a set E which can be covered by at most n disks with radii r_i such that $\sum_i r_i \leq e(2H)$. Note that all zeros of $g(z)$ are in E.

$h^n(z)Q_n^*(z)$ has leading coefficient 1 and degree $n + n^* \leq 2n$. Therefore, if the minimum distance from z to the zeros of $h^n(z)Q_n^*(z)$ is $> \delta$, then we have that $|h^n(z)Q_n^*(z)| > \left(\frac{\delta}{4e}\right)^{2n}$. From (8.7), we have that

$$|f(z) - \frac{P_n(z)}{Q_n(z)}| \leq c(\eta, f, z)\eta^n \frac{\prod_{j=1}^{2n+1} |z - \frac{1}{j}|}{\prod_{j=1}^{2n+1} |\eta - \frac{1}{j}|} \left(\frac{4e}{\delta}\right)^{2n},$$

for $\eta < |z| \leq 1$ and the minimum distance from z to the zeros of Q_n and the origin is $> \delta$. □

8.4 Cell Loss in ATM Multiplexers with Homogeneous Sources

In this section, we illustrate our approach via the computation of cell loss probability in a single ATM multiplexer in isolation and fed by a population of homogeneous two-state traffic streams. In the ATM environment, cells have the same length and, consequently, the same service time, which makes the discrete time Markov chain a natural modeling choice. Although applications of our approach are not limited to the discrete time case, we focus our discussion on discrete time models in this paper.

We model an ATM multiplexer as a discrete time server that can service c cells during one time unit. This server serves a queue with a capacity for K cells which is fed by N independent traffic sources. The arrival process associated with a source is modeled as an *Interrupted Bernoulli Process* (IBP). This process has two states: active (ON) and idle (OFF), represented by 1 and 0, respectively. In the active state, an arrival can occur with probability β (in the experiments, β is assumed to be 1). No arrivals occur while the source is in the idle state. Each of these ON-OFF sources behaves as follows. While an arrival process is in state 0, there is a probability $1-p_{00}$ that it will change to state 1 at the next time slot and a probability p_{00} that it will remain in state 0. While an arrival process is in state 1, there is a probability $1-p_{11}$ that it will transit to the idle state at the next time slot and a probability p_{11} that it will remain in state 1. When the server is busy, a maximum number of c cells will depart in each time slot. Service starts only at the beginning of each time slot.

The system can be modeled as a discrete time Markov chain with state (X_i, Y_i), where X_i is the number of cells in the queue (excluding those cells which have already been assigned for transmission) and Y_i is the number of arrival sources in the active state at the i-th time slot. We are interested in the steady state behavior $(X, Y) \triangleq \lim_{i\to\infty}(X_i, Y_i)$. Let S denote the state space. Let $\mathbf{T} = [t_{m,n;k,l}]$ be the transition matrix for this Markov chain, where $t_{m,n;k,l} = Prob[X_{i+1} = k, Y_{i+1} = l | X_i = m, Y_i = n]$. Note that the dimension of this Markov chain is $(K+1-c)*(N+1)$.

The stationary probability distribution can be obtained by solving the equation

$$\pi \mathbf{T} = \pi. \tag{8.8}$$

The cell loss probability (denoted as P_L) can then be calculated as

$$P_L = \frac{\sum_{(m,n)\in S} (m+n-K)^+ \; P[X=m, Y=n]}{\sum_{(m,n)\in S} \; m \; P[X=m, Y=n]}, \tag{8.9}$$

where $(x)^+ = \max(x, 0)$.

Equation (8.9) yields precise values of the cell loss probability. However, when the size of the model (either the buffer size K or the number of sources N) becomes large, the computational cost is prohibitively high due to the size of the state space. We now use rational interpolants to tackle this problem.

For details of algorithms and pseudo codes for calculating the rational interpolants, please refer to [8]. Here we list the major steps required to calculate the rational interpolants for $P_L(K)$:

1. Asymptotic analysis.

   Suppose

   $$\log P_L(K) \sim \theta K \qquad (K \to \infty) \tag{8.10}$$

   We calculate exponential decay rate θ for $P_L(K)$ using the algorithm proposed in [11].

2. Determine the forms of transformation and the form of approximant sequence.

   Since $P_L(K)$ is exponentially decaying, we develop approximants for the function

   $$h(K) = \log P_L(K) \tag{8.11}$$

and will use an $R_{(n+1),n}$ sequence of rational interpolants for $h(K)$, since $h(K)$ is asymptotically linear.

3. Evaluate $P_L(K)$ for small values of K (thus the corresponding values of $h(K)$ are known), by solving the Markov chain or using other available analytic methods.

4. Calculate rational interpolants $R_{(n+1),n}$ for $h(K)$. In fact, it is equivalent to calculating the $R_{n,n}$ rational interpolant sequence for $[\log P_L(K) - \theta K]$.

   We generate a sequence of rational interpolants, $R_{(n+1),n}$, with increasing orders ($n = 1, 2, \cdots$) and stop when the successive interpolants are sufficiently close in the range of K of interest.

8.4.1 Example 1

In our first example, we set $N = 50$, $p_{00} = 0.99$, $p_{11} = 0.5$, $c = 1$. The corresponding traffic intensity is

$$\rho = \frac{n(1 - p_{00})}{c(2 - p_{00} - p_{11})} = 0.98 < 1 \tag{8.12}$$

(the queue is stable). Applying techniques in [11], we obtain an exponential decay rate for the cell loss probability $P_L(K)$ of $\theta = -0.013707$ for this model.

The rational interpolants are shown in Figure 4. The first 20 points are obtained by solving the Markov chain, and the first $n+v+1$ points are then used to calculate the rational interpolant $R_{n,v}$. Over the range $[1, 1200]$, $R_{10,9}$ and $R_{9,8}$ are sufficiently close and we use $R_{10,9}$ as the final one to predict loss probability $P_L(K)$ for $K = 20, 21, \cdots, 1200$. In the figure, the solid lines are the $R_{10,9}$ and the $R_{9,8}$ (they are almost indistinguishable). To show the convergence of rational interpolants, we depicted some lower order approximants. The dashed line is $R_{2,1}$ and the dotted line is $R_{3,2}$. The differences between the approximants as the degrees increase from $R_{3,2}$ to $R_{9,8}$ are very small. In practice, $R_{4,3}$ might be sufficiently accurate. We also show (the dashdot line, in the figure) $R_{10,9}$ without using the exponential decay rate (i.e. $h(K) = P_L(K)$) for the purpose of comparison. Without the information of decay rates, the rational interpolants also yield sufficiently accurate results. However, this scheme requires additional initial points.

To verify the accuracy of the rational interpolation approach, we ran the simulation for the system with $K = 600$ and marked the simulation result with a star in Figure 4. Table 8.2 compares the numerical data of the loss probability at $K = 600$.

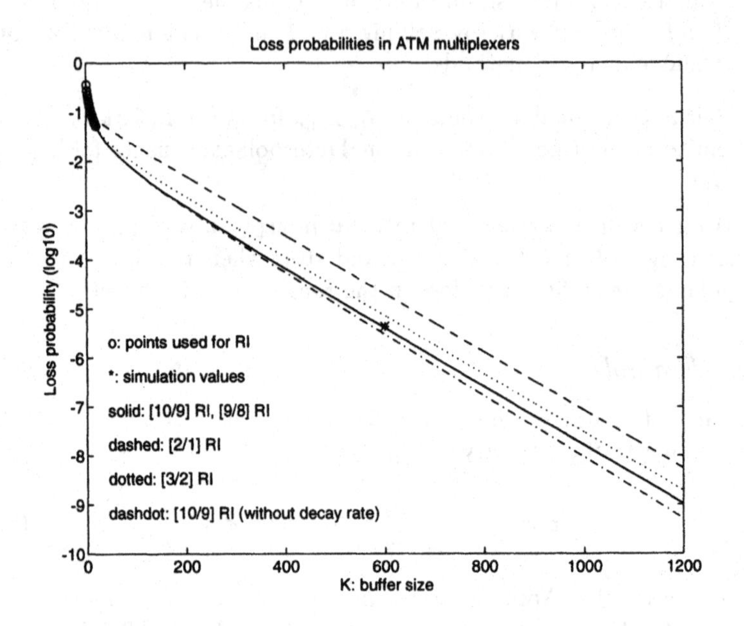

FIGURE 4. Rational interpolants in ATM multiplexers (50 sources)

	$\log_{10}$ (loss prob.)	loss prob.
$R_{10,9}$	-5.392	4.05×10^{-6}
$R_{9,8}$	-5.385	4.12×10^{-6}
Simulation 95% confidence interval		4.36×10^{-6} $(3.83 \times 10^{-6}, 4.89 \times 10^{-6})$

TABLE 8.2. Loss probabilities for ATM multiplexers (50 sources, K=600)

Guaranteed QoS	Number of supportable sessions
loss prob. $< 1 \times 10^{-7}$	$N \leq 29$
loss prob. $< 1 \times 10^{-6}$	$N \leq 39$
loss prob. $< 1 \times 10^{-5}$	$N \leq 52$
loss prob. $< 1 \times 10^{-4}$	$N \leq 67$

TABLE 8.3. Call admission control

8.4.2 Example 2: Application to Call Admission

We now consider the packet loss probability as a function of the number of sources feeding packets into it. This is of interest because it can be used to address the problem of call admission which is required to provide guaranteed QoS. We apply the rational interpolation approach to estimate this function in this section.

We have, for example, a high speed channel serving voice sessions controlled by the ATM, and consider its discrete time model. For simplicity, we assume the voice sources are two-state (ON-OFF) Markov modulated processes. It can then be modeled as an nIBP/D/c/K queue discussed in Section 2.

The call admission problem is to answer the question: what is the number of sessions that can be supported by the channel so that the cell loss probability $P_L < b$ (where b is the tolerance probability)? When the number of currently supported sessions reaches this threshold number, the admission control policy denies admission of any new call requests to the network.

We again model the ATM multiplexer as the Markov chain described in Section 2. The dimension of the Markov chain $((K+1-c)*(N+1))$ increases with the number of sessions, making the computation costs for solving the Markov chain increasingly high. We apply the rational interpolation approach to save computation costs.

For the system with $p = 0.015$, $q = 0.1$, $c = 3$, and $K = 15$, the rational interpolants are depicted in Figure 5. We chose $R_{9,9}$ as our final rational interpolant.

It is now possible to answer the question regarding guaranteed QoS. From this approximant we can say, for example, if we want the loss probability to be less than 10^{-7} we can only have less than 30 sessions. If we can tolerate the loss probability up to 10^{-6} then we can allow up to 39 sessions, etc. (See Table 8.3).

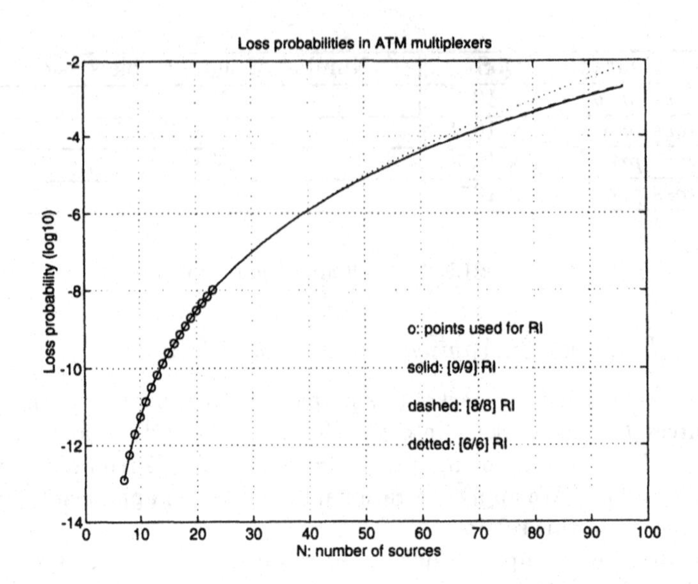

FIGURE 5. Rational interpolants for a call admission problem

8.5 Rational Interpolation Based on Simulation Results

Oftentimes the exact calculation is not feasible even for small size systems, for example in the case of ATM muliplexers with heterogneous sources. In these cases, the values at the interpolation points are obtained using simulation, and therefore contain error. This error will cause additional error in rational interpolation. It will be shown that this additional error in rational interpolation can be made arbitrarily small if the error of the values at the interpolation points is sufficiently small. Therefore, we need to make sure that the values obtained from the simulations are accurate enough. The results from the analysis of simulations using the Monte Carlo method are not applicable here because they can only provide stochastic error bounds. Therefore, analysis of simulations using the quasi-Monte Carlo method is needed in order to guarantee deterministic convergence. We will first consider the continuity of the rational interpolant of f with respect to the values of f at the interpolation points, and then we will consider the deterministic convergence of regenerative simulation.

For the continuity of the rational interpolant with respect to the values at the interpolation points, Gallucci and Jones established the following result.

Theorem 8.5.1 (Gallucci and Jones [5]) *Let* $R_{n,v}(z) = \frac{\hat{P}_n(z)}{\hat{Q}_v(z)}$ *be the* (n, v) *rational interpolant of* f, *and* $\tilde{R}_{n,v}(z)$ *be the* (n, v) *rational interpolant of*

$\tilde{f}$, both using the same set A of interpolation points $\{b_1, ..., b_{n+v+1}\}$. Let $\{y_1, ..., y_m\}$ be a set of distinct points in A, and, for $i = 1, 2, ..., m$, y_i appears r_i times in A. If $\hat{P}_n(z)$ and $\hat{Q}_v(z)$ satisfy (8.5), and the degree of $\hat{P}_n(z)$ is n, then for each $\epsilon > 0$ and compact set E such that E contains no poles of $R_{n,v}(z)$, there exists $\delta > 0$ such that

$$\max_{z \in E} |R_{n,v}(z) - \tilde{R}_{n,v}(z)| < \epsilon,$$

when

$$\max_{1 \leq k \leq r_i \text{ and } 1 \leq i \leq m} |f^{(k-1)}(y_i) - \tilde{f}^{(k-1)}(y_i)| < \delta.$$

Before we get to the deterministic convergence of regenerative simulation, we provide some material on multidimensional integration using the quasi-Monte Carlo method. More details can be found in [9, 13, 14].

Let I^m be the m-dimensional unit cube $\prod_{i=1}^m [0, 1)$, and $\bar{I}^m$ be the closed unit cube $\prod_{i=1}^m [0, 1]$. Let P be a set of points $\{\mathbf{u}_1, \mathbf{u}_2, ..., \mathbf{u}_N\} \subset I^m$. In the Monte Carlo method, we use the approximation

$$\int_{\bar{I}^m} f(\mathbf{u}) d\mathbf{u} \approx \frac{1}{N} \sum_{n=1}^{N} f(\mathbf{u}_n), \tag{8.13}$$

where $\mathbf{u}_1, \mathbf{u}_2, ..., \mathbf{u}_N$ are considered to be randomly chosen. In the quasi-Monte Carlo method, we use the same approximation except that $\mathbf{u}_1, ..., \mathbf{u}_N$ are considered to be deterministically chosen. The error of the approximation in (8.13) depends on the function f itself, and how uniformly the points $\mathbf{u}_1, \mathbf{u}_2, ..., \mathbf{u}_N$ spread over I^m. We can measure the uniformness of a set of points by using the idea of discrepancy.

Let B be an arbitrary subset of I^m. Define $A(B; P)$ to be the number of points in P which are also in B.

Definition 8.5.2 The discrepancy $D_N(P)$ of a set of points

$$P = \{\mathbf{u}_1, \mathbf{u}_2, ..., \mathbf{u}_N\} \in I^m,$$

is defined by

$$\sup_{B \in \mathcal{I}} \left| \frac{A(B; P)}{N} - \lambda_m(B) \right|,$$

where $\mathcal{I}$ is the family of all subintervals of I^m, and λ_m is the m-dimensional Lebesgue measure.

When f is Riemann integrable, the following theorem gives a deterministic error bound for the approximation in (8.13).

Theorem 8.5.3 ([9] p. 122) If f is Riemann integrable on $\bar{I}^m$, then

$$\left| \int_{\bar{I}^m} f(\mathbf{u}) d\mathbf{u} - \frac{1}{N} \sum_{n=1}^{N} f(\mathbf{u}_n) \right| \leq (1 + 2^{2m-1}) G(f, [D_N(\mathbf{u}_1, \mathbf{u}_2, ..., \mathbf{u}_N)]^{1/m}), \tag{8.14}$$

for any $\mathbf{u}_1, \mathbf{u}_2, ..., \mathbf{u}_N \in I^m$. G is defined as

$$G(f,k) = \sup_{n(p)\leq k} [U(f;p) - L(f;p)],$$

where we consider all partitions p of $\bar{I}^m$ into intervals with norm $n(p) \leq k$. The norm $n(p)$ is the largest length of all the edges of the intervals generated by p. $U(f;p)$ and $L(f;p)$ are, respectively, the upper sum and lower sum of f for the partition p.

Because f is Riemann integrable, the right-hand side of (8.14) will go to zero as $D_N \to 0$. This means that the smaller D_N is, the better the approximation will be. The study of sets of points with low discrepancy and the discrepancy of many common pseudorandom numbers can be found in [14] and references therein. Since linear congruential pseudo random numbers are the most popular ones in practice, we cite the following result.

First the linear congruential PRN is generated as follows. Choose a large positive integer m, an integer $a \in [1,m]$ with $\gcd(a,m) = 1$, an integer $c \in \{0,1,...,m-1\}$ and $y_0 \in \{0,1,...,m-1\}$. Then generate the sequence $\{x_n\}$ using

$$y_{n+1} = ay_n + c \bmod m \quad \text{for} \quad n = 0,1,...$$

$$x_n = \frac{y_n}{m} \in [0,1) \quad \text{for} \quad n = 0,1,...$$

Theorem 8.5.4 (H. Niederreiter, 1976) Let m be a prime. Then, for $1 \leq N \leq \tau$(period),

$$D_N \leq \frac{m^{1/2}}{N}\left(\frac{2}{\pi}\log m + \frac{2}{5}\right)\left(\frac{2}{\pi}\log\tau + \frac{2}{5}\right)$$

$$+\frac{(m-\tau)^{1/2}}{\tau}\left(\frac{2}{\pi}\log m + \frac{2}{5}\right) + \frac{2}{m}.$$

We now use the quasi-Monte Carlo method to obtain the deterministic convergence of regenerative simulation. Let $\{X_t : t \in D\}$ be a regenerative process with state space E. If $D = \{0,1,2,...\}$, then $\{X_t\}$ is discrete-time; if $D = [0,\infty)$, then $\{X_t\}$ is continuous-time. Let regeneration points be S_ns, the cycles be $\{X_t : S_n \leq t < S_{n+1}\}, n \geq 0$, and the cycle lengths be Y_ns. For simplicity we assume in the definition that cycle lengths cannot be zero.

Let f be a real-valued measurable function on E. What we are usually interested in is the time-average performance function $\lim_{t\to\infty} \frac{1}{t}\int_0^t f(X_u)du$ or $\lim_{t\to\infty} \frac{1}{t}\sum_{i=0}^t f(X_i)$. Define the rewards

$$R_n = \int_{S_{n-1}}^{S_n} f(X_t)dt, \quad n = 1,2,...,$$

when X_t is continuous-time; or

$$R_n = \sum_{i=S_{n-1}}^{S_n - 1} f(X_i), \quad n = 1, 2, ...,$$

when X_t is discrete-time. Note that R_ns are i.i.d. with $R_n \sim R$, where $R = \int_0^Y f(X_u)du$ when X_t is continuous-time, or $R = \sum_{i=0}^{Y-1} f(X_i)$ when X_t is discrete-time. The following theorem shows that the time-average performance function can be expressed in terms of $E[R]$ and $E[Y]$. For proof of the following theorem, see [1] p. 136.

Theorem 8.5.5 Let $\{X_t : t \in [0, \infty)\}$ be a continuous-time regenerative process with finite mean cycle length $E[Y]$. Then

$$\lim_{t \to \infty} \frac{1}{t} \int_0^t f(X_u)du = \frac{E[R]}{E[Y]}$$

with probability 1 if and only if

$$E\left[\max_{0 \le t < Y} \left| \int_0^t f(X_u)du \right| \right] < \infty.$$

The modification of the above theorem to the discrete-time case can be done in a straightforward manner.

So the purpose of a regenerative simulation is to estimate $E[R]$ and $E[Y]$. Note that Y is a special case of R when $f(x) = 1$ for all $x \in E$. Therefore, it is sufficient to consider only R.

Let the underlining probability space of R be $(\Omega, \mathcal{F}, P)$. Assume that the condition in Theorem 8.5.5 is satisfied. Note that this implies $E[|R|] < \infty$. The Monte Carlo estimate for $E[R]$ is

$$E[R] \approx \frac{1}{N} \sum_{n=1}^{N} R(\omega_n),$$

where $\omega_1, \omega_2, ..., \omega_N \in \Omega$.

In most cases in practice, the underlining probability space is as follows. Let $(\bar{I}^\infty, \prod_{i=1}^\infty \mathcal{B}([0,1]), \prod_{i=1}^\infty U)$ be a probability space, where $\bar{I}^\infty$ is an infinite-dimensional unit cube $\prod_{i=1}^\infty [0,1]$, $\mathcal{B}([0,1])$ is the class of Borel sets of $[0,1]$, and U defines a uniform distribution on $[0,1]$. For each $\mathbf{u} = (u_1, u_2, ..., u_n, ...) \in \bar{I}^\infty$, we let $\mathbf{u}^{[k]}$ be $(u_1, u_2, ..., u_k) \in \bar{I}^k$. We make the following assumptions which are true in most cases in practice:

1. R is a random variable on $(\bar{I}^\infty, \prod_{i=1}^\infty \mathcal{B}([0,1]), \prod_{i=1}^\infty U)$.

2. $\bar{I}^{\infty}$ can be written as $\cup_{m=1}^{\infty} A_m$, where A_ms are disjoint and belongs to $\prod_{i=1}^{\infty} \mathcal{B}([0,1])$. For $m = 1, 2, \ldots$, the characteristic function of A_m, denoted by $c_{A_m}(\mathbf{u})$, depends only on $\mathbf{u}^{[h(m)]}$, where $h(m)$ is a nondecreasing function whose range is the set of positive integers. For the discrete-time case, we can let A_m to be the event $\{Y = m\}$.

3. For $m = 1, 2, \ldots$,

$$R(\mathbf{u}) = g_m(\mathbf{u}^{[h(m)]}) \quad \text{when} \quad \mathbf{u} \in A_m.$$

Therefore, we can write

$$R(\mathbf{u}) = \sum_{m=1}^{\infty} g_m(\mathbf{u}^{[h(m)]}) c_{A_m}(\mathbf{u}^{[h(m)]}).$$

Therefore, $E[R]$ can be expressed as

$$\int_{\bar{I}^{\infty}} R(\mathbf{u}) d\mathbf{u},$$

where $d\mathbf{u} = \prod_{i=1}^{\infty} du_i$.

The Monte Carlo estimate for $E[R]$ is

$$\begin{aligned} E[R] &\approx \frac{1}{N} \sum_{n=1}^{N} R(\mathbf{u}_n) \\ &= \frac{1}{N} \sum_{n=1}^{N} \sum_{m=1}^{\infty} g_m(\mathbf{u}_n^{[h(m)]}) c_{A_m}(\mathbf{u}_n^{[h(m)]}). \end{aligned}$$

Note that the Monte Carlo estimate can be obtained by simulating $\{X(t) : t \in D\}$ for N cycles. $R(\mathbf{u}_n)$ can be considered as the sample of R from the nth cycle. The above estimate is not practical because it means that $h(m)$ in some cycles can be very large and, in those cycles, a lot of pseudo-random numbers are used. In practice, we limit the number of pseudo-random numbers used in one cycle by using instead the estimate $\frac{1}{N} \sum_{n=1}^{N} \hat{R}_d(\mathbf{u}_n)$, where

$$\hat{R}_d(\mathbf{u}) = \begin{cases} g_m(\mathbf{u}^{[h(m)]}) & \text{when } \mathbf{u} \in A_m \text{ and } m < d, \\ g_d(\mathbf{u}^{[h(d)]}) & \text{when } \mathbf{u} \in B_d \triangleq \cup_{m \geq d} A_m. \end{cases}$$

We can write

$$\hat{R}_d(\mathbf{u}) = \sum_{m=1}^{d-1} g_m(\mathbf{u}^{[h(m)]}) c_{A_m}(\mathbf{u}^{[h(m)]}) + g_d(\mathbf{u}^{[h(d)]}) c_{B_d}(\mathbf{u}^{[h(d)]}).$$

Therefore, the estimate is

$$\frac{1}{N} \sum_{n=1}^{N} \hat{R}_d(\mathbf{u}_n) = \frac{1}{N} \sum_{n=1}^{N} \left[\sum_{m=1}^{d-1} g_m(\mathbf{u}_n^{[h(m)]}) c_{A_m}(\mathbf{u}_n^{[h(m)]}) + g_d(\mathbf{u}_n^{[h(d)]}) c_{B_d}(\mathbf{u}_n^{[h(d)]}) \right]$$

Consider the deterministic error bound of

$$\begin{aligned}
&\left|\int_{\bar{I}^{\infty}} R(\mathbf{u})d\mathbf{u} - \frac{1}{N}\sum_{n=1}^{N} \hat{R}_d(\mathbf{u}_n)\right| \\
\leq &\left|\int_{\bar{I}^{\infty}} R(\mathbf{u})d\mathbf{u} - \int_{\bar{I}^{\infty}} \hat{R}_d(\mathbf{u})d\mathbf{u}\right| + \left|\int_{\bar{I}^{\infty}} \hat{R}_d(\mathbf{u})d\mathbf{u} - \frac{1}{N}\sum_{n=1}^{N} \hat{R}_d(\mathbf{u}_n)\right| \\
\triangleq &\ \epsilon_1 + \epsilon_2.
\end{aligned}$$

ϵ_1 is the error occurred from limiting the number of pseudo-random numbers in one cycle. Under some mild conditions, we can apply the dominated convergence theorem and make ϵ_1 as small as we want by choosing d large enough.

As for ϵ_2, we have

$$\begin{aligned}
\epsilon_2 = &\left|\sum_{m=1}^{d-1}\int_{A_m} g_m(\mathbf{u}^{[h(m)]})d\mathbf{u}^{[h(m)]} + \int_{B_d} g_d(\mathbf{u}^{[h(d)]})d\mathbf{u}^{[h(d)]} - \right. \\
&\left.\frac{1}{N}\sum_{n=1}^{N}\left[\sum_{m=1}^{d-1} g_m(\mathbf{u}_n^{[h(m)]})c_{A_m}(\mathbf{u}_n^{[h(m)]}) + g_d(\mathbf{u}_n^{[h(d)]})c_{B_d}(\mathbf{u}_n^{[h(d)]})\right]\right| \\
\leq &\sum_{m=1}^{d-1}\left|\int_{A_m} g_m(\mathbf{u}^{[h(m)]})d\mathbf{u}^{[h(m)]} - \frac{1}{N}\sum_{n=1}^{N} g_m(\mathbf{u}_n^{[h(m)]})c_{A_m}(\mathbf{u}_n^{[h(m)]})\right| + \\
&\left|\int_{B_d} g_d(\mathbf{u}^{[h(d)]})d\mathbf{u}^{[h(d)]} - \frac{1}{N}\sum_{n=1}^{N} g_d(\mathbf{u}_n^{[h(d)]})c_{B_d}(\mathbf{u}_n^{[h(d)]})\right|.
\end{aligned}$$

In most cases, $g_m c_{A_m}$ and $g_d c_{B_d}$ are usually Riemann integrable. Therefore, Theorem 8.5.3 can be applied. For any $\epsilon > 0$, we can choose d large enough so that $\epsilon_1 < \epsilon/2$. With d chosen, we then choose N large enough so that $\epsilon_2 < \epsilon/2$. Therefore we have that

$$\lim_{d\to\infty}\lim_{N\to\infty}\frac{1}{N}\sum_{n=1}^{N}\hat{R}_d(\mathbf{u}_n) = \int_{\bar{I}^{\infty}} R(\mathbf{u})d\mathbf{u}$$

in the deterministic sense.

8.5.1 *Example 1: ATM Multiplexers with Heterogeneous Traffic*

The model we consider here is similar to what we considered in the previous section, except that the ATM multiplexer is now fed by n independent heterogeneous ON-OFF sources, namely, the parameters for different ON-OFF sources, p_{00}'s and p_{11}'s, have different values. Assume the buffer size is K. In this case, the aggregate Markov chain has $K \cdot 2^n$ states. Clearly, for reasonable values of n, it is now difficult to calculate cell loss by solving the Markov chain, even for small values of K. Therefore, we resort to simulation.

As we mentioned in the introduction, the desired cell loss probability in the ATM network is very small (e.g., 10^{-9}) for moderate-size buffers. To simulate such systems involving rare events, one has to apply techniques such as the Importance Sampling (IS) to speed up simulation. [3] applied the IS technique to ATM networks, using an "exponentially twisted distribution" based on the theories of effective bandwidth and large deviations.

To verify the numerical results, we used the same models as those in [3]. Consider 16 sources with their p_{00}'s and p_{11}'s being taken as

$$(0.9, 0.9, 0.9, 0.9, 0.8, 0.8, 0.8, 0.8, 0.8, 0.7, 0.7, 0.7, 0.7, 0.6, 0.6, 0.5)$$

and

$$(0.7, 0.8, 0.6, 0.5, 0.3, 0.6, 0.6, 0.4, 0.8, 0.5, 0.9, 0.6, 0.5, 0.8, 0.6, 0.9)$$

respectively. The service capacity $c = 8$. The exponential decay rate of $P_L(K)$ is 0.34375 (see [3]).

We performed 50 simulation runs for each value of $K = 9, 10, \cdots, 20$. Each run contained 1000 regenerative cycles. Table 8.4 shows the results. These were then used to calculate the rational interpolants. The results are shown in Figure 6. The $R_{4,3}$ rational interpolant of $\log_{10} P_L(x)$ $(x \geq 8)$ is

$$-0.1493\,x + 0.4534 + \frac{0.0000104195}{-11.873 + x} + \frac{0.697539}{-5.89211 + x} - \frac{197.924}{175.935 + x}$$

Note that the pole at 11.873 does not affect the accuracy of the approximate when x takes integer values.

As Table 8.5 shows, our results match the importance sampling results obtained in [3]. For example, when $K = 20, 40, 80$, respectively, the results in [3] are $3.17 \times 10^{-4} \pm 3.2\%$, $3.76 \times 10^{-7} \pm 3.4\%$, $5.24 \times 10^{-13} \pm 3.4\%$, respectively, and the rational interpolation results are 3.21×10^{-4}, 3.85×10^{-7}, 5.57×10^{-13}, respectively.

8.5.2 *Example 2: ATM Intree Networks*

We describe an application of the rational interpolation approach to intree networks based on simulation results in this section. As an example, we consider a model with two ATM multiplexers, where the output of the first multiplexer feeds into the second one. There are a total of five ON-OFF sources, the first three of which input into the first multiplexer and the last two input into the second multiplexer. The p_{00}'s and p_{11}'s are taken to be $(0.8, 0.5, 0.7, 0.6, 0.5)$ and $(0.5, 0.6, 0.8, 0.6, 0.6)$ respectively. The service capacities are chosen to be 2 and 3 respectively. The buffer size of the first multiplexer is fixed at 25 and we vary the buffer size of the second

K	$P_L(K)$ (with 99% confidence interval)
9	$1.836218 \times 10^{-2} \pm 0.010\%$
10	$1.163464 \times 10^{-2} \pm 0.014\%$
11	$7.742995 \times 10^{-3} \pm 0.018\%$
12	$5.284334 \times 10^{-3} \pm 0.024\%$
13	$3.657093 \times 10^{-3} \pm 0.030\%$
14	$2.554200 \times 10^{-3} \pm 0.039\%$
15	$1.794637 \times 10^{-3} \pm 0.049\%$
16	$1.266396 \times 10^{-3} \pm 0.059\%$
17	$8.961004 \times 10^{-4} \pm 0.068\%$
18	$6.356303 \times 10^{-4} \pm 0.081\%$
19	$4.522467 \times 10^{-4} \pm 0.095\%$
20	$3.212790 \times 10^{-4} \pm 0.109\%$

TABLE 8.4. Cell loss probability in an ATM multiplexer

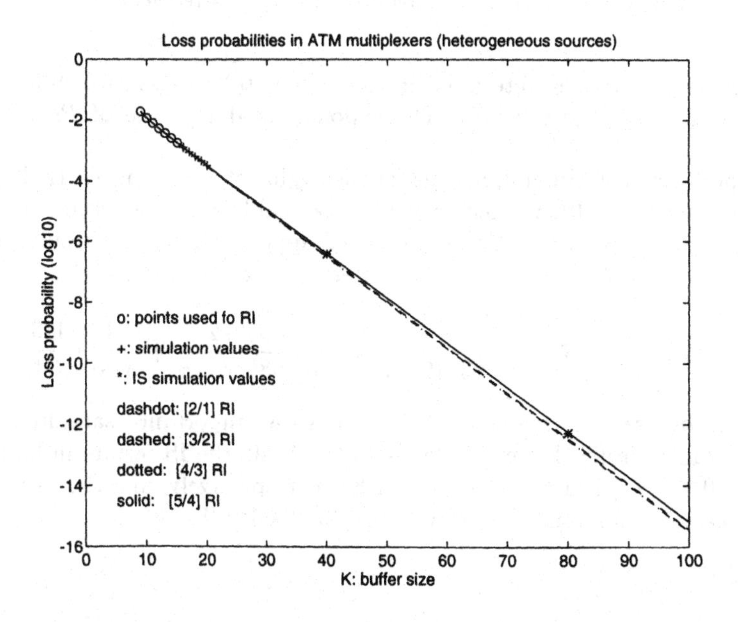

FIGURE 6. Cell loss in ATM multiplexer with heterogeneous sources

K	IS estimate in [3]	simulation in [3]	our simulation	$R_{4,3}$ estimate
10	$1.16 \times 10^{-2} \pm 2.8\%$	1.17×10^{-2}	1.163×10^{-2}	1.163×10^{-2}
15	$1.78 \times 10^{-3} \pm 3.3\%$	1.63×10^{-3}	1.795×10^{-3}	1.795×10^{-3}
20	$3.17 \times 10^{-4} \pm 3.2\%$	3.14×10^{-4}	3.213×10^{-4}	3.213×10^{-4}
40	$3.76 \times 10^{-7} \pm 3.4\%$			3.851×10^{-7}
80	$5.24 \times 10^{-13} \pm 3.4\%$			5.576×10^{-13}

TABLE 8.5. Estimates from Importance Sampling and Rational Interpolation

K	$P_L(K)$
4	$1.859231 \times 10^{-2} \pm 0.002624\%$
5	$7.965111 \times 10^{-3} \pm 0.004843\%$
6	$3.610258 \times 10^{-3} \pm 0.009431\%$
7	$1.685206 \times 10^{-3} \pm 0.017015\%$
8	$8.006919 \times 10^{-4} \pm 0.031184\%$
9	$3.850226 \times 10^{-4} \pm 0.121128\%$
10	$1.867368 \times 10^{-4} \pm 0.352957\%$
11	$9.125000 \times 10^{-5} \pm 0.458181\%$
12	$4.508183 \times 10^{-5} \pm 0.931634\%$

TABLE 8.6. Cell loss probability in an ATM intree network

multiplexer, K. We are interested in the aggregate cell loss probability in the second multiplexer $P_L(K)$. The exponential decay rate of $P_L(K)$ is 0.625.

We performed 30 simulation runs for each value of $K = 4, 5, \cdots, 12$. Each run contained 1 million regenerative cycles. Table 8.6 shows the results. These were then used to calculate rational approximants. The results are shown in Figure 7. The $R_{4,3}$ rational interpolant of $\log_{10} P_L(x)$ $(x \geq 3)$ is

$$-0.2714\,x - 3.4087 - \frac{1.25719\,10^{-6}}{-7.93119 + x} + \frac{0.700892}{-0.998966 + x} + \frac{164.463}{61.0014 + x}$$

Again, our results (see Table 8.7) match the importance sampling results obtained in [3]. For example, for $K = 25, 50$, the IS results in [3] are $5.81 \times 10^{-9} \pm 7.3\%$ and $3.57 \times 10^{-16} \pm 8.9\%$, respectively, and the rational interpolation results are 5.59×10^{-9} and 3.28×10^{-16}.

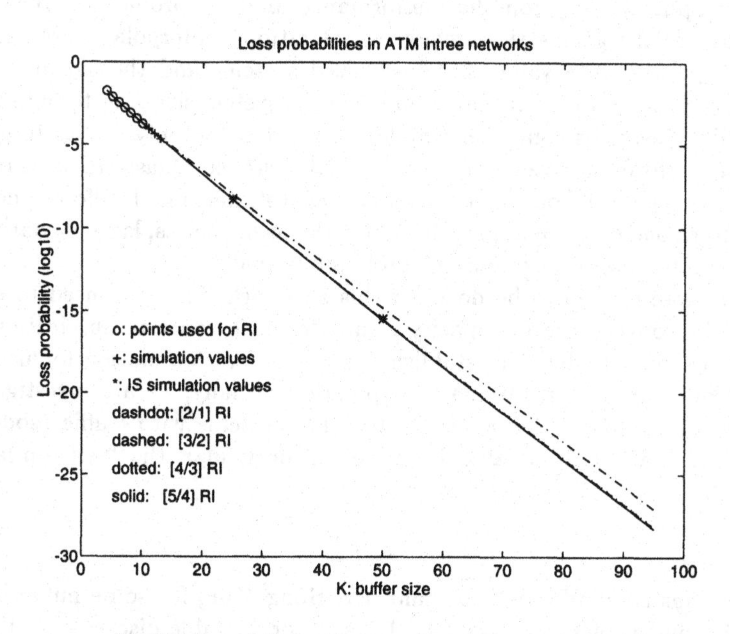

FIGURE 7. Cell loss in an ATM intree network

K	IS estimate in [3]	simulation in [3]	our simulation	$R_{4,3}$ estimate
10	1.08×10^{-4} $\pm 6.1\%$	1.87×10^{-4}	1.867×10^{-4}	1.867×10^{-4}
25	5.81×10^{-9} $\pm 7.3\%$			5.585×10^{-9}
50	3.57×10^{-16} $\pm 8.9\%$			3.278×10^{-16}

TABLE 8.7. Estimates from Importance Sampling and Rational Interpolation for an ATM intree network

8.6 Conclusion

We propose to use rational interpolants for rare-event probability and other computationally complex performance analysis problems. To take advantage of the geometric convergence of rational interpolants we need accurate performance values for down-sized systems and the asymptotic behavior of the performance function when the system size goes to infinity. For realistic systems computer simulation is often the only means to get accurate performance values for down-sized systems. Quasi-Monte Carlo analysis of the error bound for the simulation estimates is therefore a necessary component of the approach. As for the asymptotics, large deviation theory is often useful in rare-event probability analysis.

Much more remain to be done for this approach. First, we need to establish the convergence in capacity results for each concrete applications. Second, we need to further establish the theoretical foundation for using simulation estimates in rational interpolation. Finally, we are investigating the application of this approach to ATM systems with traffic models other than Markov Modulated Processes, in particular, the "self-similar" processes.

Acknowledgments: We wish to thank Dr. Hong Yang for some numerical examples and Professor Donald E. Towsley for valuable discussions. This work is supported in part by the National Science Foundation under Grant CCR-9506764 and Grant EEC-9527422, by the U.S. Army Research Office under Contract DAAH04-95-0148, by AFOSR under Contract F49620-93-1-0229DEF, and by Air Force Rome Laboratory under Contract F30602-95-C-0242.

8.7 References

[1] Asmussen, S., *Applied Probability and Queues*, John Wiley & Sons, 1987.

[2] Baker, G.A., and Graves-Morris, P., *Padé Approximants, Part I: Basic Theory*, Addison-Wesley, 1981.

[3] Baker, G. A. and Graves-Morris, P., *Padé Approximants, Part II: Extensions and Applications*, Addison-Wesley, 1981.

[4] Chang, C.S., Heidelberger, P., Juneja, S., and Shahabuddin, P. Effective Bandwidth and Fast Simulation of ATM Intree Networks, IBM Research Report RC 18586, 1992. (also in *Proc. Performance '93*, Rome, Italy, October 1993.)

[5] GALLUCCI, M.A. AND JONES, W.B., Rational Approximations Corresponding to Newton Series (Newton-Padé Approximants), *Journal of Approximation Theory*, **17**, 366–392, 1976.

[6] GONG, W.B. AND HU, J.Q., The MacLaurin Series for the $GI/G/1$ Queue, *Journal of Applied Probability*, **29**, 176–184, 1992.

[7] GONG, W.B., NANANUKUL, S., AND YAN, A., Padé Approximations for Stochastic Discrete Event Systems, *IEEE Transactions on Automatic Control*, **40**, 1349–1358, 1995.

[8] GONG, W.B. AND YANG, H., Rational Approximants for Some Performance Analysis Problems, *IEEE Transactions on Computers*, **44**, 1394–1404, 1995.

[9] HLAWKA, E., Discrepancy and Riemann Integration, *Studies in Pure Mathematics*, Edited by Mirsky, L., Academic Press, 1971.

[10] HU, J.Q., NANANUKUL, S., AND GONG, W.B., A New Approach to (s,S) Inventory Systems, *Journal of Applied Probability*, **30**, 898–912, 1993.

[11] LIU, Z., NAIN, P., AND TOWSLEY, D., Exponential Bounds for a Class of Stochastic Processes with Application to Call Admission to Networks, *Proc. 33rd Conference on Decision and Control*, 156–161, 1994.

[12] MARKUSHEVICH, A.I., *Theory of Functions of a Complex Variable: Vol. 1*, Prentice-Hall, 1965.

[13] NIEDERREITER, H., Quasi-Monte Carlo Methods and Psuedo-Random Numbers, *Bull. Amer. Math. Soc.*, **84**, 957–1041, 1978.

[14] NIEDERREITER, H., *Random Number Generation and Quasi-Monte Carlo Methods*, Society for Industrial and Applied Mathematics, 1992.

[15] NUTTALL, J., The Convergence of Padé Approximants of Meromorphic Functions, *J. Math. Anal. Appl.*, **31**, 147-153, 1970.

[16] POMMERENKE, C., Padé Approximants and Convergence in Capacity, *J. Math. Anal. Appl.*, **41**, 775-780, 1973.

[17] SALZER, H. E., Note on Osculatory Rational Interpolation, *Math. Comput.*, **16**, 486-491, 1962.

[18] STAHL, H., General Convergence Results for Rational Approximants, *Approximation Theory VI*, Edited by Chui, C.K., Schumaker, L.L. and Ward, J.D., Academic Press, 1989.

[19] WALLIN, H., The Convergence of Padé Approximants and the Size of the Power Series Coefficients, *Applicable Analysis*, **4**, 235–251, 1974.

[20] WALLIN, H., Potential Theory and Approximation of Analytic Functions by Rational Interpolation, *Lecture Notes in Mathematics*, Edited by Dold, A. and Eckmann, B., No. 747, Springer-Varlag, 1978.

[21] WOLFF, R.W., *Stochastic Modeling and the Theory of Queues*, Prentice-Hall, 1988.

[22] YANG, H., GONG, W.B. AND TOWSLEY, D.E. Efficient Algorithm for Cell Loss Probabilities in ATM multiplexers, *Proceedings of GLOBE-COM'95*, November, 1995.

[23] ZHU, Y AND LI, H The MacLaurin Expansion for a G/G/1 Queue with Markov-Modulated Arrivals and Services, *QUESTA*, pp. 125-134, 1993

9
Overloading Parallel Servers When Arrivals Join The Shortest Queue

David McDonald

ABSTRACT
We consider two parallel servers with an arrival stream of customers who join the shorter of the two servers' queues. We use the theory of Markov additive processes to prove exact asymptotic results on the mean time for the queue at the first server to attain a high level ℓ. We also give the limiting behaviour of the second queue at this hitting time as well as the steady state of the second queue when the first contains ℓ customers.

9.1 Introduction

We shall apply the theory in McDonald [9] and [10] (referred to as Parts I and II), which describes how a queue at a chosen node in a network of queues overloads. (Prepublication postscript versions are available in compressed form at anonymous@notiid.mathstat.uottawa.ca in the directory Math+Stats/people/McDonald.) The network is modeled as a Markov additive process with a boundary where the workload of the chosen node is the additive component and the queue lengths of the other nodes form the Markov chain. The boundary includes those states where the workload falls to zero.

Using the twist developed by Nummelin and Ney [13] and [14] we find the unique positive harmonic function (up to multiples), other than constant functions, for the Markov additive process. The asymptotics of this harmonic function give the rough asymptotics of the mean time for the chosen node in the network to cross some level ℓ. Performing an h-transform twists the Markov additive process into a twisted process whose workload drifts to plus infinity. The simulation of the twisted process yields the *exact* asymptotics of the mean passage time for the chosen queue. This is feasible since the time to simulate the twisted process until it crosses the level ℓ grows linearly with ℓ. The same simulation also gives the distribution of the queues at the other nodes when this rare event occurs. It also yields the steady state distribution of the other nodes when the queue at the chosen node has ℓ customers.

We shall illustrate the theory by analysing a network with two servers.

Customers arrive at nodes 1 and 2 according to two independent Poisson processes with rates α and β respectively. There is also a third independent Poisson stream with rate γ of (smart) customers who join the shorter of the two queues. The service time distributions at node 1 and node 2 are exponential with rates δ and μ respectively and the customers queue until they are served. Let (x, y) denote the number of customers waiting or being served at queue 1 and queue 2. This Markov jump process on the positive quadrant $\mathcal{S}$ is denoted by Q and has jump rates given in Figure 1.

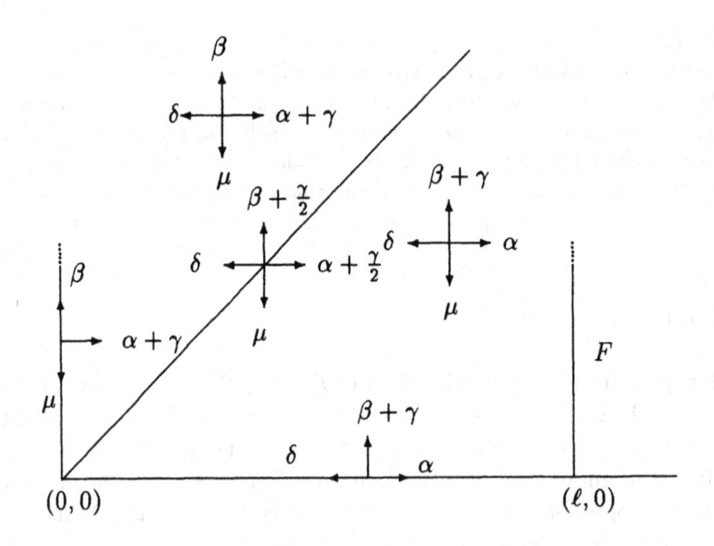

FIGURE 1. Rates for the join the shortest queue model.

Flatto and McKean [6] investigated this system when $\alpha = \beta = 0$ (i.e. no dumb customers) and $\delta = \mu$. Using analyticity arguments they showed the system was stable if the arrival rate is less than the sum of the two service rates and they obtained an exact solution for the generating function of the stationary distribution $\pi(x, y)$. Adan, Wessels and Zijm [1] used a compensation procedure to represent the stationary distribution as an infinite sum of product measures in the asymmetric case when $\mu \neq \beta$ but without dumb customers. $\gamma < \mu + \delta$ was required for stability.

Knessl, Matkowsky, Schuss and Tier [7] also studied the asymmetric case without dumb customers. Their clever heuristic procedure amounts to assuming a specific form for the asymptotic stationary distribution. The unique solution to the balance equations having this form gives the asymptotics of the stationary distribution up to an unknown scaling constant. Our asymptotic results for the diagonal case described below confirm that the asymptotic stationary distribution does have the conjectured form and our results agree with theirs in the region I, as defined in Knessl et al. [7], when x is large and y is of the same order as x. Our results on the stationary distribution allow for dumb customers and the scaling constant has a probabilistic expression which can be simulated efficiently. Consequently our theory extends the results of Knessl et al. and puts their heuristic procedure on a firm theoretical footing. On the other hand the results in Knessl et al. apply when x is large but y is small. This implies our results in the diagonal case should be extended!

Let G denote the generator of the jump process Q. Let F denote the forbidden set of points whose first coordinate is at least ℓ and let B be the complement of F. We are interested in the time $\tau_\ell \equiv \tau_F$ until the first coordinate of Q reaches some high level ℓ starting from the origin $\delta = (0,0)$.

The event rate of the jump process Q is $q = \alpha+\beta+\delta+\mu+\gamma$. If we pick the unit of time such that $q = 1$ then we can regard this jump process as the uniformization of a discrete time chain $W = (W_1, W_2)$ with the same state space and a transition kernel K given by Figure 2. Transitions outside the boundaries are simply suppressed. The transition $K((0,0),(0,0)) = \delta + \mu$ is not marked.

The discrete generator of W is $K - I$ and this is precisely G. It follows that W has the same stationary distribution π as the jump process Q. If we define $m(\vec{z}) = E_{\vec{z}}\tau_F$ to be the mean time starting from $\vec{z}$ to hit F then m satisfies:

$$\begin{cases} Gm(\vec{z}) = -1, & \vec{z} \in B; \\ m(\vec{z}) = 0, & \vec{z} \in F; \end{cases} \tag{9.1}$$

It makes no difference whether the mean hitting time m is for the Markov chain W or for its uniformization, the jump process Q. In either case m satisfies equation (9.1) so the two are the same! We shall also study the distribution of the second queue when the first reaches ℓ. This is also the same for the chain or its uniformization.

For π to exist we must first impose conditions on K to make the chain positive recurrent. Clearly $\alpha < \delta$ and $\beta < \mu$ are necessary or else either queue would be unstable even without the extra arrival stream which joins the shortest queue. Next consider a random walk on the integers with increment of $+1$ with probability $\alpha+\beta+\gamma$ and increment of -1 with probability $\mu+\delta$. If $\alpha+\beta+\gamma > \mu+\delta$ then this walk is transient to $+\infty$. If $\alpha+\beta+\gamma = \mu+\delta$ then the walk is null recurrent. We compare this to the total number of customers in both queues. The former is stochastically smaller since the queues must stay in the positive quadrant. We conclude that if $\alpha+\beta+\gamma \geq \mu+\delta$ then our network is unstable.

Assume $\alpha+\beta+\gamma < \mu+\delta$. We can show our queue must be stable using Foster's criterion (see Theorem 11.3.4 in Meyn and Tweedie [12]). Consider the total workload $v(x,y) = x+y$. First notice that if $x > 0$ and $y > 0$ then $Gv(x,y) = \alpha+\beta+\gamma-\mu-\delta < 0$. Next, for $y \geq n$,

$$G^n v(0,y) \quad = \quad E_{(0,y)}[W_1[n] + W_2[n]] - y$$

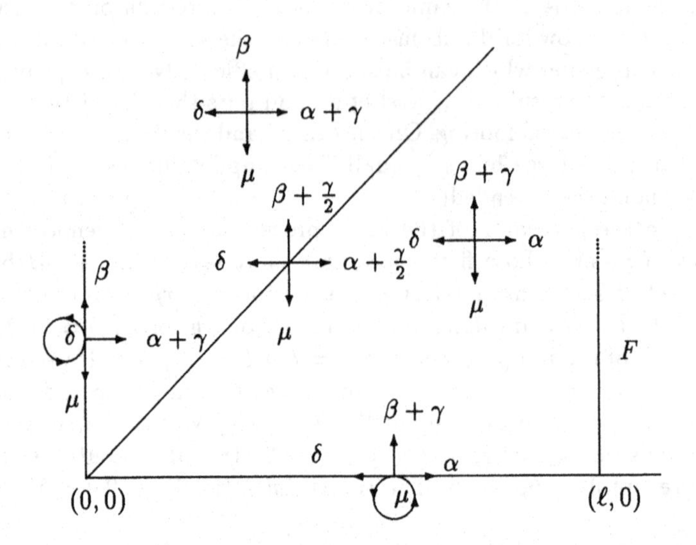

FIGURE 2. Transition probabilities of W.

$$= \quad E_{(0,y)}W_1[n] + n(\beta - \mu).$$

Since $\beta < \mu$ by hypothesis, the second term above decreases. If $\delta > \alpha + \gamma$ then $W_1[k]$ evolves like a stable discrete $M|M|1$ queue for $0 \leq k \leq n$. Hence $E_{(0,y)}W_1[n]$ is uniformly bounded in n. Consequently for n large enough, $G^n v(0, y)$ is negative.

If $\delta \leq \alpha + \gamma$ then $W_1[k]$ evolves like an unstable discrete $M|M|1$ queue and

$$W_1[n] - \sum_{k=0}^{n-1}[(\alpha + \gamma - \delta) + \delta \cdot \chi\{W_1[k] = 0\}]$$

is a sum of martingale differences. Hence,

$$\begin{aligned} E_{(0,y)}W_1[n] &= n(\alpha + \gamma - \delta) + \delta E_{(0,y)} \sum_{k=0}^{n-1} \chi\{W_1[k] = 0\} \\ &\leq n(\alpha + \gamma - \delta) + \delta \frac{1}{\alpha + \gamma - \delta} \end{aligned}$$

The last inequality follows since $E_{(0,y)} \sum_{k=0}^{n-1} \chi\{W_1[k] = 0\}$ is bounded by the expected number of visits to 0 by a discrete $M|M|1$ queue with transition kernel

$$k(x, x+1) = \alpha + \gamma, \ k(x, x-1) = \delta \cdot \chi\{x \neq 0\}, \ k(x, x) = \beta + \mu + \delta \cdot \chi\{x = 0\}.$$

Starting from $x = 0$ this expected number of visits is $(\alpha + \gamma - \delta)^{-1}$. Since $\alpha + \beta + \gamma - \mu - \delta < 0$ by hypothesis, for all n sufficiently large, $G^n v(0, y)$ is negative. Similarly for m sufficiently large, $G^m v(x, 0) < 0$ for $x \geq m$.

Let $N = \max\{n,m\}$ and let $C = \{(x,y); x,y \leq N\}$. For any (x,y) outside C define

$$-\epsilon := \max\{G^N v(x,0), G^N v(0,y), \alpha+\beta+\gamma-\mu-\delta\} < 0.$$

Let $V(x,y) = v(x,y)/\epsilon$. Define a series of stopping times τ_k such that $\tau_0 = 0$ and

$$\tau_{k+1} = \begin{cases} \tau_k + 1 & \text{if } W_1[\tau_k] > 0 \text{ and } W_2[\tau_k] > 0 \\ \tau_k + N & \text{if } W_1[\tau_k] = 0 \text{ or } W_2[\tau_k] = 0. \end{cases}$$

Let the generator of the embedded Markov chain $M[k] := W[\tau_k]$ be denoted by $\tilde{G}$. M is irreducible and for (x,y) outside C, $\tilde{G}V(x,y) \leq -1$. Consequently V is a Liapounov function for M which is bounded on C. Since M is irreducible and C is finite there exists an m such that the transition kernel $\tilde{K}$ of M satisfies $\tilde{K}^m(\vec{z},(0,0)) > \epsilon$ for all $\vec{z} \in C$. This means C is petite (as defined in Section 5.5.2 in Meyn and Tweedie [12]) so Foster's criterion is verified. Hence M is stable and the mean recurrence time to $(0,0)$ for M is finite. Finally the mean recurrence time to $(0,0)$ for W is at most N times that of M so we have that W is positive recurrent and hence stable.

The following standing assumptions ensuring stability will hold throughout the paper:

(A1) $\alpha < \delta$

(A2) $\beta < \mu$

(A3) $\alpha + \beta + \gamma < \mu + \delta$.

There are moreover two main cases; the singular case when the large deviation path to F starts by going up the y axis and the nonsingular case where the large deviation path is a straight line to F. The nonsingular case is divided into two cases: the subdiagonal case when the following condition holds

(A4)

$$\left(\frac{\alpha+\beta+\gamma}{\mu+\delta}\right)^2 < \frac{\alpha}{\delta}.$$

and the diagonal case when Condition **(A4)** fails to hold.

By the heuristic reasoning in Section 15.8 in Shwartz and Weiss [16] we see that if the most likely path to F is below the diagonal then along this path the queue at the first server is an $M|M|1$ queue with arrival rate α and service rate δ. The queue at the second server evolves independently with an arrival rate of $\beta+\gamma$ and service rate μ. If the first queue is to grow, the most likely transitions are those of a queue with the service and arrival rates reversed. Moreover the probability of reaching ℓ before emptying is *roughly* $\exp(-\ell \log(\delta/\alpha))$.

On the other hand if the most likely path is near the diagonal then in order for the first server to accumulate ℓ customers, the sum of the two queues must be around 2ℓ customers. As remarked above, the combined queue behave like an $M|M|1$ queue with arrival rate $\alpha+\beta+\gamma$ and service rate $\mu+\delta$. The probability the combined queue reaches 2ℓ customers before emptying is *roughly* $\exp\left((-2\ell \log\left((\mu+\delta)/(\alpha+\beta+\gamma)\right)\right)$. Comparing these probabilities we see that Condition **(A4)** specifies when the paths below the diagonal are most likely.

For a more detailed large deviation analysis of the likelihood of different paths to the boundary consult the forthcoming paper by Turner [18]. Our methods should be viewed as a refinement of the large deviation techniques. Once we know the the large deviation behaviour of the fluid model obtained by scaling the Markov chain the technique in this paper provides detailed information about hitting probabilities for the (unscaled) Markov chain.

The subdiagonal case is discussed in Sections 9.2 and 9.3. In Section 9.2 we consider the special case when $\beta+\gamma<\mu$. In this case when the first queue gets big the second remains in a steady state. In Section 9.3 we consider the special case when $\beta+\gamma>\mu$. In this case the second queue grows linearly with the first queue but always lags behind.

The diagonal case is discussed in Section 9.4. We consider the difference between the second and first queues. Under some conditions this difference converges in distribution when the first queue gets big. These conditions exclude singular cases.

Simulation of the numerical example given in Section 9.5 plus the heuristics given there suggest that the singular case occurs when

(A5)

$$\alpha+\gamma<\delta \text{ and } \left(\frac{\beta}{\mu}\right)\left(\frac{\alpha+\gamma}{\delta}\right)>\max\{\frac{\alpha}{\delta},\left(\frac{\alpha+\beta+\gamma}{\mu+\delta}\right)^2\}.$$

9.2 The Super Stable Case

We assume that Condition **(A4)** holds so that, as $\ell\to\infty$, the probability of hitting F on or above the diagonal is negligable. We shall first consider the case when

(B1) $\beta+\gamma<\mu$.

In such a model the service rate of the second queue is so fast that even though the smart customers all join the second queue it still remains in a steady state.

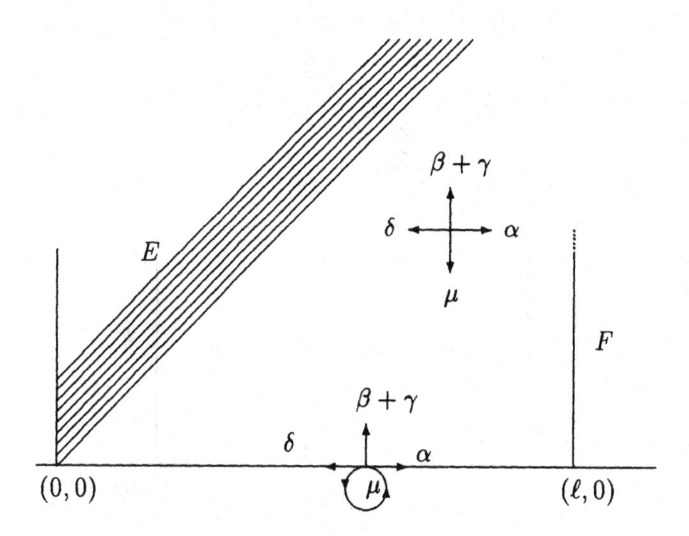

FIGURE 3. Transition probabilities of W^{∞}

9.2.1 *The Markov Additive Model*

Consider a Markov kernel K^{∞} defined on $\{(x, y) : y \geq 0\}$ in Figure 3. Define the set $E = \{(x, y) : y \geq x \geq 0\}$ which is represented by the diagonal lines in Figures 3 and 4. Denote the chain with kernel K^{∞} by W^{∞} and remark that away from E, W agrees with W^{∞}. Moreover W^{∞} can be viewed as a Markov additive process (V, Z) where the workload of the first node will be the additive process V and the queue length of the second node is represented by a Markov chain Z.

9.2.2 *The Twist*

We now introduce a natural *twist* which generalizes the associated random walk discussed in Chapter XII.4 in Feller [5]. The generating function $\hat{K}_{\theta}$ of the transition kernel of the semi-Markov process (V, Z) is given by

$$\hat{K}_{\theta}(y_1, y_2) = E(\exp(\theta \cdot (V[1] - V[0])\chi\{Z[1] = y_2\}|Z(0) = y_1).$$

For $y > 0$ this means

$$\begin{aligned} \hat{K}_{\theta}(y, y+1) &= \beta + \gamma \\ \hat{K}_{\theta}(y, y-1) &= \mu \\ \hat{K}_{\theta}(y, y) &= \delta e^{-\theta} + \alpha e^{\theta}. \end{aligned}$$

For $y = 0$ this means

$$\hat{K}_{\theta}(0, 1) = \beta + \gamma$$

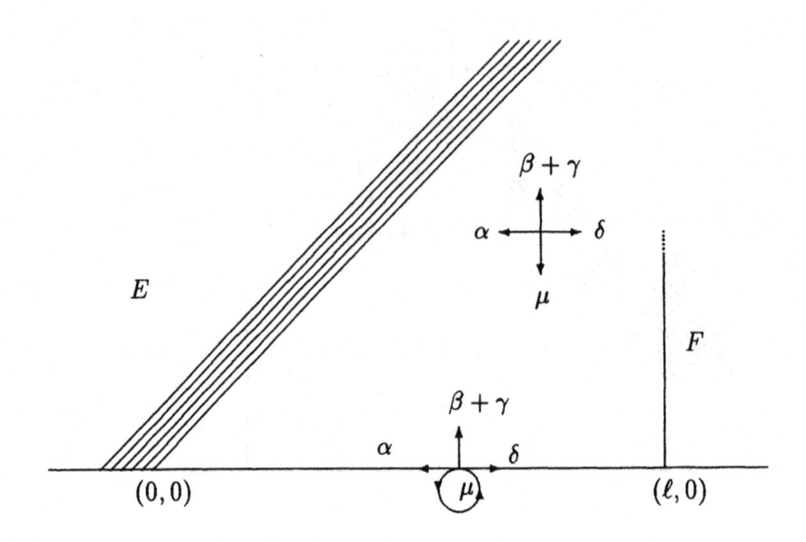

FIGURE 4. Transition probabilities of $\mathcal{W}^{\infty}$.

$$\hat{K}_{\theta}(0,0) = \mu+\delta e^{-\theta}+\alpha e^{\theta}.$$

We must find a θ and a positive eigenvector for the kernel $\hat{K}_{\theta}$ corresponding to the eigenvalue 1. We can do this by defining $r(y)=r^{y}$. We need to find an r such that $\hat{K}_{\theta}r(y)=r(y)$ for all y. For $y>0$ this means

$$(\beta+\gamma)r+\mu r^{-1}+(\delta e^{-\theta}+\alpha e^{\theta})=1.$$

For $y=0$ this means $(\beta+\gamma)r+(\mu+\delta e^{-\theta}+\alpha e^{\theta})=1$. Subtracting we find $\mu-\mu r^{-1}=0$ so $r=1$. Hence $\beta+\gamma+\mu+\delta e^{-\theta}+\alpha e^{\theta}=1$, so using the fact that $q=\beta+\gamma+\mu+\delta+\alpha=1$ we have $\alpha e^{2\theta}-(\alpha+\delta)e^{\theta}+\delta=0$. Solving this for $\exp(\theta)$ gives the solutions 1 or δ/α. Pick $\exp(\theta)$ to be the latter.

Define $a(x,y)=\exp(\theta x)r(y)=\exp(\theta x)$. a is harmonic for K^{∞} so we can define the h-transform (see Doob [4]),

$$\mathcal{K}^{\infty}(\vec{z},\vec{w})=\frac{a(\vec{w})}{a(\vec{z})}K^{\infty}(\vec{z},\vec{w}).$$

The kernel by $\mathcal{K}^{\infty}$ is the kernel of the twisted random walk $\mathcal{W}^{\infty}$ given in Figure 4. Note that with this h-transformation the jump rates α and δ have been switched. This phenomenon is well known for $M|M|1$ queues (see Shwartz and Weiss [16]).

9.2.3 Conditions on the Twisted Chain

Decompose $\mathcal{W}^\infty$ as a Markov additive process $(\mathcal{V}, \mathcal{Z})$. The Markov chain $\mathcal{Z}[n]$ has state space $y \geq 0$ and kernel $\hat{\mathcal{K}}^\infty$. For $y > 0$ this means

$$\hat{\mathcal{K}}^\infty(y, y+1) = \beta + \gamma \tag{9.2}$$
$$\hat{\mathcal{K}}^\infty(y, y-1) = \mu \tag{9.3}$$
$$\hat{\mathcal{K}}^\infty(y, y) = \delta + \alpha. \tag{9.4}$$

For $y = 0$ this means

$$\hat{\mathcal{K}}^\infty(0,1) = \beta + \gamma$$
$$\hat{\mathcal{K}}^\infty(0,0) = \mu + \delta + \alpha.$$

$\mathcal{V}[n]$ can be viewed as the (possibly negative) workload at the first node of the twisted queue. Define the sojourn time associated with the jump from state $\mathcal{Z}[n-1]$ to $\mathcal{Z}[n]$ (which corresponds to the jump from $\mathcal{W}[n-1]$ to $\mathcal{W}[n]$) to be $v[n] := \mathcal{V}[n] - \mathcal{V}[n-1]$. Hence, if $\mathcal{V}[0] = v[0]$ then $\mathcal{V}[n] = \sum_{i=0}^{n} v[i]$ describes the total accumulated workload to get to state $\mathcal{Z}[n]$ in n steps.

We must check the following conditions given in Part I and II:

(1) $\hat{\mathcal{K}}^\infty$ is a the kernel of a positive recurrent Markov chain with a stationary probability measure φ and $\sum_y r(y)^{-1}\varphi(y) < \infty$.

(2) The chain $\mathcal{W}^\infty$ is aperiodic.

(3)

$$d \equiv \sum_y \varphi(y) E_y v[0] = \sum_y \varphi(y) \sum_{x,y'} x\mathcal{K}^\infty((0,y),(x,y')) > 0.$$

The drift Condition **(3)** is easy because $d = \delta - \alpha > 0$. It is crucial that the twisted process drift toward F otherwise the theory does not apply. Condition **(1)** follows since $\varphi(y) = (1-\rho)\rho^y$ where $\rho := (\beta+\gamma)/\mu < 1$ by hypothesis. Also $r = 1$ and of course φ is a probability measure. Aperiodicity is automatic because at points $(x, 0)$ the chain $\mathcal{W}^\infty$ has null jumps. The condition **(4)** in Part I and II is vacuous for nearest neighbour walks.

Define $\tau_\ell^\infty = \min\{n \geq 1 : \mathcal{V}[n] = \ell\}$. The following is proved in Part I:

Theorem 9.2.1 *Under the Conditions* **(1-3)**, $\mathcal{Z}[\tau_\ell^\infty]$ *converges in total variation to* κ *as* $\ell \to \infty$; *that is the hitting distribution of* $\mathcal{W}^\infty$ *on* F *converges in total variation to* κ.

The probability distribution κ defined in Theorem 9.2.1 is therefore obtained by simulating the hitting distribution of $\mathcal{W}^\infty$. This is not very onerous because $\mathcal{W}^\infty$ drifts towards F. It is in general impossible to obtain κ

analytically. It does have a probabilistic interpretation however. We can construct the time reversal of $\mathcal{W}^\infty$ relative to the stationary *infinite* measure given by the product $\Pi^\infty(\vec{z}) = m(x) \times \varphi(y)$ where $\vec{z} = (x, y)$ and m is the counting measure on the integers. The kernel of the time reversal is given by

$$(\mathcal{K}^\infty)^*(\vec{w}, \vec{z}) = \frac{\Pi^\infty(\vec{z})}{\Pi^\infty(\vec{w})} \mathcal{K}^\infty(\vec{z}, \vec{w}).$$

In this case, a bit of checking shows this is precisely K^∞. In general, $(\mathcal{K}^\infty)^*$ tends to the kernel of the time reversal of K with respect to π as x tends to ∞. Moreover, in general, the distribution $\kappa(y)$ is proportional to $\varphi(y)$ times the probability the time reversal of $\mathcal{W}^\infty$ leaves $(0, y)$ and drifts of towards $-\infty$ in x without ever hitting the y axis again. In this case, the probability the increment of $\mathcal{V}$ is $+1$ or -1 doesn't depend on $\mathcal{Z}$ so $\kappa = \varphi$.

9.2.4 Asymptotics

Using the *Mean Hitting Time* Theorem in Part I and the fact that $r = 1$ we have:

Theorem 9.2.2 Under Conditions **(A1-4)** and **(B1)**,

$$(E_\delta \tau)^{-1} \sim g \exp(-\theta \ell) \text{ as } \ell \to \infty \text{ where } \exp(\theta) = \delta/\alpha$$

where

$$g \equiv \sum_{\vec{z} \in E} \pi(\vec{z}) \sum_{\vec{x} \notin E} K(\vec{z}, \vec{x}) a(\vec{x}) H(\vec{x}) \tag{9.5}$$

and $H(\vec{x})$ is the probability $\mathcal{W}^\infty$, starting at $\vec{x}$, never hits E.

The constant g can only be obtained by simulation. This is not too onerous because we only need π near the origin and as remarked above, H is not a rare event for the twisted process. The theorem shows $E_\delta \tau$ grows like $(\delta/\mu)^\ell$ since $H(\vec{x})$ is not identically equal to zero.

Let $\epsilon = 1/\ell$ and let $W^\epsilon(t) := \epsilon W(t/\epsilon)$ be our Markov process rescaled in space and time. Let the fluid limit as $\epsilon \to 0$ be $\overline{W}(t)$. In the super stable case, the fluid limit of the large deviation path to $\overline{F} := \{(1, y) : y \geq 0\}$ is along the x-axis and the hitting distribution is concentrated at the point $(1, 0)$ (see Turner [18]). In general, (see Anantharam, Heidelberger and Tsoucas [2] or Schwartz and Weiss [17]) the large deviation path to F, viewed backward in time, is the same path by which the time reversal of W with respect to π evolves, given this reverse time process starts in F. Consequently the time reversed process will drift back to $(0, 0)$ along the x-axis and may only hit E for small x. This is the essential ingredient in the *Comparison* Lemma in Part I. The probability, starting in steady state on F, of hitting E and then returning to F before returning to $(0, 0)$ must be asymptotically negligible. If by mistake we applied the results in this section to the singular case or to a case where the large deviation path is along the diagonal we see that the time reversal leaving F immediately hits E. The *Comparison* Lemma would not hold nor would the results in Parts I or II.

We can extend the *Hitting Distribution* Theorem in Part I:

Theorem 9.2.3 Under Conditions **(A1-4)** and **(B1)**, as $\ell \to \infty$,

$$P_\delta(W_2[\tau_\ell] = y \mid \tau_\ell < \tau_\delta) \to^T \quad r^{-1}(y)\kappa(y)/\left(\sum_y r^{-1}(y)\kappa(y)\right) = \varphi(y)$$

where $\to^T$ denotes convergence in total variation.

Since $r = 1$ and $\kappa = \varphi$, we conclude that when the first queue hits the level ℓ the probability the second queue has y customers is $\varphi(y) = (1-\rho)\rho^y$.

Similarly the *Steady State* Theorem in Part I becomes

Theorem 9.2.4 Under Conditions **(A1-4)** and **(B1)** we have:

$$\pi((\ell, y)) \sim e^{-\theta\ell} \sum_{\vec{z}\in E} \pi(\vec{z}) \sum_{\vec{x}\notin E} K(\vec{z}, \vec{x}) a(\vec{x}) H(\vec{x}) \frac{1}{d} \varphi(y)$$

where $\exp(\theta) = \delta/\alpha$ and $d = \delta - \alpha$.

This means the stationary measure is a product for large ℓ. It also means that the conditional distribution, $\pi((\ell, y))/\pi(\{\ell \times \{0, 1, \ldots\})$ tends asymptotically to a probability measure proportional to $\varphi(y) = (1-\rho)\rho^y$.

9.3 When the Second Queue is Not Superstable

Again we suppose Condition **(A4)** holds but we now consider the case when

(C1) $\beta + \gamma > \mu$.

In this case the second queue will explode when the first does because if it lags behind it will receive those customers who join the shortest queue. To apply theory in Part I and II we have to change our point of view. Let $E = \{(x, y) \in \mathcal{S} : y \geq x\} \cup \{(x, 0) : x \geq 0\}$ and extend $\mathcal{S}$ to be the whole plane $\mathcal{S}^\infty$ so in fact we measure the workload process for both queues and let both become negative. This gives W^∞ described by Figure 5.

According to the theory we must not twist components which become transient when the first component does but for the twist discovered above $r = 1$ so in fact we didn't twist the y component anyway. This means we can apply the theory if we let Z be a ficticious Markov chain consisting of a single state and we let V denote the vector of the queue sizes. This is still a Markov additive process with kernel K^∞ and with the twist discovered before we can still calculate $\mathcal{K}^\infty$. Both kernels are given in Figure 5.

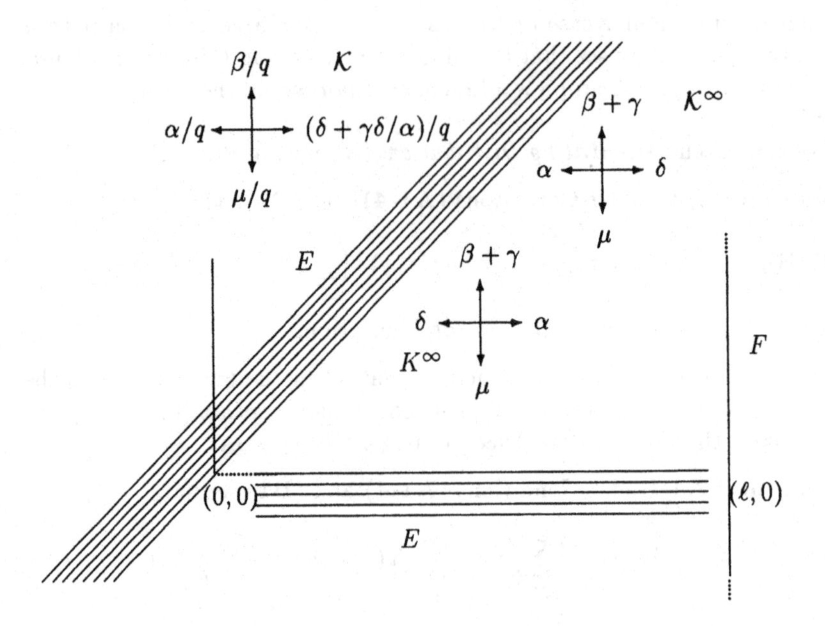

FIGURE 5. Transition probabilities of K^∞ and $\mathcal{K}^\infty$.

We can still apply the *Mean Hitting Time* Theorem to conclude the mean time until the first queue reaches ℓ grows exponentially fast as ℓ gets big.

Theorem 9.3.1 Under Conditions **(A1-4)** and **(C1)**,

$$(E_\delta \tau)^{-1} \sim g \exp(-\theta \ell) \text{ as } \ell \to \infty$$

where $\exp(\theta) = \delta/\alpha$ and where

$$g \equiv \sum_{\vec{z} \in E} \pi(\vec{z}) \sum_{\vec{x} \notin E} K(\vec{z}, \vec{x}) a(\vec{x}) H(\vec{x}) \tag{9.6}$$

where $H(\vec{x})$ is the probability $\mathcal{W}^\infty$, starting at $\vec{x}$, never hits E.

There is of course no limiting distribution for the second queue when the first reaches the level ℓ however we can apply the *Joint Bottlenecks* Theorem in Part II. Define $q(\vec{z}) \equiv q(x,y) = \sum_{\vec{w}} K(\vec{z}, \vec{w}) a(\vec{w})/a(\vec{z})$. Next define

$$\mathcal{K}(\vec{z}, \vec{w}) \equiv \frac{a(\vec{w})}{a(\vec{z})} K(\vec{z}, \vec{w}) \frac{1}{q(\vec{z})}.$$

The kernel $\mathcal{K}$ agrees with $\mathcal{K}^\infty$ outside of E. Denote the chain with kernel $\mathcal{K}$ by $\mathcal{W}$. Note that for $\vec{z} = (x,y)$ such that $x < y$, the value of $q(\vec{z})$ is $1 + \delta\gamma/\alpha - \gamma$. The kernel $\mathcal{K}$ in this region is shown in Figure 5. On the diagonal $q(x,x) = 1 + (\delta\gamma/\alpha - \gamma)/2$.

We first have to show we can couple $\mathcal{W}$ with $\mathcal{W}^\infty$. Conditions **(5)** and **(6)** in Part II could be extended to cases like this when E is a slanted boundary but it is easier just observe that $\mathcal{W}$ eventually drifts into a region where $\mathcal{K}$ agrees with $\mathcal{K}^\infty$. This is true since γ and α are greater than 0 and $\delta > \alpha$. It follows that $\mathcal{W}$ drifts to plus infinity in x with a slope less than $(\beta+\gamma-\mu)/(\delta-\alpha)$ both above and below the diagonal. It follows that eventually $\mathcal{W}$ will enter a set $I = \{(x,y) \in S : L_2 \le y \le x - L_1\}$, where L_1 and L_2 are arbitrarily large. Once inside I and away from E, $\mathcal{K}$ equals $\mathcal{K}^\infty$ so $\mathcal{W}$ can be coupled with $\mathcal{W}^\infty$. With probability arbitrarily close to one, the latter drifts to infinity away from E.

The increments of $\mathcal{V}_1$, the workload of the twisted queue at the first node, are denoted by $v_1[n] := \mathcal{V}_1[n] - \mathcal{V}_1[n-1]$. The mean drift of the twisted workload process in steady state is

$$\vec{d} \equiv (d_1, d_2) := Ev_1[1] = (\delta - \alpha, \beta + \gamma - \mu).$$

Note that the stability condition, $\alpha + \beta + \gamma < \mu + \delta$, means

$$f := \frac{\beta + \gamma - \mu}{\delta - \alpha} < 1.$$

By the *Joint Bottlenecks* Theorem, the conditional distribution of $W_2[\tau_\ell]/\ell$, given $\tau_\ell < \tau_\delta$ converges to a point measure at f. This means that when the first queue hits ℓ the second queue is also large but still only the fraction f of the first. This gives the following theorem,

Theorem 9.3.2 If Conditions **(A1-4)** and **(C1)** are satisfied then almost surely $W_2[\tau_\ell]/\ell \to d_2/d_1 = (\beta+\gamma-\mu)/(\delta-\alpha)$ as $n \to \infty$.

Note that we must extend the *Joint Bottlenecks* Theorem in Part II to show the twisted process satisfies $\mathcal{W}[\tau_\ell]/\ell \to (1, d_2/d_1)$. The key idea is to show that ultimately the twisted process lies entirely in the region where the second queue is the shortest; that is $\mathcal{W}$ ultimately leaves E and can therefore be coupled with $\mathcal{W}^\infty$.

The conclusion of Theorem 9.3.2 is consistent with large deviation theory. The fluid limit of the large deviation path to F is precisely along the line with slope $(\beta+\gamma-\mu)/(\delta-\alpha)$. The results above apply because the fluid limit of the time reversal leaving F retraces this line and thus avoids E.

9.4 The Diagonal Case

In addition to the stability conditions, let us now assume Condition **(A4)** fails to hold. In fact we assume

(D1)

$$\left(\frac{\alpha+\beta+\gamma}{\mu+\delta}\right)^2 > \frac{\alpha}{\delta}.$$

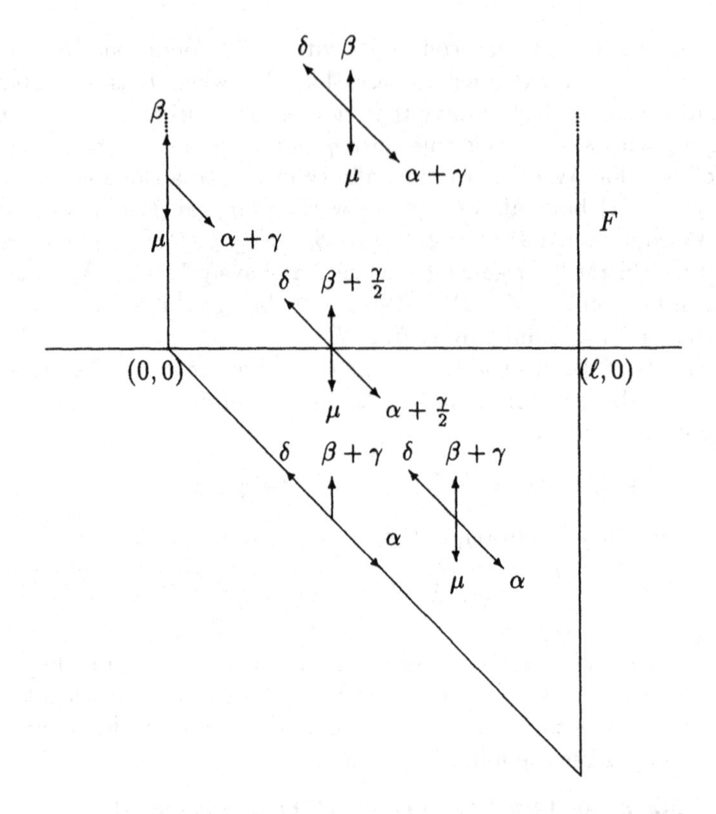

FIGURE 6. The rates are homogeneous in x except at the boundary E.

The fluid limit of the large deviation path to F is on the diagonal consequently the x-axis can be neglected and included in E.

9.4.1 The Markov Additive Model

Let's consider a new Markov chain composed of the length of the first queue and the difference of the second queue minus the first. Redefine the state space $\mathcal{S} = \{(x, y) : x \geq 0, y \geq -x\}$. The jump rates are given in Figure 6. This new model has the advantage that the jump rates are homogeneous in x away from the new boundary $E := \{(0, y) : y \geq 0\} \cup \{(x, -x) : x \geq 0\}$. This is of course a stable process and the stationary probability of a point (x, y) such that $x \geq 0$ and $y \geq -x$ is $\pi(x, x + y)$. As before, we pick the unit of time such that the event rate is 1 so the jump process is the uniformization of a chain W with the same state space. The transition kernel K is obtained by suppressing the transitions into E as in the super stable case. Since the Markov jump process is the uniformization of W it follows that they have the same stationary distribution.

If we remove the boundary E we have the Markov chain W^{∞} with

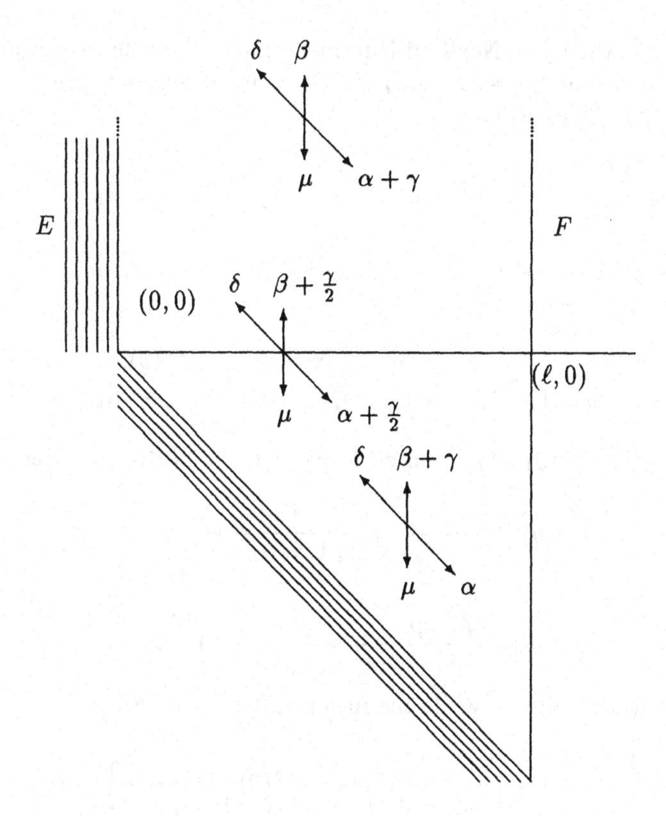

FIGURE 7. The transitions of K^{∞}.

Markov kernel K^{∞} defined in Figure 7. Remark that, away from E, W agrees with W^{∞}. Moreover W^{∞} can be viewed as a Markov additive process (V, Z) where the workload of the first node will be the additive process V and the second coordinate representing the difference in the queue lengths will be the Markov chain Z.

The chain Z has transition probabilities given, for $y > 0$, by

$$\begin{aligned}\hat{K}^{\infty}(y, y+1) &:= p_1 \equiv \beta + \delta \\ \hat{K}^{\infty}(y, y-1) &:= q_1 \equiv \alpha + \mu + \gamma;\end{aligned}$$

for $y < 0$, by

$$\begin{aligned}\hat{K}^{\infty}(y, y+1) &:= q_2 \equiv \beta + \gamma + \delta \\ \hat{K}^{\infty}(y, y-1) &:= p_2 \equiv \alpha + \mu;\end{aligned}$$

and for $y = 0$, by

$$\begin{aligned}\hat{K}^{\infty}(0, 1) &:= p \equiv \beta + \gamma/2 + \delta \\ \hat{K}^{\infty}(0, -1) &:= q \equiv \alpha + \mu + \gamma/2.\end{aligned}$$

To apply the theory in Ney and Nummelin [13] [14] we need to suppose that $\delta + \beta < \alpha + \mu + \gamma$ so $\rho_1 := p_1/q_1 < 1$ and $\delta + \beta + \gamma > \alpha + \mu$ so $\rho_2 = p_2/q_2 < 1$. This means

(D2) $\delta + \beta < 1/2$

(D3) $\alpha + \mu < 1/2$.

Under these conditions the chain with kernel $\hat{K}^\infty$ has stationary distribution $\hat{\pi}^\infty$ given by reversibility;

$$\begin{aligned}\hat{\pi}^\infty(n+1)q_1 &= \hat{\pi}^\infty(n)p_1, \hat{\pi}^\infty(1)q_1 = \hat{\pi}^\infty(0)p\\ \hat{\pi}^\infty(-n-1)q_2 &= \hat{\pi}^\infty(-n)p_2, \hat{\pi}^\infty(-1)q_2 = \hat{\pi}^\infty(0)q\end{aligned}$$

so $\hat{\pi}^\infty(n) = \rho_1^{n-1}\hat{\pi}^\infty(0)(p/q_1)$ and $\hat{\pi}^\infty(-n) = \rho_2^{n-1}\hat{\pi}^\infty(0)(q/q_2)$. Hence,

$$\hat{\pi}^\infty(0)\left(\frac{p}{q_1}\frac{1}{1-\rho_1} + \frac{q}{q_2}\frac{1}{1-\rho_2} + 1\right) = 1$$

so

$$\hat{\pi}^\infty(0) = \left(\frac{p}{q_1 - p_1} + \frac{q}{q_2 - p_2} + 1\right)^{-1}.$$

The drift to the left is δ while the mean drift to the right is

$$\alpha + \gamma\sum_{n=1}^{\infty}\hat{\pi}^\infty(n) + \frac{\gamma}{2}\hat{\pi}^\infty(0) = \alpha + \gamma\hat{\pi}^\infty(0)\frac{p}{q_1 - p_1} + \frac{\gamma}{2}\hat{\pi}^\infty(0).$$

Consequently, the walk W^∞ has negative mean drift if

$$\alpha + \gamma\hat{\pi}^\infty(0)\frac{p}{q_1 - p_1} + \frac{\gamma}{2}\hat{\pi}^\infty(0) < \delta$$

or

$$\gamma\left(\frac{1-(\alpha+\mu+\gamma/2)}{2(\alpha+\mu+\gamma)-1} + \frac{1}{2}\right) < (\delta - \alpha)(\hat{\pi}^\infty(0))^{-1}.$$

Simplifying this means

$$\frac{\gamma}{2}\left(\frac{1+\gamma}{2(\alpha+\mu+\gamma)-1}\right) < (\delta-\alpha)\left(\frac{1-(\alpha+\mu+\gamma/2)}{2(\alpha+\mu+\gamma)-1} + \frac{\alpha+\mu+\gamma/2}{1-2(\alpha+\mu)} + 1\right)$$

or

$$\frac{\gamma}{2}\left(\frac{1+\gamma}{2(\alpha+\mu+\gamma)-1}\right) < (\delta-\alpha)\frac{\gamma(1+\gamma)}{(2(\alpha+\mu+\gamma)-1)(1-2(\alpha+\mu))}.$$

Cross multiplying and simplifying this gives $1-2\mu < 2\delta$ or $\alpha+\beta+\gamma < \mu+\delta$. Since this condition must hold under Condition **(A3)** for the original chain W to be stable we conclude the mean drift is negative provided **(D2-3)** hold.

Under Conditions **(D2-3)** it is easy to check that the chain Z is geometrically ergodic using Lyapounov functions (see Section 15.5.1 in Meyn and Tweedie [12]). Since the additive processes is a nearest neighbour random walk it follows that Condition **(M2)** in Part II holds (see the remark there about conditional Laplace transforms). This means a unique twist exists. In the next section we calculate the twist directly.

9.4.2 The Twist

Again the generating function $\hat{K}_\theta$ of the transition kernel of the semi-Markov process (V, Z) is given by

$$\hat{K}_\theta(y_1, y_2) = E(\exp(\theta \cdot (V[1] - V[0])\chi\{Z[1] = y_2\}|Z(0) = y_1).$$

For $y > 0$,

$$\begin{aligned} \hat{K}_\theta(y, y+1) &= \beta + \delta e^{-\theta} \\ \hat{K}_\theta(y, y-1) &= \mu + (\alpha + \gamma)e^{\theta} \end{aligned}$$

For $y < 0$,

$$\begin{aligned} \hat{K}_\theta(y, y+1) &= (\beta + \gamma) + \delta e^{-\theta} \\ \hat{K}_\theta(y, y-1) &= \alpha e^{\theta} + \mu. \end{aligned}$$

For $y = 0$,

$$\begin{aligned} \hat{K}_\theta(0, 1) &= (\beta + \gamma/2) + \delta e^{-\theta} \\ \hat{K}_\theta(0, -1) &= \mu + (\alpha + \gamma/2)e^{\theta}. \end{aligned}$$

We must find a θ and a positive eigenvector for the kernel $\hat{K}_\theta$ corresponding to the eigenvalue 1. If Conditions **(D2-3)** hold then for every value θ there is a corresponding eigenvalue $\exp(\Lambda(\theta))$ where in fact we are looking for θ so $\Lambda \equiv \Lambda(\theta) = 0$. In general we guess the form of the eigenfunction to be $r(y) = r_1^y$ for $y \geq 0$ and $r(y) = r_2^y$ for $y < 0$. We need to find an $r(y)$ such that $\hat{K}_\theta r(y) = \exp(\Lambda)r(y)$. For $y > 0$ this means

$$(\beta + \delta e^{-\theta})r_1 + (\mu + (\alpha + \gamma)e^{\theta})r_1^{-1} = \exp(\Lambda). \tag{9.7}$$

For $y < 0$ this means

$$(\beta + \gamma + \delta e^{-\theta})r_2 + (\alpha e^{\theta} + \mu)r_2^{-1} = \exp(\Lambda). \tag{9.8}$$

For $y = 0$ this means

$$(\beta + \gamma/2 + \delta e^{-\theta})r_1 + (\mu + (\alpha + \gamma/2)e^{\theta})r_2^{-1} = \exp(\Lambda). \tag{9.9}$$

Note that, if $\Lambda(\theta) = 1$ then from the above,

$$(\beta + \delta e^{-\theta})(\mu + (\alpha + \gamma)e^{\theta}) \leq 1/4 \text{ and } (\alpha e^{\theta} + \mu)(\beta + \gamma + \delta e^{-\theta}) \leq 1/4.$$

Solve the first two equations for r_1 and r_2^{-1} in terms of θ and $\Lambda(\theta)$. There are two solutions for each equation:

$$\begin{aligned}r_1(\theta) &= \frac{\exp(\Lambda)\pm\sqrt{\exp(2\Lambda)-4(\beta+\delta e^{-\theta})(\mu+(\alpha+\gamma)e^{\theta})}}{2(\beta+\delta e^{-\theta})},\\ r_2^{-1}(\theta) &= \frac{\exp(\Lambda)\pm\sqrt{\exp(2\Lambda)-4(\alpha e^{\theta}+\mu)(\beta+\gamma+\delta e^{-\theta})}}{2(\alpha e^{\theta}+\mu)}.\end{aligned}$$

Call these solutions $r_1^- < r_1^+$ and $r_2^- < r_2^+$. Since we want $\exp(\Lambda(\theta))$ to be the Perron-Frobenius eigenvalue of $\hat{K}_\theta$ we pick $r_1(\theta) = r_1^-(\theta)$ and $r_2(\theta) = r_2^-(\theta)$ to be consistent with the fact that $\Lambda(0) = 0$. This is true if and only if Conditions **(D2-3)** hold.

Now substitute into (9.9) and solve for $\Lambda(\theta)$ in terms of θ. Under Conditions **(D2-3)** there must exist a θ such that $\Lambda(\theta) = 0$. In general the function $\Lambda(\theta)$ is convex and $\Lambda(0)$=0. In this case $\Lambda(\theta) \to \infty$ as $\theta \to \infty$. This follows by substituting into (9.9). This gives $\exp(\Lambda)$ as the sum of two terms

$$\frac{(\beta+\gamma/2+\delta e^{-\theta})}{2(\beta+\delta e^{-\theta})}\left(\exp(\Lambda)-\sqrt{\exp(2\Lambda)-4(\beta+\delta e^{-\theta})(\mu+(\alpha+\gamma)e^{\theta})}\right)$$

and

$$\frac{(\mu+(\alpha+\gamma/2)e^{\theta})}{2(\alpha e^{\theta}+\mu)}\left(\exp(\Lambda)-\sqrt{\exp(2\Lambda)-4(\alpha e^{\theta}+\mu)(\beta+\gamma+\delta e^{-\theta})}\right).$$

Now divide through by $\exp(\Lambda)$ and let $\theta \to \infty$. It is clear that the above sum can be one only if $\exp(\Lambda(\theta))$ grows like $\exp(\theta/2)$.

To be sure there is another root we now show the derivative of $\Lambda(\theta)$ at $\theta = 0$ is negative. First recall that $r_1(0) = 1$, $r_2(0) = 1$. Now derive (9.7), (9.8) and (9.9) implicitly.

$$\begin{aligned}\dot{\Lambda}(0) &= (\beta+\delta)\dot{r}_1-\delta+(\alpha+\gamma)-(\mu+\alpha+\gamma)\dot{r}_1\\ \dot{\Lambda}(0) &= -\delta+(\beta+\gamma+\delta)\dot{r}_2+\alpha-(\alpha+\mu)\dot{r}_2\\ \dot{\Lambda}(0) &= -\delta+(\beta+\frac{\gamma}{2}+\delta)\dot{r}_1+(\alpha+\frac{\gamma}{2})-(\mu+\alpha+\frac{\gamma}{2})\dot{r}_2\end{aligned}$$

Solving the first two equations for $\dot{r}_1$ and $\dot{r}_2$ and substituting into the last we get

$$\begin{aligned}&(\dot{\Lambda}(0)-(\alpha-\delta))\left[1+\frac{\mu+\alpha+\gamma/2}{\beta+\delta-\alpha-\mu+\gamma}-\frac{\beta+\gamma/2+\delta}{\beta+\delta-\mu-\alpha-\gamma}\right]\\ &= \frac{\gamma}{2}-\frac{\gamma(\beta+\gamma/2+\delta}{\beta+\delta-\mu-\alpha-\gamma}=-\frac{\gamma}{2}\frac{1+\gamma}{(2(\beta+\delta)-1)}.\end{aligned}$$

Simplifying we get

$$(\dot{\Lambda}(0)-(\alpha-\delta))\frac{\gamma(\gamma+1)}{(\beta+\delta-\mu-\alpha)^2-\gamma^2} = -\frac{\gamma}{2}\frac{1+\gamma}{(2(\beta+\delta)-1)}$$

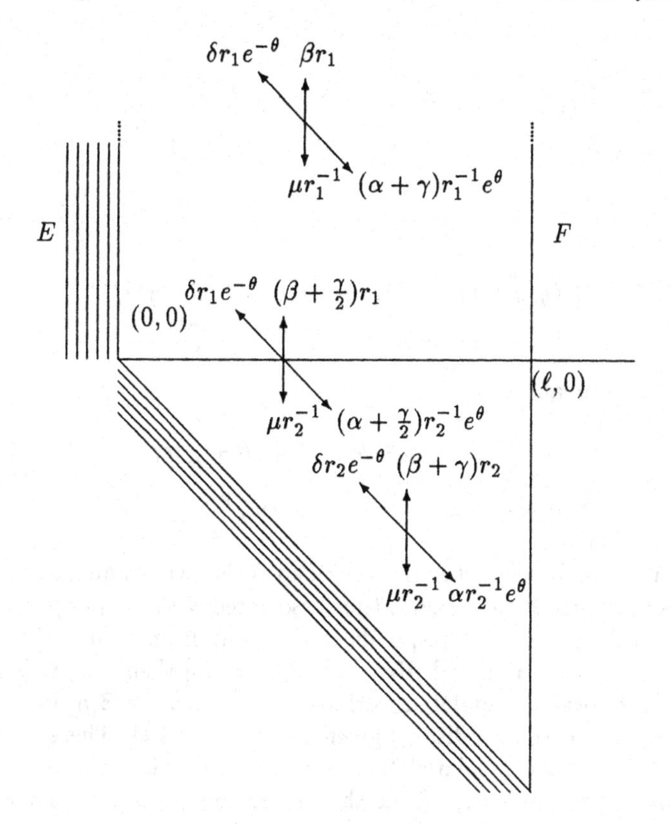

FIGURE 8. The transitions of $\mathcal{K}^{\infty}$.

or

$$\begin{aligned}\dot{\Lambda}(0) &= (\alpha-\delta)) + \frac{(2(\beta+\delta)-1+\gamma)^2-\gamma^2}{2(2(\beta+\delta)-1)} \\ &= 1/2-\mu-\delta<0\end{aligned}$$

by Condition **(A3)**. Hence we will always have another root which we still denote by θ.

For any point $\vec{z}=(x,y)$ define $a(\vec{z}) \equiv a(x,y) = \exp(\theta x) r(y)$. a is harmonic for K^{∞} so we can define the h-transform

$$\mathcal{K}^{\infty}(\vec{z},\vec{w}) = \frac{a(\vec{w})}{a(\vec{z})} K^{\infty}(\vec{z},\vec{w}).$$

The kernel $\mathcal{K}^{\infty}$ of the twisted random walk $\mathcal{W}^{\infty}$ is given in Figure 8.

9.4.3 Conditions on the Twisted Chain

Decompose $\mathcal{W}^{\infty}$ as a Markov additive process $(\mathcal{V},\mathcal{Z})$. The Markov chain $\mathcal{Z}[n]$ has the integers as a state space and has kernel $\hat{\mathcal{K}}$. For $y>0$ this means

$$\begin{aligned}\hat{\mathcal{K}}(y,y+1) &:= p_1 \equiv \beta r_1 + \delta r_1 e^{-\theta}\\ \hat{\mathcal{K}}(y,y-1) &:= q_1 \equiv \mu r_1^{-1} + (\alpha+\gamma) r_1^{-1} e^{\theta}.\end{aligned}$$

For $y < 0$ this means

$$\begin{aligned}\hat{\mathcal{K}}(y,y+1) &:= q_2 \equiv \delta r_2 e^{-\theta} + (\beta+\gamma) r_2\\ \hat{\mathcal{K}}(y,y-1) &:= p_2 \equiv (\alpha e^{\theta} + \mu) r_2^{-1}.\end{aligned}$$

For $y = 0$ this means

$$\begin{aligned}\hat{\mathcal{K}}(0,1) &:= p \equiv \delta r_1 e^{-\theta} + (\beta + \frac{\gamma}{2}) r_1\\ \hat{\mathcal{K}}(0,-1) &:= q \equiv (\alpha e^{\theta} + \mu) r_2^{-1} + \frac{\gamma}{2} r_2^{-1} e^{\theta}.\end{aligned}$$

Again $\mathcal{V}[n]$ can be viewed as the workload of the twisted queue at the first node and again define the sojourn time associated with the jump from state $\mathcal{Z}[n-1]$ to $\mathcal{Z}[n]$ (which corresponds to the jump from $\mathcal{W}[n-1]$ to $\mathcal{W}[n]$) to be $v[n] := \mathcal{V}[n] - \mathcal{V}[n-1]$. Hence, if $\mathcal{V}[0] = v[0]$ then $\mathcal{V}[n] = \sum_{i=0}^{n} v[i]$ describes the total accumulated workload to get to state $\mathcal{Z}[n]$ in n steps.

We must check the conditions given in Part I and II. There is a small catch. Condition **(2)** is violated! This is really no problem however because we could modify the kernel K at the outset to stay put or make a null transition with probability 1/2, say. The periodicity disappears but the stationary distribution, the hitting distribution and the twist all remain the same. The only difference is that it takes twice as long, on average, to hit the boundary. Consequently we have ignored this periodicity throughout this section.

$\hat{\mathcal{K}}$ is a the kernel of a positive recurrent Markov chain with a stationary probability measure φ if

$$\rho_1 := \frac{p_1}{q_1} = \frac{\beta r_1 + \delta r_1 e^{-\theta}}{\mu r_1^{-1} + (\alpha+\gamma) r_1^{-1} e^{\theta}} = r_1^2 \frac{\beta + \delta e^{-\theta}}{\mu + (\alpha+\gamma) e^{\theta}} < 1$$

and

$$\rho_2 := \frac{p_2}{q_2} = \frac{(\alpha e^{\theta} + \mu) r_2^{-1}}{\delta r_2 e^{-\theta} + (\beta+\gamma) r_2} = r_2^{-2} \frac{(\alpha e^{\theta} + \mu)}{\delta e^{-\theta} + (\beta+\gamma)} < 1.$$

Now, with $\Lambda(\theta) = 0$, substitute the expressions for $r_1^2(\beta + \delta e^{-\theta})$ obtained from (9.7) and $r_2^{-2}\mu$ from (9.8) into the above. This gives the equivalent necessary conditions for the existence of φ:

$$\frac{r_1}{\mu + (\alpha+\gamma) e^{\theta}} < 2 \text{ and } \frac{r_2^{-1}}{\beta + \gamma + \delta e^{-\theta}} < 2.$$

These two conditions follow if and only if $r_1 = r_1^-$ and $r_2 = r_2^-$. First

$$\frac{r_1}{\mu + (\alpha+\gamma)e^\theta} = \frac{1 - \sqrt{1 - 4(\beta + \delta e^{-\theta})(\mu + (\alpha+\gamma)e^\theta)}}{2(\beta + \delta e^{-\theta})(\mu + (\alpha+\gamma)e^\theta)}.$$

Since $(\beta + \delta e^{-\theta})(\mu + (\alpha+\gamma)e^\theta) \leq 1/4$ the above expression is always less than 2. Next

$$\frac{r_2^{-1}}{\beta + \gamma + \delta e^{-\theta}} = \frac{1 - \sqrt{1 - 4(\mu + \alpha e^\theta)(\beta + \gamma + \delta e^{-\theta})}}{2(\alpha e^\theta + \mu)(\beta + \gamma + \delta e^{-\theta})}.$$

and again the above expression is always less than 2 if and only if we have $(\alpha e^\theta + \mu)(\beta + \gamma + \delta e^{-\theta}) < 1/4$.

Now consider the condition that $\sum_y r(y)^{-1}\varphi(y) < \infty$ also follows. Just substitute

$$\begin{aligned} r_1^{-1}\rho_1 &= r_1 \frac{\beta + \delta e^{-\theta}}{\mu + (\alpha+\gamma)e^\theta} = \frac{1 - \sqrt{1 - 4(\beta + \delta e^{-\theta})(\mu + (\alpha+\gamma)e^\theta)}}{2(\mu + (\alpha+\gamma)e^\theta)} \\ &= \frac{1 - \sqrt{1 - 4vu}}{2u} < 1 \end{aligned}$$

where we let $u := \mu + (\alpha+\gamma)e^\theta$ and $v := \beta + \delta e^{-\theta}$. If $1 - 2u \leq 0$ or $\mu + (\alpha+\gamma)e^\theta \geq 1/2$ then the above condition is automatic. If however, $\mu + (\alpha+\gamma)e^\theta < 1/2$, the last inequality boils down to $u + v < 1$ and is equivalent to the following condition:

(D4) If $\mu + (\alpha+\gamma)e^\theta < 1/2$ then $(\alpha+\gamma)e^\theta + \delta e^{-\theta} < (\alpha + \gamma + \delta)$.

Next,

$$\begin{aligned} r_2\rho_2 &= r_2^{-1}\frac{\alpha e^\theta + \mu}{\beta + \gamma + \delta e^{-\theta}} \\ &= \frac{1 - \sqrt{1 - 4(\alpha e^\theta + \mu)(\beta + \gamma + \delta e^{-\theta})}}{2(\beta + \gamma + \delta e^{-\theta})} \\ &= \frac{1 - \sqrt{1 - 4ts}}{2s} < 1 \end{aligned}$$

where $s = \beta + \gamma + \delta e^{-\theta}$ and $t = \alpha e^\theta + \mu$. If $1 - 2s \leq 0$ or $\beta + \gamma + \delta e^{-\theta} \geq 1/2$ then the above condition is automatic. If however, $\beta + \gamma + \delta e^{-\theta} < 1/2$ then the last inequality boils down to $s + t < 1$ or equivalently $\alpha e^\theta + \delta e^{-\theta} < \alpha + \delta$. We therefore have the condition:

(D5) If $\beta + \gamma + \delta e^{-\theta} < 1/2$ then $\alpha e^\theta + \delta e^{-\theta} < \alpha + \delta$.

The probability distribution φ is calculated just like $\tilde{\pi}^\infty$ where we just change the definitions of p_1, q_1, p_2, q_2, p, q, ρ_1 and ρ_2 as above. Hence $\varphi(n) = \rho_1^{n-1}\varphi(0)(p/q_1)$ and $\varphi(-n) = \rho_2^{n-1}\varphi(0)(q/q_2)$ and

$$\varphi(0) = \left(\frac{p}{q_1 - p_1} + \frac{q}{q_2 - p_2} + 1\right)^{-1}.$$

Finally we have to check the drift Condition **(3)**. If Conditions **(D2-3)** hold then by the *Drift* Lemma in Part I it is sufficient to check that the drift of V is negative and this we did in the previous section. Alternatively just calculate:

$$\begin{aligned} d &\equiv \sum_y \varphi(y) E_y v_1[0] = \sum_y \varphi(y) \sum_{x,y'} x\mathcal{K}^\infty((0,y),(x,y')) > 0. \\ &= \varphi(0)\left(\frac{p}{q_1 - p_1}[(\alpha+\gamma)r_1^{-1}e^\theta - \delta r_1 e^{-\theta}] + (\alpha + \frac{\gamma}{2})r_2^{-1}e^\theta - \delta r_1 e^{-\theta}\right) \\ &\quad + \varphi(0)\left(\frac{q}{q_2 - p_2}[\alpha r_2^{-1}e^\theta - \delta r_2 e^{-\theta}]\right). \end{aligned}$$

In fact d must be positive since $\dot{\Lambda}(0) < 0$ which means $\dot{\Lambda}(\theta) > 0$.

9.4.4 Asymptotics

We again use the asymptotic results in Parts I and II:

Theorem 9.4.1 If Conditions **(A1-3)** and **(D1-5)** hold then

$$(E_\delta \tau)^{-1} \sim g \exp(-\theta \ell) \text{ as } \ell \to \infty.$$

θ is the unique positive value making $\Lambda(\theta) = 0$, and

$$g \equiv \sum_{\vec{z} \in E} \pi(\vec{z}) \sum_{\vec{x} \notin E} K(\vec{z}, \vec{x}) a(\vec{x}) H(\vec{x}) \sum_y r^{-1}(y)\kappa(y), \tag{9.10}$$

where $H(\vec{x})$ is the probability $\mathcal{W}^\infty$, starting at $\vec{x}$, never hits E and where κ is the limiting hitting distribution of $\mathcal{W}^\infty[\tau_\ell]$ given $\tau_\ell < \tau_E$.

The constants $\pi(\vec{z})$ can be obtained by simulation of W while H can be obtained by simulating $\mathcal{W}^\infty$.

If by mistake we applied this analysis to a singular case, the large deviation path to F would contain a segment on the vertical axis connected either to a straight line segment to F above the diagonal (see the numerical example in Section 9.5) or to a segment going to the diagonal followed by a segment on the diagonal to F (see Turner [18] for this sort of trajectory). The first trajectory could not minimize the action if it hit F above the diagonal so either way the large deviation trajectory to the boundary would hit F near the diagonal. By Conditions **(D2-3)**, the time reversal having kernel $(\mathcal{K}^\infty)^*$ drifts to $-\infty$ hugging the x-axis. This means W^* hugs the diagonal as it returns from (ℓ, ℓ). This in turn means the large deviation path to F is indeed the diagonal if the conditions of the theorem hold.

We can extend the *Hitting Distribution* Theorem in Part I:

Theorem 9.4.2 Under the conditions of Theorem 9.4.1 above, as $\ell \to \infty$,

$$P_\delta(W_2[\tau_\ell] = y \mid \tau_\ell < \tau_\delta) \to^T \quad r^{-1}(y)\kappa(y) / \left(\sum_y r^{-1}(y)\kappa(y)\right)$$

where $\rightarrow^T$ denotes convergence in total variation.

Similarly the *Steady State* Theorem in Part I becomes:

Theorem 9.4.3 Under the conditions of Theorem 9.4.1 above, as $\ell \rightarrow \infty$, we have:

$$\pi((\ell,\ell+y)) \sim e^{-\theta\ell}\sum_{\vec{z}\in E}\pi(\vec{z})\sum_{\vec{x}\notin E}K(\vec{z},\vec{x})a(\vec{x})H(\vec{x})\frac{1}{d}r^{-1}(y)\varphi(y).$$

In the diagonal case, the two servers effectively pool their efforts. It is reasonable to guess that the the rough asymptotics of the steady state probability $\pi(\ell,\ell)$ should be the same as the the rough asymptotics of the steady state probability an $M|M|1$ queue with service rate $\delta+\mu$ and arrival rate $\alpha+\beta+\gamma$ has 2ℓ customers in it. The latter probability has rough asymptotics given by $((\alpha+\beta+\gamma)/(\delta+\mu))^{2\ell}$. From Theorem 9.4.3 $\exp(-\theta\ell)$ gives the rough asymptotics of $\pi(\ell,\ell)$. This implies that that $\exp(-\theta)=((\alpha+\beta+\gamma)/(\delta+\mu))^2$.

To check this, first substitute the expressions $r_1(\theta)$ and $r_2^{-1}(\theta)$ with $\Lambda(\theta)=1$ into (9.9). After simplifying (9.9) becomes $c=a\sqrt{A}+b\sqrt{B}$ where

$$\begin{aligned}
c &= \frac{\gamma}{2}\left((\alpha e^{\theta}+\mu)+(\beta+\delta e^{-\theta})e^{\theta}\right)\\
a &= (\alpha e^{\theta}+\mu)(\beta+\gamma/2+\delta e^{-\theta})\\
b &= (\beta+\delta e^{-\theta})(\mu+(\alpha+\gamma/2)e^{\theta})\\
A &= 1-4(\beta+\delta e^{-\theta})(\mu+(\alpha+\gamma)e^{\theta})\\
B &= 1-4(\alpha e^{\theta}+\mu)(\beta+\gamma+\delta e^{-\theta}).
\end{aligned}$$

Now we have to check that

$$(c^2-a^2A-b^2B)^2-4a^2b^2AB=0$$

when we substitute $\exp(-\theta)=((\alpha+\beta+\gamma)/(\delta+\mu))^2$ and we impose $\gamma=1-\alpha-\beta-\delta-\mu$.

This symbolic calculation was checked using *Mathematica*©. Since the intervening expansion generates thousands of terms it is not recommended for hand calculation. The *Mathematica* code is available in the file where Parts I and II are found.

The product $\exp(-\theta\ell)r^{-1}(y)\varphi(y)$ agrees with formula (13) in Knessl et al. [7]. Also, the conditional distribution, $\pi((\ell,\ell+y))/\pi(\{\ell\times\{0,1,\ldots\})$ tends asymptotically to $r^{-1}(y)\varphi(y)$.

9.5 Numerical Results and Heuristics

In what follows we take $\alpha=0.05$, $\delta=0.2$, $\mu=0.4$, $\beta=.15$ and $\gamma=0.2$. Note that Conditions **(A1-3)** and **(D1-5)** hold (**(D4-5)** are vacuous).

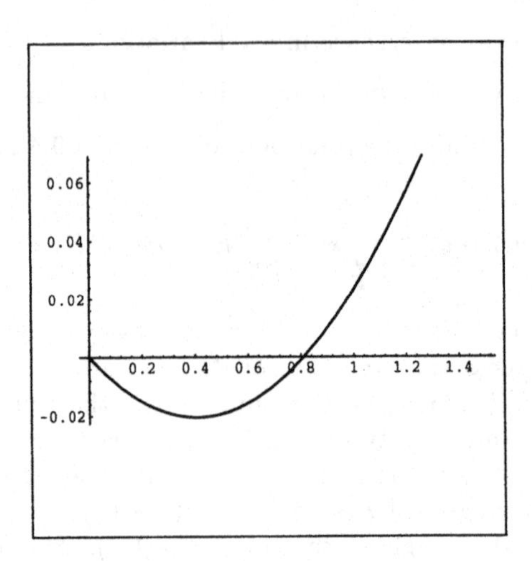

FIGURE 9. Plot of $\Lambda(\theta)$ against θ.

Note, however, that Condition **(B1)** also holds. This shows how important it is to first use large deviation theory to decide on the global behaviour (in this case the likely paths follow the diagonal). Figure 9 was produced by *Mathematica* code (available with Parts I and II)). We use *Mathematica* to solve for $\Lambda(\theta)$, $r_1(\theta)$ and $r_2(\theta)$ using the Equations (9.7), (9.8) and (9.9). We then plot $\Lambda(\theta)$ against θ in Figure 9. Note that the curve has two intercepts at $\theta = 0$ and $\theta = 0.81093$. This means that $\exp(0.81093\ell)$ gives the rough asymptotics for the mean time for the first queue to reach a level ℓ. Note that $\exp(-0.81093) = ((\alpha+\beta+\gamma)/(\mu+\beta))^2$ as predicted.

Now take $\alpha = .0$, $\beta = .1$, $\gamma = .1$, $\mu = .2$ and $\delta = .6$. Since $\alpha = .0$ this might be a diagonal case but Condition **(D2)** fails! Now, consider the rare event of climbing up the y-axis to $\ell(1-(\mu-\beta)/(\delta-(\gamma+\alpha)))$ followed by a straight line large deviation in x and y to (ℓ,ℓ). Large deviation paths like this one were discovered by Anantharam, Heidelberger and Tsoucas [2] for Jackson networks. The probability this occurs before a return to $\delta = (0,0)$ is of order

$$\left(\frac{\beta}{\mu}\right)^{\ell-\ell(\mu-\beta)/(\delta-(\alpha+\gamma))}\left(\frac{\beta}{\mu}\right)^{\ell(\mu-\beta)/(\delta-(\alpha+\gamma))}\left(\frac{\alpha+\gamma}{\delta}\right)^{\ell} > \left(\frac{\alpha+\beta+\gamma}{\mu+\beta}\right)^{2\ell}$$

so this rare event is more likely than a large deviation up the diagonal. This is the basis of the conjectured Condition **(A5)**.

This is a singular case and our theory doesn't apply. On the other hand, simulation of this example seems to show that the above probability,

$$\left(\frac{\beta}{\mu}\right)^{\ell}\left(\frac{\alpha+\gamma}{\delta}\right)^{\ell},$$

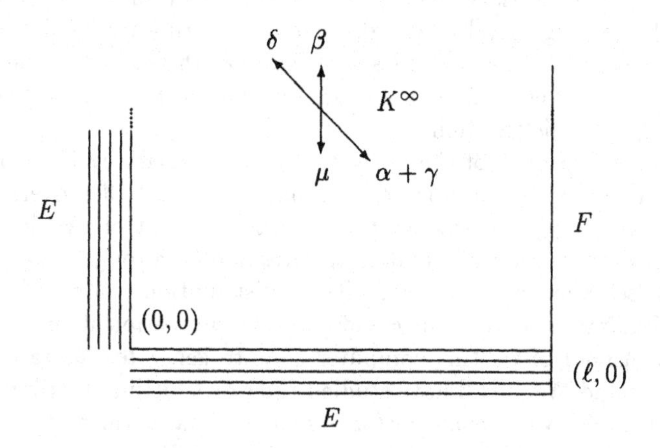

FIGURE 10. Suppose the large deviation path is a straight line F.

does give the rough asymptotics of the probability of reaching F before returning to $(0,0)$. Since the unscaled large deviation path hits F close to (ℓ,ℓ) it doesn't seem unreasonable that there might be a limiting hitting distribution on F even in this case!

Suppose the large deviation path to F was a straight line above the diagonal to F. We could then define E to be the union of the y axis and the diagonal. After making the change of variable $x \to x$ and $y \to y-x$, as in Figure 6, we get Figure 10. Now, to apply the *Joint Bottlenecks* Theorem in Part II, we can only twist in the horizontal direction. This has the effect of reversing the arrows with weights δ and $\alpha+\gamma$ in Figure 10. However this twisted process surely drifts down and hits E so such a picture is impossible. We conclude it is not possible to have a straight large deviation path to F above the diagonal. If $\alpha+\gamma>\delta$, no large deviation path above the diagonal could have a segment on the y axis because of the drift in x. Since straight lines to F are excluded by the above reasoning, $\alpha+\gamma>\delta$ must yield a nonsingular case. This accounts the rest of the conjectured Condition (**A5**).

9.6 Conclusion

The join the shortest queue model studied here provides an excellent test for the theory in Parts I and II. Another applications is [11]. Also Labrèche and McDonald [8] have applied these methods to Jackson networks. Our theory is seen to be a refinement of the vastly more general theory of large deviations. Unlike large deviation theory which only gives rough asymptotics, our theory gives the exact asymptotics of the mean time for the first queue to fill to a high level ℓ. We also get the distribution of the second queue at this hitting time and the steady state of the second queue when the first queue has ℓ customers in it. Large deviation theory only gives the fluid limit of these distributions.

The exact asymptotics of the mean time to hit a high level has proved useful in [11] where one studies the probability of a buffer overload in an ATM switch. Since in practice the constant in front of the dominant exponential term can be very small, estimates based on rough asymptotics are poor. Also, knowing the exact hitting distribution could prove very useful for simulation because one could start the simulation at the moment of overload instead of wasting computer time to generate this rare event first. Shwartz and Weiss [17] have studied the exact hitting distribution in a two server model where some customers come in pairs which split to join each queue. This model is also studied in Parts I and II. Exact asymptotics (called tight limits) have also been studied with a similar technique in Sadowsky and Szpankowski [15] in the case of a $GI|G|n$ queue.

The asymptotic expressions involve constants like g which cannot be given explicitly but they may be obtained quickly by simulating the twist process $\mathcal{W}^\infty$ which is transient towards F. Similar twist processes are widely used for importance sampling. In particular one could use importance sampling to produce an estimate $\hat{p}$ of the probability p of hitting F before returning to $\delta = (0,0)$. Using the fact that $E_\delta\tau \sim m/p$ where m is the mean return time by W to δ we could get an estimate for the mean time for W to reach F.

This is not a sensible approach. First of all one wants an unbiased estimator of $E_\delta\tau$ not of p. Most importantly however, why go to the trouble of proving the importance sampling estimator $\hat{p}$ has optimally small variance when one can prove a limit theorem as we have above. It's far better to spend one's time estimating g once and for all rather than providing a procedure which must be repeated when ℓ changes. If one can find a harmonic function one should prove a limit theorem! On the other hand, in most practical problems one can only guess an approximately optimal twist so importance sampling is the only option. There is, moreover, no counterpart in our theory for operations like turning the importance sampling on and off, as discussed in Chang, Heidelberger, Juneja and Shahabuddin [3].

It is impossible to find a harmonic function on the entire state space so we must create a boundary E including δ and find a function which is harmonic for K outside E. If, outside E, the kernel K agrees with the kernel K^∞ of a Markov additive process then general theory provides a recipe for finding a harmonic function.

For the theory to work, the large deviation path to F must avoid the boundary we pick! A complete large deviation analysis of the join the shortest queue model is not available at present. The first numerical example shows our theory won't determine whether the large deviation path is on the diagonal or is subdiagonal. There are also examples where

$$\left(\frac{\beta}{\mu}\right)\left(\frac{\alpha+\gamma}{\delta}\right) > \frac{\alpha}{\delta} > \left(\frac{\alpha+\beta+\gamma}{\mu+\delta}\right)^2$$

so we won't be able to decide between a singular case and a subdiagonal case unless the conjectured condition for singularity **(A5)** is true. Nevertheless Turner's methods [18] plus our theory do provide some information. **(A1-3)** and **(D1-5)** are (almost) rigorous conditions for the diagonal case. We still need to show that even in a singular case the large deviation path hits F near (ℓ, ℓ).

Once the large deviation theory is settled one should be able to obtain a generalization of the join the shortest queue problem by adding in feedback so that with some probability a customer leaving either queue feeds back into the system. A more difficult problem would be to increase the number of servers. One can also try a similar analysis of the serve the longest queue model (see Shwartz and Weiss [16]). Singular cases like the second numerical example where the large deviation path is not a straight line need to be studied. Finally, the search for harmonic functions is related to the study of the Martin boundary so one might hope to see new and interesting Markov chains whose rare events can be studied using these techniques.

Acknowledgments: I thank the Center for Applied Probability at Columbia University for providing the opportunity to publish this work. I also thank Stephen Turner and Bob Foley for their helpful comments. I am also grateful to both referees for their suggestions, corrections and encouraging words. Finally I thank Karl Sigman and David Yao for accepting several final versions with good humour.

9.7 References

[1] Adan, I. J., Wessels, J. and Zijm, W. H. M., Analysis of the asymmetric shortest queue problem, *Queueing Systems Theory Appl.*, **8**, 1–58, 1991.

[2] Anantharam, V., Heidelberger, P. and Tsoucas, P., Analysis of rare events in continuous time Markov chains via time reversal and fluid approximation, *IBM Research Report RC 16280 (#71858)*, 1990.

[3] Chang, C-S., Heidelberger, P., Juneja, S. and Shahabuddin P., Effective bandwidth and fast simulation of ATM intree networks, *Performance Evaluation.*, **20**, 45–65, 1994.

[4] DOOB, J. L., Discrete potential theory and boundaries, *J. Math. and Mech.*, **8**, 433–458, 1959.

[5] FELLER, W., *An Introduction to Probability Theory and Its Applications*, John Wiley & Sons, New York, 1971.

[6] FLATTO, L. AND MCLEAN, H. P., Two Queues in Parallel, *Communications in Pure and Applied Math.*, **30**, 255–263, 1977.

[7] KNESSL, C., MATKOWSKY, B. J., SCHUSS Z. AND TIER, C. Two parallel queues with dynamic routing, *IEEE Trans. Commun.*, **COM-34**, 1170-1175, 1986.

[8] LABRÈCHE, N. AND MCDONALD, D., Large deviations of the total backlog in a queueing network, *manuscript*, 1995.

[9] MCDONALD, D., Asymptotics of first passage times for random walk in a quadrant, *submitted Ann. Applied Probab.*, 1995.

[10] MCDONALD, D., Asymptotics of first passage times for random walk in a quadrant II, *submitted Ann. Applied Probab.*, 1995.

[11] MCDONALD, D., Buffer overload probabilities for a Markov-modulated fluid model, *manuscript*, 1996.

[12] MEYN, S.P. AND TWEEDIE, R.L., *Markov Chains and Stochastic Stability*, Springer Verlag, New York, 1993.

[13] NEY, P. AND NUMMELIN, E., Markov Additive Processes I. Eigenvalue Properties and Limit Theorems, *Ann. Probab.*, **15**, 561–592, 1987a.

[14] NEY, P. AND NUMMELIN, E., Markov Additive Processes II. Large Deviations, *Ann. Probab.*, **15**, 593–609, 1987b.

[15] SADOWSKY, J. AND SZPANKOWSKI, W., The probability of large queue lengths and waiting times in a heterogeneous multiserver queue I: tight limits. *Adv. Appl. Prob.*, **27**, 532–566, 1995.

[16] SHWARTZ, A. AND WEISS, A., *Large deviations for performance analysis*, Chapman and Hall, 1994.

[17] SHWARTZ, A. AND WEISS, A., Induced rare events: analysis via large deviations and time reversal, *Adv. Appl. Prob.*, **25**, 667-689, 1993.

[18] TURNER, S., Resource pooling via large deviations, *manuscript*, 1996.

10
Rare Events in the Presence of Heavy Tails

Søren Asmussen

ABSTRACT We consider random walks and queues where one or more of the underlying distributions are heavy-tailed, more precisely subexponential (e.g., the tails are regularly varying, DFR Weibull type or lognormal type). This contrast the standard light-tailed assumptions (for example in large deviations theory) where exponential moments are required. The problem is to derive asymptotic expressions for rare events like large random walk maxima or large cycle maxima, large waiting times or large sojourn time in a network, or excursions above a large level, and also to study the behaviour of the process leading to the occurence of the rare event. The results support the folklore that in the heavy-tailed case, rare events occur as consequence of one big jump

10.1 Introduction

The present paper is a survey of some recent results on rare events when the underlying probability distributions are heavy-tailed (exponential moments do not exist): what are their probabilities and given the occurence of the rare event, what was the process behaviour leading to it?

These problems are fairly well understood in the light-tailed case and a considerable body of theory has been developed. The setting is typically that of large deviations theory (e.g. Cottrell *et al.* [22], Bucklew [19] or Dembo & Zeitouni [24]) which provides an asymptotic expression for the rare event probability as well as a description of the path of the process prior to the rare event (though it is often a problem that the large deviations results are not very explicit in their form). The intuitive picture is that rare events occur as a consequence of a build-up over a period where the underlying parameters change, typically by exponential family transformations, and that the change which is seen in the sample path over that period is a change of drift.

The intuition in the heavy-tailed case is completely different: rare events occur as a consequence of one big jump, and apart from that, the process evolves in its typical way. Some classical references supporting this intuition are Durrett [25] and Anantharam [2]. We provide here a number of further examples in a somewhat different vein.

The framework in most of the paper is that of random walk–like processes in discrete or continuous time. In the discrete time case, $\{S_n\}_{n=0,1,2,\dots}$ is a random walk, $S_n = X_1 + \cdots + X_n$ where $X_1, X_2, \dots$ are i.i.d. with common distribution F with mean $\mu < 0$, and $\{W_n\}_{n=0,1,2,\dots}$ is the reflected version (Lindley process; GI/G/1 waiting time process) given by $W_0 = 0$,

$$W_n = (W_{n-1} + X_n)^+ = S_n - \min_{k=0,\dots,n} S_k;$$

a random variable with the limiting stationary distribution is denoted by W_∞ and has the same distribution as $M = \max_{0,1,\dots} S_n$. It is a classical fact (Embrechts & Veraverbeke [26] and references there) that in the heavy–tailed case,

$$\mathbb{P}(W_\infty > x) = \mathbb{P}(M > x) \approx \mathbb{P}(\tau(x) < \infty) \approx \frac{1}{|\mu|} \int_x^\infty \overline{F}(dy)dy , \quad (10.1)$$

where $\tau(x) = \inf\{n : S_n > x\}$. In continuous time, we only look at the special case where $\{S_t\}_{t \geq 0}$ is a compound process with a drift term: at Poisson times with rate β, an upwards jump with distribution B occurs, and in between jumps, $\{S_t\}$ drifts downwards at a unit rate. We assume throughout $\rho = \beta \int_0^\infty x B(dx) < 1$. The process $\{S_t\}$ can be identified with the classical compound Poisson insurance risk claim surplus process and the reflected version $W_t = S_t - \inf_{0 \leq v \leq t} S_v$ with the M/G/1 virtual waiting time process.

The paper is organised as follows. We start in Section 2 by a particular rare events problem, large values within a cycle. Both the light–tailed and the heavy–tailed case are surveyed in order to highlight the intrinsic differences. In Section 3, we give the formal definition of heavy–tailed distributions and some of their analytical properties. Sections 4–7 are devoted to further instances of heavy–tailed behaviour. In particular, we look at extreme values (Section 4), convergence rates (Section 5), excursions (Section 6) and random walk maxima (Section 7). In Section 8, we show by an example (large sojourn times in Jackson networks) how the results developed in the random walk setting may lead to conjectures for more general models, and Section 9 contains some concluding discussion, in particular of simulation aspects.

10.2 Cycle Maxima

We consider the discrete time reflected random walk W_n. The process is regenerative with generic cycle

$$C = \min\{k = 1, 2, \dots : W_k = 0\} = \min\{k = 1, 2, \dots : S_k \leq 0\} .$$

The object of study is the tail of the cycle maximum M_C defined as $\max_{k=0,\dots,C-1} W_k$.

10.2.1 Light Tails

Let $\hat{F}[s] = \mathbb{E}e^{sX}$ denote the m.g.f. of F. Assuming the existence of sufficiently many exponential moments, the assumption $\mu < 0$ ensures the existence of a solution $\gamma > 0$ to the equation $\hat{F}[\gamma] = 1$. We write $\tilde{F}(dx) = e^{\gamma x} F(dx)$ and let $\tilde{\mathbb{P}}$, $\tilde{\mathbb{E}}$ refer to the case where the random walk has increments governed by $\tilde{F}$ rather than F (the *associated random walk*, cf. Feller [27] or [4] Ch. 12; a crucial feature is that its drift is positive rather than negative as for the original random walk). The likelihood ratio up to time n is $L_n = d\mathbb{P}/d\tilde{\mathbb{P}}\Big|_{1,\ldots,n} = e^{-\gamma S_n}$.

Letting $\sigma = \min\{\tau(x), C\}$, $\xi(x) = S_{\tau(x)} - x$ (the overshoot), we get

$$\begin{aligned}
\mathbb{P}(M_C > x) &= \mathbb{P}(S_\sigma > x) = \tilde{\mathbb{E}}\left[L_\sigma; S_\sigma > x\right] \\
&= e^{-\gamma x}\tilde{\mathbb{E}}\left[e^{-\gamma\xi(x)}; \tau(x) < C\right] \\
&\approx e^{-\gamma x}\tilde{\mathbb{E}}\left[e^{-\gamma\xi(x)}; C = \infty\right] \\
&= K e^{-\gamma x} \qquad (10.2)
\end{aligned}$$

where $K = \lim_{x\to\infty} \tilde{\mathbb{E}}e^{-\gamma\xi(x)} \cdot \tilde{\mathbb{P}}(C = \infty)$. This result is the classical one in the light-tailed case and was obtained by Iglehart [34] (see also Bartfai [17]).

An easy extension of the calculation provides also a description of the behaviour of the process given $M_C > x$ (Asmussen [3]). To this end, let $\hat{F}_n(x) = n^{-1}\sum_1^n I(X_k \le x)$ be the empirical distribution function. From $\|\hat{F}_n - \tilde{F}\| \to 0$ $\tilde{\mathbb{P}}$-a.s. (supremum norm), we get

$$\begin{aligned}
&\mathbb{P}(M_C > x,\, \|\hat{F}_{\tau(x)} - \tilde{F}\| > \epsilon) \\
&= e^{-\gamma x}\tilde{\mathbb{E}}\left[e^{-\gamma\xi(x)}; S_\sigma > x,\, \|\hat{F}_{\tau(x)} - \tilde{F}\| > \epsilon,\, \tau(x) < C\right] \\
&\le e^{-\gamma x}\tilde{\mathbb{P}}\left(\|\hat{F}_{\tau(x)} - \tilde{F}\| > \epsilon\right) = o\left(e^{-\gamma x}\right),
\end{aligned}$$

$$\mathbb{P}(\|\hat{F}_{\tau(x)} - \tilde{F}\| > \epsilon \mid M_C > x) \approx \frac{o(e^{-\gamma x})}{Ke^{-\gamma x}} \to 0\,.$$

I.e., the behaviour up to the upcrossing of level x is as if the increment distribution changed from F to $\tilde{F}$, and the main dramatic feature we see in the sample path is a change of drift causing the upcrossing to occur as consequence of a build-up while this happens.

10.2.2 Heavy Tails

For the heavy-tailed case, the analogue of (10.2) is

Proposition 10.2.1 $\mathbb{P}(M_C > x) \approx \mathbb{E}C \cdot \overline{F}(x)$, $x \to \infty$.

The idea of the proof of Asmussen [6] is the following. Choose a level, say $x/2$, to roughly distinguish between small and big jumps, and let $p_1(x) = \mathbb{P}(S_{\tau(x)-1} < x/2, \tau(x) < C)$ (the upcrossing occurs as consequence of a big jump), $p_2(x) = \mathbb{P}(x/2 \le S_{\tau(x)-1} < x, \tau(x) < C)$ (the upcrossing occurs as consequence of a possibly small jump). One then shows that the number $N_1(x)$ of upcrossings of level x within the cycle which occurs from a level $< x/2$ is approximately a $0-1$ r.v., so that $p_1(x) \approx \mathbb{E}N_1(x)$. Estimation of a heavy–tailed integral yields $\mathbb{E}N_1(x) \approx \mathbb{E}C \cdot \overline{F}(x)$ so that it remains to show that $p_2(x)$ is negligible.

This is carried out by a level crossing argument which relies crucially on the available steady–state asymptotics (10.1). Let $D(x)$ be the steady–state probability of $W_n > x, W_{n+1} \le x$, $D_C(x)$ the expected number of times the event $W_n > x, W_{n+1} \le x$ occurs within a cycle and D the limit as $x \to \infty$ of the expected number of downcrossings ($S_n > -x, S_{n+1} \le -x$) of level $-x$ of $\{S_n\}$. Using (10.1), simple calculations yield $D(x) \approx \mu_-/|\mu| \cdot \overline{F}(x)$ where $\mu_- = \mathbb{E}X_-$. To get the asymptotics of $D_C(x)$, note first that

$$\lim_{x \to \infty} \frac{\overline{F}(x+y)}{\overline{F}(x)} = 1 \text{ uniformly in } y \le y_0 \tag{10.3}$$

(that is, if a heavy–tailed r.v. jumps over x, then it jumps in fact much higher). Further, D can be seen to be $\mu_-/|\mu|$ so that asymptotically we have $p_1(x)D + p_2(x) \le D_C(x)$. Inserting

$$p_1(x) \approx \mathbb{E}C \cdot \overline{F}(x), \quad D_C(x) = \mathbb{E}C \cdot D(x) \approx \mathbb{E}C \cdot \mu_-/|\mu| \cdot \overline{F}(x)$$

and dividing by $\overline{F}(x)$ yields $p_2(x) = o(p_1(x))$.

A closer inspection of the argument reveals that indeed the level $S_{\tau(x)-1}$ before the big jump is not only bounded by $x/2$ but is in fact $O(1)$.

10.3 Heavy–Tailed Distributions

Let G be a distribution on $(0, \infty)$ and let $Y_1, Y_2, \ldots$ be i.i.d. with common distribution G. The definition of G to be heavy–tailed requires inevitably that $\mathbb{E}e^{sY} = \infty$ for all $s > 0$ (which follows from (10.3)). Some standard examples are

Regularly varying tails $\overline{G}(x) = L(x)/x^\alpha$ where $L(x)$ is slowly varying (e.g. Pareto– or stable distributions);

DFR Weibull distributions $\overline{G}(x) = e^{-x^\beta}$ where $\beta < 1$;

The lognormal distribution $Y = e^{\sigma U + \mu}$ where $U \sim N(0,1)$.

Note that only moments of order $\alpha - 1 - \epsilon$ exist for the regularly varying case, whereas the Weibull– and lognormal distributions have moments of all orders (the lognormal distribution represents, however, a borderline case by not being determined by its moments).

An established common framework is that $G \in \mathcal{S}$, the subexponential class introduced by Chistyakov [21] and defined by the requirement

$$\lim_{x\to\infty} \frac{\mathbb{P}(Y_1 + Y_2 > x)}{\mathbb{P}(Y_1 > x)} = 2 . \tag{10.4}$$

Since for any distribution G, $\mathbb{P}(\max(Y_1, Y_2) > x) \approx 2\mathbb{P}(Y_1 > x)$, this formal definition contains the intuition behind heavy tails: the only way the sum can get large is by one of the summands getting large (in contrast, in the light–tailed case both summands are large if the sum is so).

Analytic estimates are often facilitated by restrictiong to the class S^* (only slightly smaller than the full subexponential class), defined by Klüppelberg [36] by the requirement

$$\int_0^{x/2} \overline{G}(y)\frac{\overline{G}(x-y)}{\overline{G}(x)}dy \to \int_0^{\infty} \overline{G}(y)dy \; ; \tag{10.5}$$

this is naturally motivated by (10.3) but note that one cannot apply dominated convergence.

The class $\mathcal{S}$ splits naturally into two, the class of distributions with a regularly varying tail and the rest. Another convenient classification is according to extreme value theory. Recall that $G \in MDA(H)$ (maximum domain of attraction) if there exist constants a_n, b_n such that $a_n(\max(Y_1, \ldots, Y_n) - b_n) \xrightarrow{\mathcal{D}} H$. Since subexponential distributions have unbounded support, $H(x)$ must either be the Frechet law $\Phi_\alpha(x) = e^{-x^{-\alpha}}$ or the Gumbel law $\Lambda(x) = e^{-e^{-x}}$. Goldie & Resnick [30] showed that if $G \in \mathcal{S}$, then it only requires a few smoothness conditions for G to be either in $MDA(\Phi_\alpha)$ or in $MDA(\Lambda)$. A further fundamental result (a sharpening of (10.3)) is that then there exist constant $\gamma(x) \to \infty$ (which can be taken as $\mathbb{E}[Y - x \mid Y > x]$) such that

$$\mathbb{P}\left(\frac{Y-x}{\gamma(x)} > y \mid Y > x\right) = \frac{\overline{G}(x+\gamma(x)y)}{\overline{G}(x)} \to \mathbb{P}(V_\alpha > y) , \tag{10.6}$$

where V_α is Pareto when $G \in MDA(\Phi_\alpha)$ ($\alpha < \infty$) and exponential when $G \in MDA(\Lambda)$ ($\alpha = \infty$). This fact is sometimes the crucial one rather than extreme value behaviour.

The *integrated tail*, occuring, e.g., in (10.1), is defined as $\int_x^\infty \overline{G}(y)dy$. It is important to note that a distribution with the same tail behaviour as the integrated tail of G inherits many of the properties of G, e.g. being subexponential, in S^*, in $MDA(H)$ (then with the samme $\gamma(\cdot)$) etc. In fact, the precise conditions for many results are in terms of the integrated tail rather than G. We shall not state what precisely is needed (also not concerning the role of S^*) since in practice this makes little difference.

10.4 Extreme Values

We return to reflected random walks and consider first the extremal index θ. Letting $W_\infty^{(1)}, W_\infty^{(2)}, \ldots$ be i.i.d. copies of W_∞, the intuition between the extremal index is that $\overline{W}_n = \max_{k \le n} W_k$ grows like $\max_{k \le \theta n} W_\infty^{(k)}$ so that extreme value behaviour of the dependent sequence $\{W_n\}$ can be reduced to the i.i.d. case when $\theta > 0$. This is in fact what happens in the light–tailed case, see Iglehart [34] and Rootzén [40]. However, comparing (10.1) and Proposition 10.2.1 show that the tail of the stationary distribution is heavier than that of the cycle maximum in the heavy–tailed case, and general results from extreme value theory (Rootzén [40]) yield

Proposition 10.4.1 The extremal index of $\{W_n\}$ is $\theta = 0$.

Since $\theta = 0$, comparison with the stationary distribution is useless, but it holds in general that the extremal behaviour is like that of $n/\mathbb{E}C$ i.i.d. copies of M_C. Now Proposition 10.2.1 shows that the asymptotics of M_C is like that of $\mathbb{E}C$ i.i.d. copies of the increment X, and so we are lead to

$$\overline{W}_n \approx \overline{X}_n = \max_{k \le n} X_k. \tag{10.7}$$

In fact (Asmussen [6]):

Proposition 10.4.2 For any sequence $\{u_n\}$ and any $\tau \in [0, \infty]$, $\mathbb{P}(\overline{W}_n \le u_n) \to e^{-\tau}$ if and only if $\mathbb{P}(\overline{X}_n \le u_n) \to e^{-\tau}$.

Here by general theory ([38]), $\mathbb{P}(\overline{X}_n \le u_n) \to e^{-\tau}$ is equivalent to $n\overline{F}(u_n) \to \tau$, and if $F \in MDA(H)$ with constants a_n, b_n, we can take $u_n = x/a_n + b_n$, $\tau = -\log H(x)$. Further:

Proposition 10.4.3 $\overline{W}_n - \overline{X}_n \xrightarrow{\mathcal{D}} W_\infty^{(1)} + W_\infty^{(2)}$.

Due to the asymptotic exponentiality of hitting times of rare events, we expect large values to occur at approximately Poisson times. Define a point process N_n on $[0, 1]$ by

$$N_n(A) = \frac{1}{\gamma(u_n)} \# \left\{ k \le n : \frac{k}{n} \in A, \, W_k > u_n \right\}.$$

Proposition 10.4.4 Assume that $n\overline{F}(u_n) \to \tau$ and that (10.6) holds with $G = F$. Then $N_n \xrightarrow{\mathcal{D}} N$ where N is a compound Poisson process with intensity τ and the compounding distribution being that of $V_\alpha / |\mu|$.

10.5 Convergence Rates

Some of the intuition when the extremal index is > 0 is that the process can be split into blocks which are not too far separated and roughly independent. Thus when $\theta = 0$, it may be expected that the convergence rate is slow.

In the light-tailed case, the rate of convergence to stationarity is exponential, see e.g. [4] pp. 95, 261. The only available precise result in the heavy-tailed case is the following (Asmussen & Teugels [14]):

Proposition 10.5.1 For the M/G/1 queue with $\overline{B}(x) = L(x)/x^\alpha$, it holds for any fixed x that

$$\mathbb{P}(W_\infty \le x) - \mathbb{P}(W_t \le x) \approx \frac{\rho \mathbb{P}(W_\infty \le x)}{(\alpha-1)(1-\rho)^\alpha} \cdot \frac{L(t)}{t^{\alpha-1}}, \quad t \to \infty$$

Previous results are coupling bounds obtained by Thorisson [43] (see also Højgaard & Møller [33]); Proposition 10.5.1 shows that these are off by approximately one power of t. The analogoue of Proposition 10.5.1 for reflected random walks or general subexponential distributions is open.

10.6 Excursions

Consider the M/G/1 workload process. An excursion above level x at time t is initiated by a jump from a level $W_{t-} < x$ (the pre-level) to a level $W_t > x$ ($W_t - x$ is the overshoot) and lasts until the next downcrossing of level x. On Fig. 1, there are three such excursions.

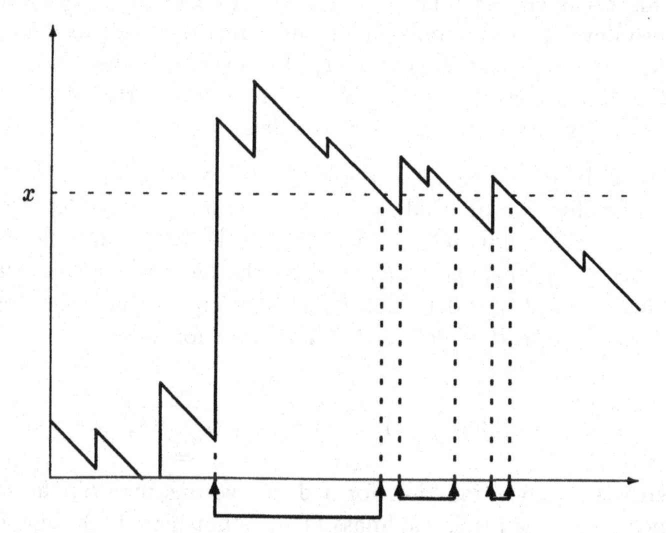

Figure 1: M/G/1 excursions

Motivated by a problem raised by M.F. Neuts, Asmussen & Klüppelberg [11] investigated the asymptotic form of such excursions under stationary conditions in the heavy–tailed case. In the light–tailed case, the overshoot and hence (by the Markov property) the whole excursion has a limit as $x \to \infty$. However, in the heavy–tailed case, asymptotically two types occur with frequencies $1-\rho$, ρ: the ones with a big jump (on Fig. 1 there is one, marked by big arrows), starting from a pre–level which is $O(1)$ (in fact it has the stationary distribution) and has an overshoot converging to ∞; and the ones with a mall jump (two marked by small arrows on the figure), starting from a pre–level close to x (in fact, $x - W_{t-}$ is $O(1)$) and has an overshoot converging in distribution. It can be seen also that small excursions occur only as aftereffects of big ones.

10.7 Random Walk Maxima and Ruin Times

10.7.1 The Tail Probability $\mathbb{P}(M > x) = \mathbb{P}(W_\infty > x) = \mathbb{P}(\tau(x) < \infty)$

The approximation (10.1) is basic in the area, and so we shall comment upon its proof.

1. An extremely short proof of (10.1) proceeds by letting $Z(x) = \mathbb{P}(M > x)$, $z(x) = \mathbb{P}(X_1 > x) = \overline{F}(x)$. Then conditioning upon $X_1 = y$ yields

$$Z(x) \;=\; z(x) + \int_{-\infty}^{x} Z(x-y)F(dy) \;.$$

This is a Wiener–Hopf equation, for which asymptotic properties have been developed in Asmussen [5], and direct reference to results of that paper immediately yields (10.1). However, obviously this approach relies on a considerable body of theory and furthermore, it is not easy to extend to more general models.

2. The only direct rigorous proof of (10.1) we know of utilizes ladder heights. Recall that C is the descending ladder epoch and let $C_+ = \tau(0) = \inf\{n > 0 : S_n > 0\}$ be the ascending one, $G_-(x) = \mathbb{P}(S_C \le x)$, $G_+(x) = \mathbb{P}(S_{C_+} \le x)$ the ladder height distributions. Since $\mu < 0$, G_+ is defective, $\|G_+\| < 1$, and we have the well-known representation (the Pollaczek–Khintchine formula)

$$\mathbb{P}(M > x) \;=\; (1 - \|G_+\|)\sum_{n=1}^{\infty} \overline{G_+^{*n}}(x) \tag{10.8}$$

with the convention that for a defective measure H, the tail $\overline{H}(x)$ means $H(x,\infty)$ (i.e., the mass at ∞ is not included). The argument then proceeds in two steps:

(a) First one shows that

$$\overline{G_+}(x) \approx \frac{1}{|m_-|}\int_x^\infty \overline{F}(y)dy \ , \tag{10.9}$$

where $|m_-|$ is the mean of G_-. To this end, let

$$R_+(A) = \mathbb{E}\sum_0^{C_+-1} I(S_n \in A)$$

be the pre–C_+ occupation measure so that

$$\overline{G_+}(x) = \int_{-\infty}^0 \overline{F}(x-y)R_+(dy) \ . \tag{10.10}$$

By a standard time–reversion argument, R_+ equals the renewal measure $U_- = \sum_0^\infty G_-^{*n}$ which in turn for large negative y is asymptotically $dy/|m_-|$. The contribution to the integral from $[-K,0]$ is $O(\overline{F}(x))$ which is negligible compared to $\int_{-\infty}^{-K}\overline{F}(x-y)dy$ which in turn is of the same order of magnitude as $\int_{-\infty}^0 \overline{F}(x-y)dy = \int_x^\infty \overline{F}(y)dy$. This easily yields (10.9).

(b) By (10.9) and the subexponential property of the integrated tail,

$$\begin{aligned}\overline{G_+^{*n}}(x) &= \|G_+\|^n\overline{\left(\frac{G_+}{\|G_+\|}\right)^{*n}}(x)\\ &\approx \|G_+\|^n\frac{n}{|m_-|\cdot\|G_+\|}\int_x^\infty \overline{F}(y)dy \ .\end{aligned}$$

In fact, $\overline{G_+^{*n}}(x)$ is dominated by $\|G_+\|^n C_\epsilon(1+\epsilon)^n\int_x^\infty \overline{F}(y)dy$ for any ϵ (e.g. [15]). Taking $(1+\epsilon)\|G_+\| < 1$, dominated convergence yields

$$\begin{aligned}\mathbb{P}(M > x) &\approx \frac{1-\|G_+\|}{|m_-|}\int_x^\infty \overline{F}(y)dy\sum_{m=1}^\infty n\|G_+\|^{n-1}\\ &= \frac{1}{|m_-|(1-\|G_+\|)}\int_x^\infty \overline{F}(y)dy \ .\end{aligned}$$

But by Wiener–Hopf theory, $|m_-|(1-\|G_+\|) = |\mu|$.

3. It is tempting to try to simplify the argument by approximating $\mathbb{P}(M > x)$ by $\int_{-\infty}^x \overline{F}(x-y)R(dy)$ where $R(A) = \mathbb{E}\sum_0^\infty I(S_n \in A)$ is the full occupation measure of the random walk. If this is permissible and $\int_{-K}^x$ can be ignored, the fact that $R(dy)$ for large negative

y is asymptotically $dy/|\mu|$ would yield (10.1) just as in (a). This approach would have the potential of extension to more general models. However, the difficulty is not only to make the details rigorous but that the approach leads to a wrong answer: $\int_{-\infty}^{x} \overline{F}(x-y)R(dy)$ is the expected number of upcrossings of level x which asymptotically behaves like $\mu_-/|\mu| \cdot \mathbb{P}(M > x)$ rather than $\mathbb{P}(M > x)$.

4. Heuristically, (10.1) can be derived by using the regenerative representation of $\{W_n\}$ combined with the intuition that the first upcrossing of level x within a cycle occur by a jump from a small value, cf. Section 2.2. This would yield

$$\begin{aligned}
\mathbb{P}(M > x) &= \frac{1}{\mathbb{E}C}\mathbb{E}\sum_{n=1}^{C-1} I(W_n > x) \\
&= \frac{\mathbb{P}(M_C > x)}{\mathbb{E}C}\mathbb{E}\left[\sum_{n=\tau(x)}^{C-1} I(W_n > x) \mid \tau(x) < C\right] \\
&\approx \overline{F}(x)\int_x^\infty \frac{F(dy)}{\overline{F}(x)} R[x-y,\infty) \\
&= \int_x^\infty F(dy)\frac{y-x}{|\mu|} = \frac{1}{|\mu|}\int_x^\infty \overline{F}(y)dy
\end{aligned}$$

The difficulty is to make rigorous that the upcrossing occurs by a jump from a small value, and in fact, the argument for this in Section 2.2 relied on that (10.1) had already been established.

It should be noted that numerical evidence (e.g. Abate *et al.* [1] shows that (10.1) may be a quite poor approximation unless x is extremely large. Asmussen & Binswanger [7] point out an approximation based upon ideas of Hogan [32] which is better for small or moderately large values of x.

Markov–modulated extensions of (10.1) appear in Asmussen *et al.* [8], Asmussen & Højgaard [9] and Jelenkovic & Lazar [35], with the final expression for the constant in front of the integral being given in [35]. These results might suggest the further extension to general stationary input. However, counterexamples can be given even for regenerative input showing that some further structure is required.

10.7.2 The Ruin Time

We next address the question of how the unrestricted random walk $\{S_n\}$ attains a large value x. The asymptotic form of the probability of this event is well known, cf. (10.1), and the heavy tail intuition predicts exceedances of level x to occur as consequence of one big jump. Thus, what remains is to find the asymptotic distribution of the time of the big jump and to make precise that the random walk behaves in its typical way except for this one big jump. The answer to the first question is the following (Asmussen & Klüppelberg [10]).

Proposition 10.7.1 Assume that $F \in MDA(H)$. Then the conditional distribution of $\tau(x)/\gamma(x)$ given $\tau(x) < \infty$ converges weakly to that of $V_\alpha/|\mu|$.

Proof. We give a heuristical argument different from that of [10]. In view of (10.1) and (10.6) (recall that the integrated tail has the same $\gamma(\cdot)$ as F), it must be shown that

$$\mathbb{P}(\infty > \tau(x) > z/|\mu|) \approx \mathbb{P}(M > z) \tag{10.11}$$

where $z = z(x) = y\gamma(x)$; to this end, note that $\mathbb{P}(M > z) \approx \mathbb{P}(V_\alpha > y)\mathbb{P}(M > x)$. Define

$$\tau_-(z) = \inf\{n : S_n < -z\}, \quad \xi(z) = z - S_{\tau_-(z)},$$

$$K(x,v) = \mathbb{P}(\xi(z) \le v), \quad A(x) = \{\tau(x) > \tau_-(z)\} .$$

Since $(\tau_-(z)/z, \xi(x)) \xrightarrow{\mathcal{D}} (1/|\mu|, K(\infty,\cdot))$ for some distribution $K(\infty,\cdot)$ and $\mathbb{P}(A(x)) \to 1$, the same convergence in distribution holds given $A(x)$, and so in (10.11), we can heuristically replace the l.h.s. by $\mathbb{P}(\infty > \tau(x) > \tau_-(z))$. But clearly,

$$\mathbb{P}(\infty > \tau(x) > \tau_-(z)) = \mathbb{P}(A(x)) \int_0^\infty \mathbb{P}(M > x + z + v) K(x, dv) .$$

Here $\mathbb{P}(A(x)) \ge \mathbb{P}(\tau(x) = \infty) \to 1$. Since $\mathbb{P}(M > x+z+v)/\mathbb{P}(M > x+z)$ is bounded by 1 and converges to 1 uniformly on bounded intervals, it follows that

$$\begin{aligned}
&\mathbb{P}(\infty > \tau(x) > \tau_-(z)) \\
&\approx 1 \cdot \mathbb{P}(M > x + z) \int_0^\infty \frac{\mathbb{P}(M > x + z + v)}{\mathbb{P}(M > x + z)} K(x, dv) \\
&\approx \mathbb{P}(M > x + z) \int_0^\infty K(\infty, dv) = \mathbb{P}(M > x + z).
\end{aligned}$$

□

Note that the Pareto– or exponential limit in Proposition 10.7.1 contrasts the normal limit in the light–tailed case, a classical result appearing already in Segerdahl [41].

10.7.3 Conditioned Limit Theorems

Next consider the problem of formulating and showing that the random walk behaves in its typical way except for the big jump. We state the result for the continuous time insurance risk process $\{S_t\}$ only where we obtain particularly sharp results in terms of total variation convergence. An illustration is given on Fig. 2; we want to describe the distribution of $\{S_t\}_{0\le t\le \tau(x)}$ given $\tau(x) < \infty$.

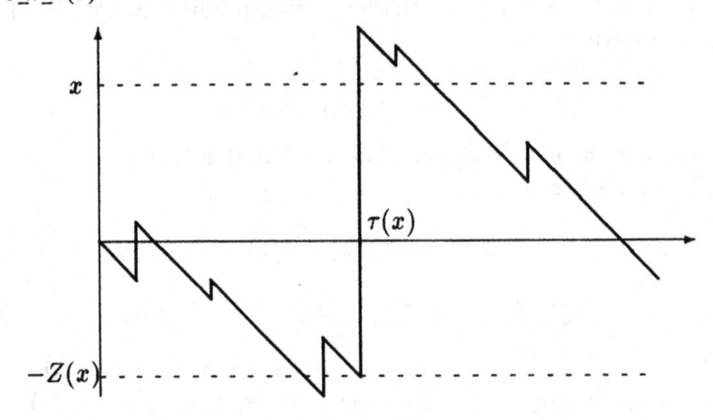

Figure 2: Ruin for the risk process

Define $-Z(x) = S_{\tau(x)-}$ (the value just before ruin), $B^{(x)}(y) = B(x+y)/\overline{B}(x)$ (the overshoot distribution over x), $\delta(z) = \sup\{t : S_t = -z\}$ (the last downcrossing of $-z$),

$$\mathbb{P}^{(x)} = \mathbb{P}(\cdot \mid \tau(x) < \infty), \quad \mathbb{P}^{(x,z)} = \mathbb{P}(\cdot \mid \tau(x) < \infty, Z(x) = z)$$

and let $\|\cdot\|$ denote total variation distance between probability measures.

Theorem 10.7.2 (a) $\|\mathbb{P}^{(x)}(Z(x) \in \cdot) - B^{(x)}\| \to 0$;
(b) $g(u, Z(u)) \to 0$ *in* $\mathbb{P}^{(x)}$*-probability where*

$$g(x,z) = \left\|\mathbb{P}^{(x,z)}\left(\{S_t\}_{0\le t<\tau(x)}\right) - \mathbb{P}\left(\{S_t\}_{0\le t<\delta(z)}\right)\right\| \to 0 .$$

That is, given ruin occurs from level $-z$, the conditioned process behaved up to that as the unconditioned process up to its last downcrossing of $-z$.

The tools used in the proof of Theorem 10.7.2 are path decompositions and excursion theory for Markov processes (combined, of course, with subexponential asymptotics). For example, the following description of a ladder segment is crucial. Given the first upcrossing of level 0 occurs from $-z$, the process behaviour up to that is as follows: take a process $\left\{\tilde{S}_t\right\}$ distributed as $\{S_t\}$ and run $\left\{-\tilde{S}_t\right\}$ in forwards time starting from $-z$ until it hits 0. The distribution of this finite sample path is then the same as the ladder segment of $\{S_t\}$ with the time reversed. See Fig. 3. This follows from general results of Fitzsimmons [28] on the relation between the excursions for a given Markov process and its classical dual (duality is here w.r.t. Lebesgue measure), which are used at a number of further occasions in the proof. The result on the ladder segment alone can also be derived from Asmussen & Schmidt [13].

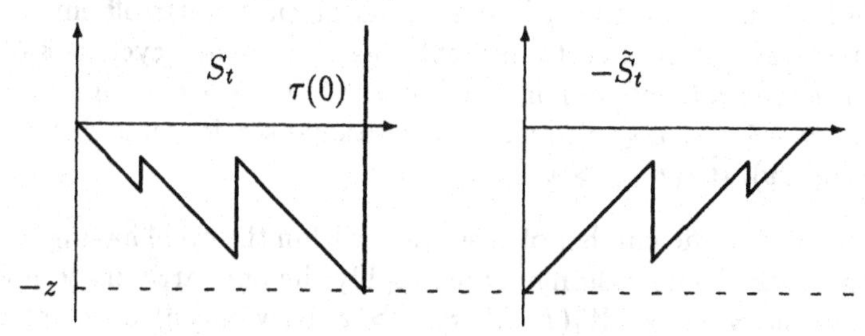

Figure 3: A ladder segment and its reverse

The results for general discrete–time random walks are of just the same spirit but mathematically somewhat weaker, being given in terms of weak convergence of the process suitably normalized. The reason that the results for the insurance risk process are stronger is mainly the downwards skipfree property of the process.

10.8 Heuristics for Jackson Networks

Consider a Jackson network with K nodes, Poisson input at rate λ_i and service time distribution B_i at node i, and Markovian routing: a customer leaving node i goes to no $_{ij}$ and exits the system w.p. $1 - p_{i1} - \cdots - p_{iK}$.

The throughput equations are

$$\gamma_i = \lambda_i + p_{1i}\gamma_1 + \cdots + p_{Ki}\gamma_K, \; i = 1, \ldots, K.$$

It is well known that (assuming for simplicity that each B_i has a strictly positive density on $(0, \infty)$) that the necessary and sufficient stability condition is that the unique positive solution satisfies $\gamma_i \mu_i < 1$, where $\mu_i = \int_0^\infty x B_i(dx)$, for all i, see e.g. Sigman [42], Chang *et al.* [20], Dai [23] and Baccelli & Foss [16]; note, however, that some of the literature on the stability of queueing networks (e.g. Borovkov [18], Down & Meyn [39]) does not apply in the present setting because exponential tails is a crucial assumption. We let D_n denote the sojourn time of the nth customer and D_∞ the corresponding steady–state r.v. Few explicit results on the distribution of D_∞ are available, and motivated from this, we shall here heuristically derive a tail approximation for the heavy–tailed case. To this end, we assume that $B_K \in \mathcal{S}$ and that $\overline{B}_i(x) = o\left(\overline{B}_K(x)\right)$, $i = 1, \ldots, K-1$ (the discussion is readily generalized to the case where several B_i have a subexponential tail of the same order of magnitude). We proceed in the following steps:

1. Based upon Section 2.2, it seems reasonable to assume that sojourn times $> x$ within a cycle occur as a consequence of one customer at node K experiencing a service time $> x$. Here a cycle C is defined as the first emptiness time of the network given it starts off empty. The expected number of customers entering node K in a cycle is $\mathbb{E}C \cdot \gamma_K$. Hence the probability of one having service time $> x$ is approximately $p(x) = \mathbb{E}C \cdot \gamma_K \overline{B}_K(x)$, and his conditional service time distribution is $B_K(dy)/\overline{B}_K(x)$, $y > x$.

2. Let $N(x)$ be the number of customers within the cycle having sojourn times $> x$. These customers are roughly the ones arriving to node K while the workload $W_t(K)$ there is $> x$. Now output does not occur from node K while the long service time is being processed, and so the input rate to node K during that time is roughly

$$\gamma_K^{(K)} = \lambda_K + p_{1K}\gamma_1^{(K)} + \cdots + p_{K-1,K}\gamma_{K-1}^{(K)},$$

where the $\gamma_i^{(K)}$ are solutions to the reduced throughput equations

$$\gamma_i^{(K)} = \lambda_i + p_{1i}\gamma_1^{(K)} + \cdots + p_{K-1,i}\gamma_{K-1}^{(K)}, \quad i = 1, \ldots, K-1$$

(note that $\gamma_i^{(K)} < \gamma_i$ for all i). Hence

$$\begin{aligned} \mathbb{E}N(x) &\approx p(x)\int_x^\infty \frac{y-x}{1-\gamma_K^{(K)}\mu_K}\frac{F(dy)}{\overline{B}_K(x)} \\ &= \mathbb{E}C\frac{\gamma_K}{1-\gamma_K^{(K)}\mu_K}\int_x^\infty \overline{B}_K(y)dy\,. \end{aligned}$$

3. By regenerative process theory, we arrive at

$$\mathbb{P}(D_\infty > x) = \frac{\mathbb{E}N(x)}{\mathbb{E}C} \approx \frac{\gamma_K}{1-\gamma_K^{(K)}\mu_K}\int_x^\infty \overline{B}_K(y)dy\,. \qquad (10.12)$$

10.9 Concluding Remarks

1. Many of the results presented in this paper has recently been generalised (Asmussen [6]) to storage processes $\{W_t\}$ (evolving as the M/G/1 workload process except that the decrease between jumps occurs at rate $r(x)$ at level x rather than it is linear corresponding to $b = 0$). In particular, this provides tail asymptotics for $\mathbb{P}(W_\infty > x)$ which is of considerable interest in insurance risk in view of a dual

interpretation in terms of ruin probabilities. For a regularly varying tail and $r(x) = a + bx$, this was earlier obtained by Klüppelberg & Stadtmüller [37].

2. An interesting aspect of the description of heavy–tailed behaviour is its implication for simulation algorithms, say we want to determine $\ell(x) = \mathbb{P}(W_\infty > x) = \mathbb{P}(M > x)$ or $\ell(x) = \mathbb{P}(M_C > x)$ by simulation; since approximations in the heavy–tailed area are often not very good, simulation may be required even in quite simple settings. Simulating from the conditional distribution given the rare event is optimal but not feasible (see e.g. Asmussen & Rubinstein [12]). However, conditioned limit results for the process behaviour given the rare event may lead to ideas of how to simulate from a probability distribution $\tilde{F}$ not too different from the true conditional distribution. This idea is classical in the light–tailed case, where simulating using the exponentially twisted distribution $\tilde{F}$ has proved extremely efficient.

   In the simulation context, the relevant measure of efficiency is relative error. This is defined as $\epsilon(x) = Var(Z(x))/\ell(x)^2$ where $Z(x)$ is the unbiased ($\mathbb{E}Z(x) = \ell(x)$) r.v. generated by the simulation (in practice, N replications are generated, the empirical mean is used as point estimator and a confidence interval is given using the empirical variance). E.g. in the setting of the cycle maximum M_C in Section 2.1 (light tails), $Z(x) = L_\sigma I(S_\sigma > x)$ and $\epsilon(x)$ remains bounded as $x \to \infty$. This is the best performance observed in any realistic example of rare events simulation and has motivated to consider $Z(x)$ optimal if $\epsilon(x)$ grows substantially slower than $\ell(x)^{-1}$ (e.g., the definition in [12] requires that $\epsilon(x) = O(|\log \ell(x)|^p)$ for some p).

   In the light–tailed case, many examples of such optimal simulation estimators are known and are typically based upon exponential twisting, possibly in a Markov–modulated context. The situation in the heavy–tailed case is far more open. The only example we know of an estimator which is optimal in the above sense has been produced by Asmussen & Binswanger [7] for the very special case of the M/G/1 queue with regularly varying service time distribution; the idea is a conditional Monte Carlo method involving the order statistics. For more general cases, the above description of how rare events occur certainly leads to many ideas on how to simulate a twisted process (typically having one or a few big jumps rather than overall twisted parameters) which asymptotically looks like the optimal conditioned process given the rare event. However: this criterion does not apriori ensure good properties of $\epsilon(x)$, and there are in fact examples in the simulation literature (Glasserman [29]) showing that this is not always the case. Thus, the area of rare events simulation in the presence of heavy tails seems only in its early stages.

10.10 References

[1] Abate, J., Choudhury, G.L. & Whitt, W., Waiting-time tail probabilities in queues with long-tail service-time distributions, *Queueing Systems*, **16**, 311–338, 1994.

[2] Anantharam, V., How large delays build up in a $GI/GI/1$queue, *Queueing Systems*, **5**, 345–368, 1988.

[3] Asmussen, S., Conditioned limit theorems relating a random walk to its associate, with applications to risk reserve processes and the GI/G/1 queue, *Adv. Appl. Probab.*, **14**, 143-170, 1982.

[4] Asmussen, S., *Applied Probability and Queues*, Wiley, 1987.

[5] Asmussen, S., A probabilistic look at the Wiener-Hopf equation. Submitted, 1995.

[6] Asmussen, S., Subexponential asymptotics for stochastic processes: extremal behaviour, stationary distributions and first passage probabilities. Submitted, 1996.

[7] Asmussen, S. & Binswanger, K., Ruin probability simulation for subexponential claims. Manuscript, University of Lund and ETH Zürich, 1996.

[8] Asmussen, S., Fløe Henriksen, L. & Klüppelberg, C., Large claims approximations for risk processes in a Markovian environment, *Stoch. Proc. Appl.*, **54**, 29–43, 1994.

[9] Asmussen, S. & Højgaard, B., Ruin probability approximations for Markov-modulated risk processes with heavy tails. To appear in *Theor. Stoch. Proc.*, 1995.

[10] Asmussen, S. & Klüppelberg, C., Large deviations results for subexponential distributions, with applications to insurance risk. To appear in *Stoch. Proc. Appl.*, 1997.

[11] Asmussen, S. & Klüppelberg, C., Stationary M/G/1 excursions in the presence of heavy tails. To appear in *J. Appl. Probab.*, March 1997.

[12] Asmussen, S. & Rubinstein, R.Y., Steady-state rare events simulation and its complexity properties, *Advances in Queueing: Models, Methods & Problems* (J. Dshalalow ed.), 429–466. CRC Press, Boca Raton, Florida, 1995.

[13] Asmussen, S. & Schmidt, V., Ladder height distributions with marks, *Stoch. Proc. Appl.*, **58**, 105–119, 1995.

[14] ASMUSSEN, S. & TEUGELS, J.L., Convergence rates for M/G/1 queues and ruin problems with heavy tails, *J. Appl. Probab.*, 1997.

[15] ATHREYA K.B. & NEY, P., *Branching Processes*, Springer, Berlin, 1972.

[16] BACCELLI, F. & FOSS, S.G., Stability of Jackson–type queueing networks, *Queueing Sytems*, **17**, 5–72, 1994.

[17] BARTFAI, P., Limes superior Sätze für die Wartemodelle, *Stud. Sci. Math. Hung.*, **5**, 317–325, 1970.

[18] BOROVKOV, A.A., Limit theorems for queueing networks, *Th. Prob. Appl.*, **31**, 413–427, 1986.

[19] BUCKLEW, J.A., *Large Deviation Techniques in Decision, Simulation and Estimation*, Wiley, New York, 1990.

[20] CHANG, C.S., THOMAS, J.A., & KIANG, S.H., On the stability of open networks: a unified approach by stochastic dominance, *Queueing Systems.*

[21] CHISTYAKOV, V.P., A theorem on sums of independent random variables and its application to branching random processes, *Th. Prob. Appl.*, **9**, 640–648, 1964.

[22] COTTRELL, M., FORT, J.–C. & MALGOUYRES, G., Large deviations and rare events in the study of stochastic algorithms, *IEEE Trans. Aut. Control*, **AC–28**, 907–920, 1983.

[23] DAI, J.G., On positive Harris recurrence of multiclass queueing networks: a unified approach via fluid limit models, *Ann. Appl. Probab* **5**, 49–77, 1995.

[24] DEMBO, A. & ZEITOUNI, O., *Large Deviation Techniques*, Jones & Bartlett, Boston, 1993.

[25] DURRETT, R., Conditioned limit theorems for random walks with negative drift, *Z. Wahrscheinlichkeitsth. verw. Geb.*, **52**, 277–287, 1980.

[26] EMBRECHTS, P. & VERAVERBEKE, N., Estimates for the probability of ruin with special emphasis on the possibility of large claims, *Insurance: Mathematics and Economics*, **1**, 55-72, 1982.

[27] FELLER, W., *An Introduction to Probability Theory and its Applications* **II** (2nd ed.), Wiley, New York, 1971.

[28] FITZSIMMONS, P.J., On the excursions of Markov processes in classical duality, *Probab. Th. Rel. Fields*, **75**, 159-178, 1987.

[29] GLASSERMAN, P., & KOU, S.–G., Analysis of an importance sampling estimator for tandem queues, *ACM TOMACS*, **5**, 22–42, 1995.

[30] GOLDIE, C.M. & RESNICK, S.I., Distributions that are both subexponential and in the domain of attraction of an extreme–value distribution, *J. Appl. Probab.*, **20**, 706–718, 1988.

[31] HEIDELBERGER, P., Fast simulation of rare events in queueing and reliability models, *ACM TOMACS* **6**, 43–85, 1995. Preliminary version published in *Performance Evaluation of Computer and Communications Systems*. Springer Lecture Notes in Computer Science, **729**, 165–202, 1995.

[32] HOGAN, M.L., Comment on "Corrected diffusion approximations in certain random walk problems", *J. Appl. Probab.*, **23**, 89–96, 1986.

[33] HØJGAARD, B. & MØLLER, J.R., Convergence rates in matrix–analytic models, *Stochastic Models*, **12**, 1996.

[34] IGLEHART D.L., Extreme values in the GI/G/1 queue, *Ann. Math. Statist.*, **43**, 627-635, 1972.

[35] JELENKOVIC, P.R. & LAZAR, A.A., Subexponential asymptotics of a network multiplexer. Manuscript, Center for Telecommunications Research, Columbia University, New York, 1995.

[36] KLÜPPELBERG, C. Subexponential distributions and integrated tails. *J. Appl. Probab.*, **25**, 132–141, 1988.

[37] KLÜPPELBERG, C. & STADTMÜLLER, U. Ruin probabilities in the presence of heavy tails and interest rates. Submitted, 1995.

[38] LEADBETTER, R., LINDGREN, G. & ROOTZÉN, H., *Extremes and Exceedances of Stationary Processes.* Springer–Verlag, 1983.

[39] MEYN, S.P. & DOWN, D., Stability of Jackson–type queueing networks, *Ann. Appl. Probab*, **4**, 124–148, 1994.

[40] ROOTZÉN, H., Maxima and exceedances of stationary Markov chains, *Adv. Appl. Probab.*, **20**, 371-390, 1988.

[41] SEGERDAHL, C.–O., When does ruin occur in the collective theory of risk? *Skand. Aktuar Tidsskr.*, 22–36, 1955.

[42] SIGMAN, K., The stability of open queueing networks, *Stoch. Proc. Appl.*, **35**, 11–25, 1990.

[43] THORISSON, H., The queue GI/G/1: Finite moments of the cycle variable and uniform rates of convergence, *Stoch. Proc. Appl.*, **19**, 85–99, 1985.

11 A Network Multiplexer with Multiple Time Scale and Subexponential Arrivals

Predrag R. Jelenković and Aurel A. Lazar

ABSTRACT Real-time traffic processes, such as video, exhibit multiple time scale characteristics, as well as subexponential first and second order statistics. We present recent results on evaluating the asymptotic behavior of a network multiplexer that is loaded with such processes.

11.1 Introduction

One of the key features in Asynchronous Transfer Mode (ATM) based broadband networks is statistical multiplexing (SMUX). Most of the multiplexed entities are calls originating from various sources. In order to operate properly, each of these calls has to satisfy some quality of service requirements (QOS). QOS requirements are usually bounds on performance measures characterizing the dynamic behavior of the multiplexed traffic. The most basic model of a SMUX is an infinite buffer single server queue with a work conserving scheduler. The fundamental performance measure is the queue length distribution ($\mathbb{P}[Q > x]$). Therefore, it is of utmost importance to have feasible procedures for calculating this distribution under reasonable assumptions on the arrival processes.

Numerous investigations have shown that the arrival processes (sources) that arise in ATM networks (like voice and video) have a very complex statistical structure; an especially troublesome characteristic is the high statistical dependency (e.g., see [25, 30]). Modeling of this high dependency usually leads to analytically very complex statistical characteristics, typically making the associated evaluation of the queue length distribution intractable. However, because of the stringent QOS requirements in ATM, only the tail of the queue length distribution in the domain of very small probabilities is needed. This has motivated researchers to investigate possible approximations of the asymptotic behavior of the queue length distribution. This is the main subject of our presentation.

More formally, given an infinite buffer single-server queue, let $A = \{A_t, t \geq 0\}$, $C = \{C_t, t \geq 0\}$, be two discrete time, stationary, and ergodic processes (on a probability space $(\Omega, \mathcal{F}, \mathbb{P})$); A_t represents the amount of arrivals to the queue at time t, and C_t is the server capacity at time t. Then, for any initial random variable Q_0, the following (Lindley's) equation

$$Q_{t+1} = (Q_t + A_t - C_t)^+ \tag{11.1}$$

completely defines the queue length process $\{Q_t, t \geq 0\}$. Queues of this type represent a natural model for ATM multiplexers. According to the classical result of Loynes [31], if $\mathbb{E}A_t < \mathbb{E}C_t$ (and $\{A_t, C_t, t \geq 0\}$ are stationary and ergodic) $\{Q_t\}$ couples with the unique stationary solution $\{Q_t^s\}$ of the recursion (11.1) for any initial condition Q_0; in particular $\mathbb{P}[Q_t \geq x] \to \mathbb{P}[Q_0^s \geq x]$ as $t \to \infty$ (for simplicity we will refer to Q_t^s simply as Q). In what follows the difference between the arrival and the service process $\{X_t \stackrel{def}{=} A_t - C_t, t \geq 0\}$ will be called the *queue increment process*.

Stationarity and ergodicity comprise the general framework for our current exposition, and will be assumed in the rest of the paper. We will see that under different assumptions on the distribution of the queue increment process, the queue length asymptotics may exibit very different behavior. Two major probabilistic categories of assumptions are the *exponential* (Cramér) and *subexponential*. Informally, the exponential category is represented with random variables whose moment generating functions are finite in some positive neighborhood of zero, whereas the subexponential category consists of random variables whose m.g.f.s are infinite on the positive real axis. This paper is organized according to this categorization.

In the first part of the paper (section 11.2) we examine the exponential asymptotic queueing behavior in the presence of multiple time scales. We demonstrate that in this case the dominant (or so called Equivalent Bandwith) multiplexer approximation may be very inaccurate. To try to alleviate this problem, in section 11.2.1 we present an asymptotic expansion approach for approximating all queue length probabilities for the case of structured Markovian multiple time scale (decomposable) arrivals. In section 11.2.2 we prove that for arrival processes that spend long-tailed (random) time in their high activity states, the Equivalent Bandwith (EB) constant does not depend on slow time scale statistics and is equal to the case when processes stay in high activity states all of the time.

In the second part of the paper (section 11.3) we discuss the problem of approximating the queue length probabilities under subexponential (non Cramér) assumptions. We first give precise definitions and some intuition behind the modeling of real time processes using subexponential statistics. Some very recent asymptotic results for arrival processes with both subexponential marginals and subexponetial autocorrelation function are summarized in section 11.3.1. The paper is concluded in section 11.4.

11.2 Multiple Time Scale Arrivals

Very often, arrival processes that arise in modern communication networks exhibit a multiple time scale structure. A typical example is Variable Bit Rate (VBR) video traffic. This traffic consists of ATM cells, that, when grouped together, correspond to slices; slices are the building blocks for frames, and finally, a large number of frames form scenes [30]. Each of these VBR video building blocks (cells, slices, frames, scenes) belong to different time scales, and are characterized by different statistics. Furthermore, on an even larger time scale these building blocks form calls with their own statistics. The call statistics themselves may change according to the time of the day. Thus, from this brief analysis, we see that there is a wide spectrum of time scales that are involved in modeling flows in broadband networks. The total range is from a few nanoseconds (ns) to a few hours ($1hour = 3.6\ 10^{12} ns$). In this section we will attempt to answer some questions on the queue length asymptotics in the presence of multiple time scale arrivals and Gärtner-Ellis (Cramér) assumptions.

Using the Theory of Large Deviations (see [37]), under general assumptions of the Gärtner-Ellis (Cramér) type, one can show that

$$\lim_{x \to \infty} -\frac{\log \mathbb{P}[\mathrm{Q} > \mathrm{x}]}{\mathrm{x}} = \theta^*, \tag{11.2}$$

for some positive constant θ^*, called the *asymptotic decay rate* (or the equivalent bandwidth constant) [7, 17]. Also, in some cases, like finite Markov arrival and service processes, the following stronger result holds: $\mathbb{P}[\mathrm{Q} > \mathrm{x}] \sim \alpha e^{-\theta^* \mathrm{x}}$ as $\mathrm{x} \to \infty$, where α and θ^* are positive constants; θ^* is the same as in (11.2). For simple arrival processes (like On-Off Markov sources) it turns out that the constant α is of the order one. This led many authors to believe that the simple approximation $\mathbb{P}[\mathrm{Q} > \mathrm{x}] \approx e^{-\theta^* \mathrm{x}}$ holds; this approximation is commonly referred to as [9] the effective bandwidth (EB) approximation (sometimes it is also called the dominant root approximation). Following this result admission control policies based on the concept of effective bandwidth have been developed; see [7, 16, 18, 17, 27].

However, as discussed in [9], the EB approximation may often be very inaccurate. This is usually the case when many sources (N) are multiplexed; under this assumption it was shown in [9] that $\alpha \approx e^{-\gamma N}$ for some constant γ. A more formal analysis of the multiplexing of a large number of sources and an improvement of the EB approximation is given in [14, 15]. Complementing the work done in [9], in [19] we have shown that EB approximation may be very inaccurate in the presence of multiple time scale arrivals. Similar observations of inaccuracy of the EB approximation in the presence of multiple time scales (in the context of nearly decomposable Markov-modulated arrivals) were independently obtained in [35].

From a mathematical point of view, the inaccuracy of the EB approximation is due to fact that two processes that are "close" in the distribution *sense may be far apart in the cumulant sense.* Recall that a family of processes is said to converge in distribution if *all finite dimensional* distributions of these processes converge in distribution. On the other hand, the asymptotic decay rate constant θ^* is completely determined by the cumulant function $\varphi(\theta)$ which is a functional of the *whole (infinite dimensional)* arrival process. Therefore, convergence in distribution of a family of processes does not necessarily imply convergence of their cumulant functions.

A simple numerical example with two state Markov-modulated arrivals that illustrate the preceding comments (on the disagreement of the two convergence concepts) is given in [19]. The structure of the example is as follows. The modulating chain is assumed to have two states (say $\{1,2\}$) with transition probabilities $p_{21} = \epsilon, p_{12} = o(\epsilon)$; when in state $i = 1,2$ the source is producing i.i.d. arrivals $Y(j)$ such that in state 1 the source is producing stochastically smaller arrivals than in state 2. Since, $p_{12} = o(\epsilon)$ it is easy to see that the arrival process converges in distribution to the stochastically smaller process $Y(1)$, as $\epsilon \to 0$. However, its cumulant function converges to the cumulant function of the stochastically larger process $Y(2)$ (a formal argument that justifies this can be found in the proof of Theorem 2.1). For different values of the paramenter ϵ queue probabilities are presented in Figure 1 (solid lines). We can see that as $\epsilon \to 0$ the queue distribution converges to the queue distribution when $A \equiv Y(1)$ (represented by the common steep decline of the three solid lines on Figure 1), but the tail always decays as if the arrival process is the stochastically larger process $A \equiv Y(2)$ (parallel lines). Also note that the EB approximation (dashed line on the same figure) is off by orders of magnitude from the true probabilities. (For more details and more examples see [19].)

This idea was exploited in greater generality in [19], where, for a family of arrival processes $A^\epsilon, \epsilon > 0$, we give sufficient conditions under which the queue length distribution satisfies the following extension of the logarithmic asymptotic relation (11.2)

$$\lim_{\epsilon \to 0} \overline{\underline{\lim}}_{x \to \infty} \frac{-\log \mathbb{P}[Q^\epsilon \geq x]}{x} = \theta^*,$$

for some $\theta^* > 0$; symbol $\overline{\underline{\lim}}$ denotes that either $\overline{\lim}$ or $\underline{\lim}$ is taken. Using this result we have shown, under strict stability conditions, that the asymptotic decay rate of an ATM multiplexer *does not depend on the slow time scale statistics* (larger time units). However, the rate at which the queue length distribution decreases for small buffer sizes could be much larger than the asymptotic decay rate. This implies that an equivalent bandwidth admission control policy (based only on θ^*) may *significantly underutilize the system resources*, and that the slower timescales can be very important here.

In the same paper it was experimentally confirmed that the histogram of the queue distribution obtained by statistical multiplexing of 6 parts of the

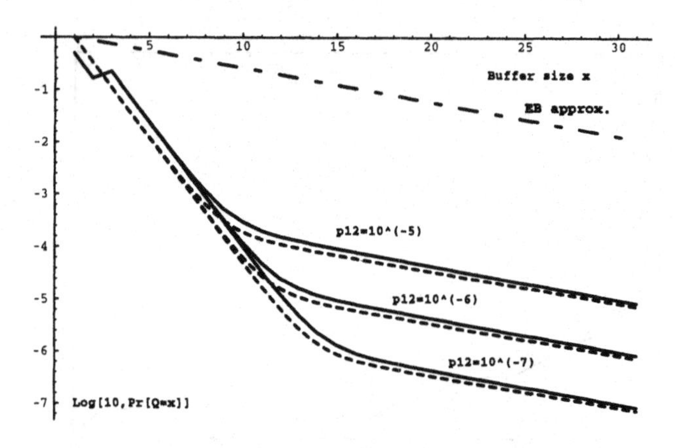

FIGURE 1. Graph of $\log_{10} \mathbb{P}[Q^{\epsilon} = x]$ from example 1.

Star Wars video sequence has a *"polygonal shape"* (multiple decay rates), typical for multiple time scale models, see Figure 2.

11.2.1 Queueing Analysis

Overall, as shown in [19], the EB approximation may give very inaccurate results. Also, as pointed out in [9], the exact asymptotic single exponential approximation may be poor (the authors suggested a procedure for approximating the queue length distribution with three exponentials). For that reason we have investigated a perturbation theory based approach for approximating all queue probabilities in the presence of multiple time scale (nearly decomposable) [20] arrival processes (a comprehensive treatment of a discrete time queue with multiple time scale arrivals can be found in [21]).

In that work we developed a *recursive* asymptotic expansion method for approximating the queue length distribution and investigated the radius of convergence of the queue asymptotic expansion series. The analysis focused on "small" to "moderate" buffer sizes under the conditions of strictly stable multiple time scale arrivals. For a class of examples we *analytically* determined the radius of convergence using methods of linear operator theory. We also gave general sufficient conditions under which the radius converges to zero; this showed roughly what situations have to be avoided for the proposed method to work (well). We combined the asymptotic expansion method with the EB approximation, and gave an approximation procedure for the buffer probabilities for all buffer ranges. The procedure was tested on extensive numerical examples. We illustrate this procedure in the following numerical example.

The asymptotic expansion approximation with $k = 0, 3$ expansion terms

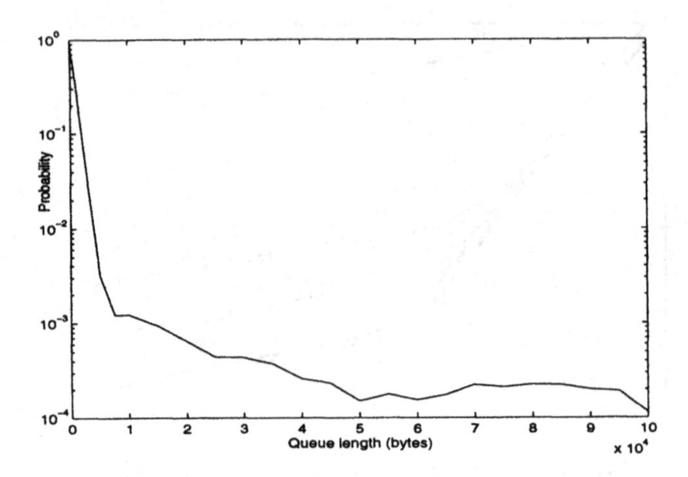

FIGURE 2. Queue length distribution for multiplexing 6 parts of the Star Wars video sequence on the slice level. The total length of the multiplexed sequence is 1,000,000 slices ($\approx$ 23 min).

for multiplexing 8 heterogeneous On-Off sources is displayed in Figure 3 (capacity of the server was taken to be $C_t \equiv 1$). The combination of asymptotic expansion and EB approximation is plotted in Figure 4. We see that the transition between the two approximations is smooth. Therefore, although we have no error estimate in the EB domain, from the smoothness of transition, we can expect that the approximation is excellent in the EB domain as well. This *smoothness of fit* can be used as a heuristic criterion for the overall accurateness of the approximation.

Let us now compare this approximative method with the classical exact z-transform inversion. In order to obtain the exact solution, one must find the inverse of $64X64$ z-transform matrix, then find 63 roots of the characteristic polynomial in the unit circle; use these roots to obtain boundary probabilities, and, at last, find the inverse z-transform of the queue z-transform. We were not able to complete even the first step, i.e., finding the z-transform matrix inverse after 24 hours, after which we stopped the program. (Computation was attempted with Mathematica 2.2 on a (150MHz, 64M RAM, 100M virtual memory)) SGI machine. However, using the same environment Mathematica + SGI, we obtained a three term expansion approximation in less than an hour. This clearly shows the efficacy of the asymptotic expansion method.

As all of the calculations were done with Mathematica 2.2, which is known to be slow for intensive numerical problems, we expect that the asymptotic expansion method when implemented in C will produce much faster results. Therefore, we predict that this method will be very useful for large practical problems that often appear in the fine tuning of ATM

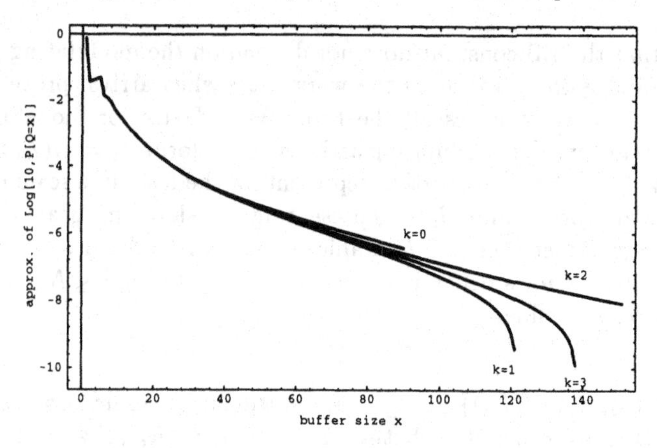

FIGURE 3. Approximate "probabilities" obtained by using k expansion terms.

admission controllers.

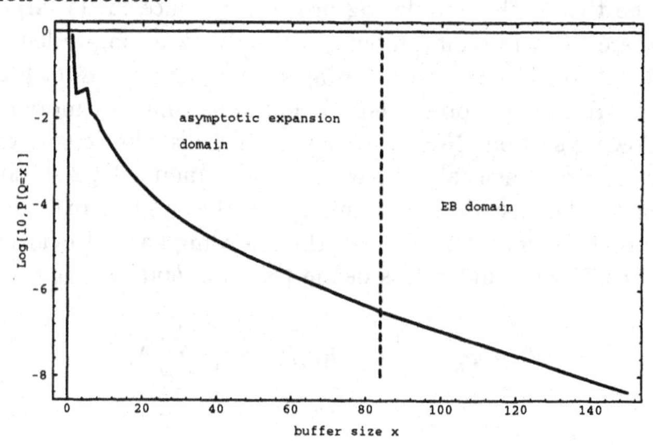

FIGURE 4. Total queue distribution approximation, obtained by combination of the asymptotic expansion method and EB approximation.

We next present a class of arrivals that also cannot be well approximated with an EB approximation. Unlike the case of nearly decomposable Markov arrivals, for which the asymptotic expansion method for approximating queue probabilities is available, approximating the queue probabilities for this type of arrivals remains an open problem.

11.2.2 Arrivals with Long-tailed High Activity Periods

In this section we will examine arrival processes that stay in their high activity states for a long-tailed (random) period of time. For these arrivals we prove that the EB constant does not depend on the modulating process statistics, and is the same as in the worst case when arrival processes are in their high activity states all the time. As a device for modeling time scales, we will consider modulating arrivals of the form $A_t = Y_t(B_t)$, where intuitively Y_t can be thought of as representing the fast time scale changes in the arrival process, and B_t as representing the slow time scale changes.

More formally, let there be N traffic sources all modeled as stationary, ergodic, discrete time stochastic processes $\{A_t^i, t \geq 0, 1 \leq i \leq N\}$. For each i, $1 \leq i \leq N$, we define

$$A_t^i \stackrel{def}{=} Y_t^i(B_t^i),$$

where $\{Y^i(1), \ldots, Y^i(K_i)\}, K_i \geq 1$ are stationary ergodic processes that are stochastically ordered such that $Y^i(j) \leq_{st} Y^i(K_i), 1 \leq j < K_i$ (for stochastic ordering see [34], or [4], chapter 4). Further, the processes $B^i = \{B_t^i, t \geq 0\}$ are stationary, ergodic, discrete time process with a finite state space $\mathcal{S}_i = \{1, \ldots, K_i\}$. All processes $Y^i(j)$, and B^i are assumed independent of each other.

We assume that each modulating process B^i, once in its largest state K_i, stays there for a random amount of time with a long-tailed distribution (see Definition 11.8 in the following section). Examples of long-tailed distributions are Pareto, some Weibull, and lognormal; for more examples see the following section. Note that long-tailed distributions decay more slowly than any exponential; in particular, the moment generating function of a long-tailed distribution is infinite on the positive real axis.

In order to state our result we need the following Large Deviations Theory definitions. For $\theta \in \mathbb{R}^+$ let us define the *cumulant function*

$$\varphi_n(\theta) \equiv \varphi_n^A(\theta) \stackrel{def}{=} \frac{1}{n} \log \mathbb{E}\{\exp[\theta \sum_{t=1}^{n} A_t]\}, \tag{11.3}$$

where $n \geq 1$, and let

$$\varphi(\theta) \stackrel{def}{=} \lim_{n\to\infty} \varphi_n(\theta). \tag{11.4}$$

Furthermore, let us define

$$D \stackrel{def}{=} \{\theta \geq 0 \ : \ \varphi(\theta) < \infty\},$$

and make usual Large Deviation **Assumptions:**

A1 $\varphi(\theta)$ is strictly convex on D,

A2 $\varphi(\theta)$ is differentiable for all $\theta \in D$.

For the sake of simplicity, in this section we will assume that the server capacity process $C_t \equiv c \in \mathbb{R}^+$. This is frequently the case in communication networks.

Theorem 11.2.1 Assume that for each i and $\theta \in \mathbb{R}^+, \varphi^{Y^i(K_i)}(\theta)$ exist, and satisfies conditions **A1** and **A2** with $\sum_{i=1}^{N}(\varphi^{Y^i(K_i)})'(0) < c$. Also, let $\pi_{K_i} = \mathbb{P}[B_t^i = K_i] > 0$ for all i, and assume that the residual time of staying in the high activity state K_i is long-tailed. Then

$$\lim_{x\to\infty} -\frac{\log \mathbb{P}[Q > x]}{x} = \theta_N^*, \tag{11.5}$$

where θ_N^* is the equivalent bandwidth constant. This constant, if it exists, is the positive solution of the equation $\sum_{i=1}^{N} \varphi^{Y^i(K_i)}(\theta) = \theta c$; if the positive solution of this equation does not exists, we set $\theta_N^* = \infty$.

Proof. Let T_{K_i} be the residual time of staying in state K_i; From the assumption that T_{K_i} is long-tailed and Lemma 11.3.4 it follows that

$$\lim_{n\to\infty} \frac{\log \mathbb{P}[T_{K_i} > n]}{n} = 0. \tag{11.6}$$

Now the theorem will follow from Theorem 3.9 in [7] if we prove that the cumulant function of the arrival process i is equal to the cumulant function of the process $Y^i(K_i)$, i.e., when the arrival process is in its highest activity period K_i all the time.

First, from the stochastic ordering of $Y^i(k), 1 \le k \le K_i$, and Strassen's theorem (see section 4.2.3 [4]) it follows that $Y^i(k)$ can be constructed on the same probability space, such that $Y^i(j) \le Y^i(K_i), 1 \le j < K_i$, holds along each sample path. From this, it follows that

$$\overline{\varphi}^{A^i}(\theta) \le \varphi^{Y^i(K_i)}(\theta). \tag{11.7}$$

The lower bound follows from

$$\begin{aligned}
&\underline{\lim}_{T\to\infty} \frac{1}{T} \log \mathbb{E}\left[e^{\theta \sum_{t=1}^{T} A_t^i}\right] \\
&\ge \underline{\lim}_{T\to\infty} \frac{1}{T} \log \left(\pi_{K_i}^i \mathbb{P}[T_{K_i} > T]\mathbb{E}\left[e^{\theta \sum_{t=1}^{T} Y_t^i(K_i)}\right]\right) \\
&= \underline{\lim}_{T\to\infty} \frac{\log \mathbb{P}[T_{K_i} > T]}{T} + \varphi^{Y^i(K_i)}(\theta) = \varphi^{Y^i(K_i)}(\theta),
\end{aligned}$$

where the last equality follows from (11.6). This proves that $\varphi^{A^i}(\theta) = \varphi^{Y^i(K_i)}(\theta)$, and the assertion of the theorem follows.□

This theorem is illustrated in the following numerical example.

Example 11.2.2 *General on-off source.* Let (the modulating process) B_t be a $\{0, 1\}$ valued process whose dynamics are described as follows. When in state zero (off), B_t stays there for a geometrically distributed random time

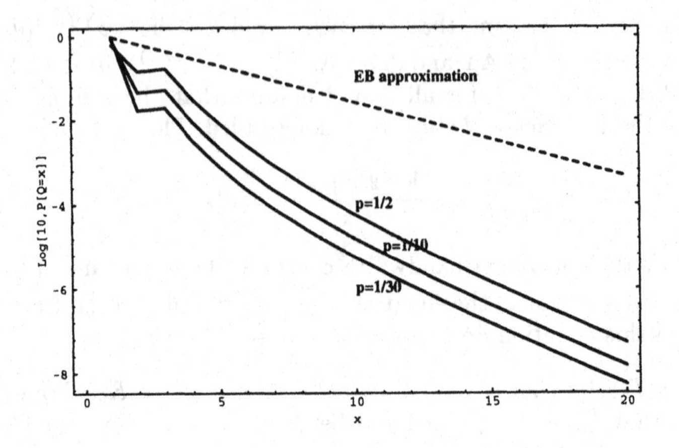

FIGURE 5. Graph of $\log_{10} \mathbb{P}[Q = x]$ from Example 1 for different values of T_{off} period parameter p.

T_{off}, $\mathbb{P}[T_{off} = k] = (1-p)^{k-1}p, k \in \mathbb{N}$. When in state one (on) the process stays there for a generally distributed random time T_{on} (independent of T_{off}). When in state zero the source is not producing anything ($Y(0) \equiv 0$), and while in state one the source is producing i.i.d. arrivals with distribution $\mathbb{P}[Y(1) = 2] = 1 - \mathbb{P}[Y(1) = 0] = a$. Assume that the capacity of the server is $c = 1$. (Due to slightly simpler boundary conditions, all numerical examples in this paper were done for the recursion $Q_{t+1} = (Q_t - 1)^+ + A_t$; this recursion is asymptotically equivalent to (11.1).) From Theorem 11.2.1 it follows that as long as T_{on} is long-tailed the equivalent bandwidth constant is independent of the distribution of T_{on} and T_{off} (p). This is numerically illustrated for different values of p in Figure 5; other parameters are taken to be $a = 2/5$, $\mathbb{P}[T_{on} \geq k] = k^{-3}, 1 \leq k \leq 80$, $\mathbb{P}[T_{on} \geq k] = 0, k > 80$. From the figure we can observe that the queue length probabilities are decreasing as p decreases. This is intuitively obvious since off periods are getting larger. Also all the graphs eventually become parallel, as predicted by the previous theorem.

Although this example (as well as all the other examples in this paper) shows that the EB approximation is (too) conservative, it does not have to be so in general. In the context of the on-off source model we have observed that the EB approximation is too optimistic when the distribution of the on period decays faster than an exponential. Although we were not able to theoretically formulate this observation in greater generality, we believe it to be equally important. We illustrate this insight numerically. Take $\mathbb{P}[T_{on} \geq k] = e^{-(k-1)^2}, k = 1, 2, 3, 4, 5$, $\mathbb{P}[T_{on} \geq k] = 0, k > 5$. Queue probabilities for $p = 1/10, a = 4/5$ are presented in Figure 6. We see that

EB approximation is too optimistic. For more examples and details see [21]. Informally, if the input process "doesn't look exponential" the queue output is not exponential either.

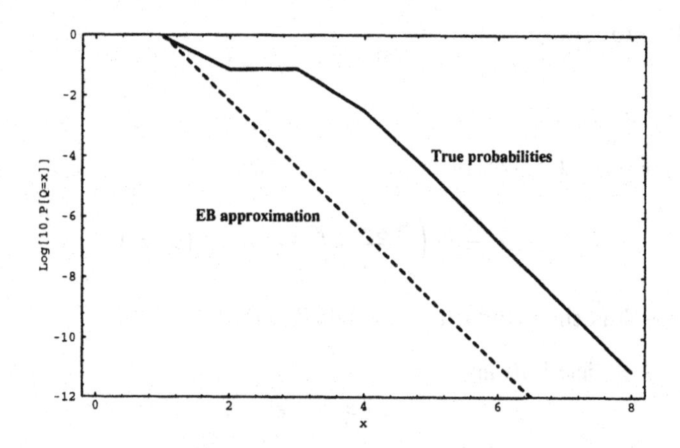

FIGURE 6. Comparison between the equivalent bandwidth approximation and the true probabilities for the case when the distribution of the on period decreases faster than an exponential.

11.3 Subexponential Arrivals

In this section our goal is to examine the asymptotics of the queue length distribution when the Cramér type conditions are replaced by subexponential assumptions. The two largest non Cramér families of distributions are long-tailed and subexponential distributions.

Definition 11.3.1 *A distribution function F on $[0, \infty)$ is called* long-tailed *($F \in \mathcal{L}$) if*

$$\lim_{x \to \infty} \frac{1 - F(x - y)}{1 - F(x)} = 1, \quad y \in \mathbb{R}. \tag{11.8}$$

Definition 11.3.2 *A distribution function F on $[0, \infty)$ is called* subexponential *($F \in \mathcal{S}$) if*

$$\lim_{x \to \infty} \frac{1 - F^{*2}(x)}{1 - F(x)} = 2, \tag{11.9}$$

where F^{*2} denotes the 2-nd convolution of F with itself, i.e., $F^{*2}(x) = \int_{[0,\infty)} F(x - y) F(dy)$.

The class of subexponential distributions was first introduced by Chistakov [8]. The definition is motivated by the simplification of the asymptotic analysis of the convolution tails. Some examples of distribution functions in $\mathcal{S}$ are:

(I) the Pareto family

$$F(x) = 1 - (x - \beta + 1)^{-\alpha},$$

$x > \beta > 0, \alpha > 0$.

(II) the lognormal distribution

$$F(x) = \Phi\left(\frac{\log x - \mu}{\sigma}\right), \quad \mu \in \mathbb{R}, \sigma > 0,$$

where Φ is the standard normal distribution.

(III) Weibull distribution

$$F(x) = 1 - e^{-x^{\beta}},$$

for $0 < \beta < 1$.

(IV)

$$F(x) = e^{-x(\log x)^{-a}},$$

for $a > 0$. This class was proven to be subexponential in [33].

(V) Benktander Type I distribution [28]

$$F(x) = 1 - cx^{-a-1}x^{-b\log x}(a + 2b\log x),$$

$a > 0, b > 0$, and c appropriately chosen.

(V) Benktander Type II distribution [28]

$$F(x) = 1 - cax^{-(1-b)}\exp\{-(a/b)x^{b}\},$$

$a > 0, 0 < b < 1$, and c appropriately chosen.

The general relation between $\mathcal{S}$ and $\mathcal{L}$ is the following.

Lemma 11.3.3 *(Athrey and Ney, [3])* $\mathcal{S} \subset \mathcal{L}$.

The following lemma [8] clearly shows that for long-tailed distributions Cramér type conditions are not satisfied.

Lemma 11.3.4 *If* $F \in \mathcal{L}$ *then* $(1 - F(x))e^{\alpha x} \to \infty$ *as* $x \to \infty$, *for all* $\alpha > 0$.

An extensive treatment of subexponential distributions (and further references) can be found in Cline [11, 12].

Before we proceed any further, let us try to understand some of the basic properties of the sequence $\{X_n, n \geq 1\}$ of subexponentially distributed i.i.d. random variables. One of the main sample path characteristics of subexponential distributions follows from its definition [8], and that is

$$\mathbb{P}[X_1 + X_2 + \cdots + X_n > x] \sim n\mathbb{P}[X_1 > x], \qquad (11.10)$$

as $x \to \infty$. This means that a sum of subexponential random variables exceeds a large value x by having one of them excede this value x; in terms of the appearance of the sample path of a sequence of subexponential random variables, we note that the sequence exhibits isolated peaks.

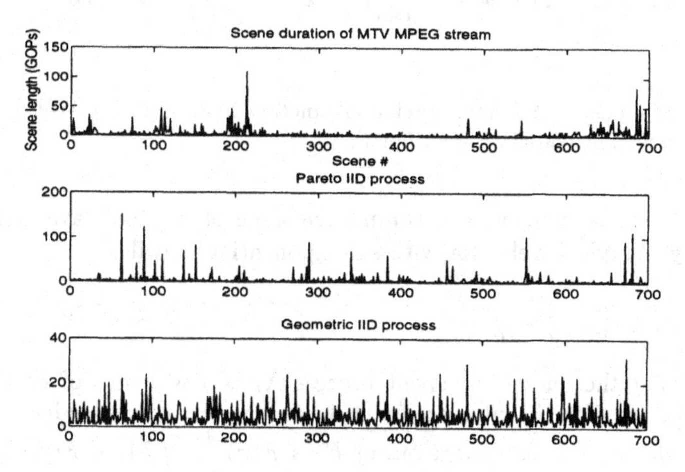

FIGURE 7. MPEG scene length duration (top); i.i.d. sample paths generated with the Pareto (middle) and geometric distribution (bottom).

Such a sample path behavior characterizes the scene lengths of video streams coded using the MPEG standard. Figure 7 shows a sequence of scene length durations (top), and for rough comparison, the sample paths generated by i.i.d. processes with Pareto (middle) and geometric distribution (bottom). Clearly, the scene length duration process has a subexponential character, as does the Pareto process, where the large peaks tend to be isolated in time, as suggested by (11.10). This is unlike the case of the geometrically distributed process. (For the description of MPEG data and the definition of scenes see [24].)

In terms of video traffic, subexponentiality can also manifest itself in the time-dependent (autocorrelation) structure. As shown in Figure 8, the autocorrelation function of MPEG video (17 streams multiplexed) matches the (subexponential) Pareto function $f(t) = \beta/t^{\alpha}$, for $\alpha = 0.513, \beta = 1.195$.

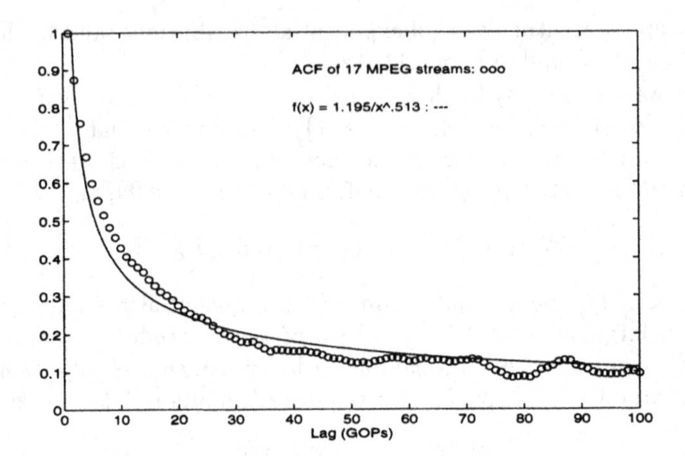

FIGURE 8. Modeling the autocorrelation function of MPEG video with an appropriate (subexponential) Pareto function.

In the next section, we will summarize some of the tools available for analyzing the queue behavior with subexponential arrivals.

11.3.1 Queueing Analysis

Assume that the queue increment process X_t is a sequence of i.i.d. random variables with distribution function F, and A_t is independent of C_t. Further, denote the *integrated tail* of F as $\hat{F}(x) \stackrel{def}{=} \int_x^\infty [1 - F(t)]dt$, and define by $F_1(x) = m^{-1}(1 - \hat{F}(x))$, where $m = \hat{F}(0)$. Similarly, in the rest of the paper for any d.f. G, we define its corresponding $\hat{G}(x)$ and $G_1(x)$. Then the following result on the waiting time distribution asymptotics of the $GI/GI/1$ queue holds (see Veraverbeke [36]). Let K be the d.f. of A_t.

Theorem 11.3.5 **(i)** *$F_1 \in \mathcal{S} \Longleftrightarrow K_1 \in \mathcal{S}$ and* $\lim_{x\to\infty} \frac{\hat{F}(x)}{\hat{K}(x)} = 1$.

(ii) *If $K, K_1 \in \mathcal{S}$, then*

$$\mathbb{P}[\mathrm{Q_t > x}] \sim \frac{1}{\mathbb{E}\mathrm{C_t} - \mathbb{E}\mathrm{A_t}} \int_x^\infty \mathbb{P}[\mathrm{A_t > u}]\mathrm{du}, \quad \text{as } \mathrm{x} \to \infty. \qquad (11.11)$$

This theorem was first proved in [32]; in [36] the same result was shown using a random walk technique. Some of the first applications of long-tailed distributions in queueing theory were made by Cohen [13], and Borovkov [6] for functions of regular variations [26, 5]. Recent results on long-tailed and subexponential asymptotics of a $GI/GI/1$ are given in [1, 38]. (Also, in [1] further motivation is given for the application of long-tailed distributions to communication networks.)

The assumption that $K, K_1 \in \mathcal{S}$ in the theorem above can be replaced by an assumption on K only. (Note that $K \in \mathcal{S}$ does not necessarily imply that $K_1 \in \mathcal{S}$.) This has been investigated in [28].

Definition 11.3.6 *$F \in \mathcal{S}^*$ if*

$$\int_0^x \frac{\overline{F}(x-y)}{\overline{F}(x)}\overline{F}(y)dy \to 2m_F < \infty, \quad \text{as } x \to \infty,$$

where $m_F = \int_0^\infty yF(dy)$.

This class has the property that $\mathcal{S}^* \subset \mathcal{S}$, and that $F \in \mathcal{S}^* \Rightarrow F_1 \in \mathcal{S}$. Sufficient conditions for $F \in \mathcal{S}^*$ can be found in [29], where it was explicitly shown that lognormal, Pareto, and certain Weibull distributions are in $\mathcal{S}^*$.

An extension of Theorem 11.3.5 was investigated in [2]. In that paper the authors established the subexponential asymptotics of a Markov-modulated M/G/1 queue. However the constant of proportionality was left in a complex form. Full extension of Theorem 11.3.5 to Markov-modulated G/G/1 queues was given in [23] (preliminary results were reported in [22]), where it was proved that the queue length asymptotics are invariant under Markov modulation. A precise statement of this result follows.

Let $\{J_t\}$ be a stationary irreducible aperiodic Markov chain with a finite state space E (say with N elements) and transition matrix P, and let $\{X_t\}$ be a sequence of real valued random variables. A stationary Markov process $\{(J_t, X_t)\}$ on $E \times \mathbb{R}$ whose transition distribution depends only on the first coordinate is called a Markov-modulated random walk (MMRW). This process is completely defined by its transition matrix measure $F_{ij}(B) = \mathbb{P}[\mathrm{J}_1 = \mathrm{j}, \mathrm{X}_1 \in \mathrm{B} | \mathrm{J}_0 = \mathrm{i}]$, and $F = \{F_{ij}\}$ (note that $\|F\| = F((-\infty, \infty)) = P$). Let $\{(J_t^r, X_t^r)\}$ denote the associated reversed process. This process is determined by the set of transition measures $F_{ij}^r(B) = \mathbb{P}[\mathrm{J}_0 = \mathrm{j}, \mathrm{X}_1 \in \mathrm{B} | \mathrm{J}_1 = \mathrm{i}]$, with $F^r = \{F_{ij}^r\}$ being the corresponding transition matrix measure.

Let (J_t, A_t) and (J_t, C_t) be two MMRWs such that A_t and C_t are conditionally independent given J_{t-1}, J_t; $\{A_t\}$ and $\{C_t\}$ are arrival and service processes, respectively. Let K and D be the corresponding transition measures for these MMRWs, i.e., $K = \{K_{ij}\} = \{\mathbb{P}[\mathrm{A}_1 \in \mathrm{B}, \mathrm{J}_1 = \mathrm{j} | \mathrm{J}_0 = \mathrm{i}]\}$, and $D = \{D_{ij}\} = \{\mathbb{P}[\mathrm{C}_1 \in \mathrm{B}, \mathrm{J}_1 = \mathrm{j} | \mathrm{J}_0 = \mathrm{i}]\}$; the reversed transition measure for the arrival process is $K^r = \{K_{ij}^r\} = \{\mathbb{P}[\mathrm{A}_1 \in \mathrm{B}, \mathrm{J}_0 = \mathrm{j} | \mathrm{J}_1 = \mathrm{i}]\}$, $B \in \mathcal{B}(\mathbb{R})$. For any (matrix) measure H, we denote $\overline{H}(x) = H(x, \infty)$. Then the following theorem holds [23].

Theorem 11.3.7 *Let $\lim_{x\to\infty} \overline{K^r}(x)/\overline{H}(x) = W$, as $x \to \infty$, $W = \{W_{ij}\}, W_{ij} \in [0, \infty)$, $H(x) \in \mathcal{L}, H_1(x) \in \mathcal{S}$ (or $H \in \mathcal{S}^*$), with at least one $W_{ij} > 0$. If $\mathbb{E}\mathrm{C}_\mathrm{t} > \mathbb{E}\mathrm{A}_\mathrm{t}$, and P $(= \|K\| = \|D\|)$ is irreducible and aperiodic, then,*

$$\frac{1}{\hat{H}(x)}\overline{Q}(x) \to \frac{1}{\mathbb{E}\mathrm{C}_\mathrm{t} - \mathbb{E}\mathrm{A}_\mathrm{t}} e\pi We, \quad \text{as } x \to \infty, \tag{11.12}$$

where $\overline{Q}(x)$ *is a column vector with its ith component equal to* $\mathbb{P}[Q_t > x|J_t = i]$. *In particular,*

$$\mathbb{P}[Q_t > x] \sim \frac{1}{\mathbb{E}C_t - \mathbb{E}A_t} \int_x^\infty \mathbb{P}[A_t > u]du, \quad \text{as } x \to \infty. \qquad (11.13)$$

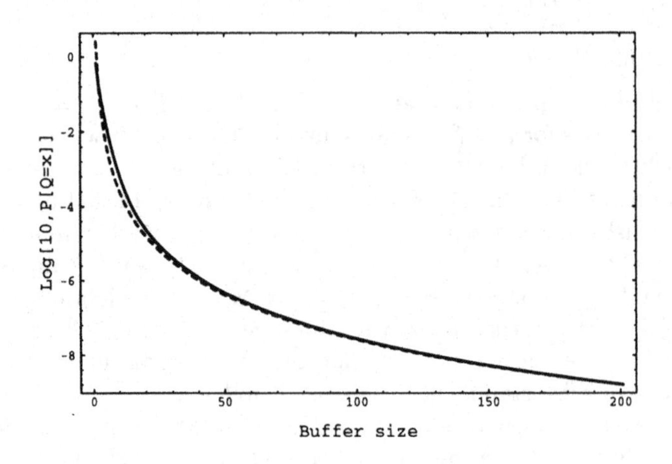

FIGURE 9. Graph of $\log_{10} \mathbb{P}[Q = i]$ versus buffer size i from Example 2; the solid line represents the true probabilities, and the dashed line represents the approximation $2.603/i^{-4}$.

An illustration of the preceding theorem is given in the following numerical example.

Example 11.3.8 Consider a constant server queue with $C_t = 1$ and two state (e.g. $\{0, 1\}$) Markov-modulated arrivals (source). The transition probabilities for the modulating Markov chain are $p_{01} = 1/3, p_{10} = 3/4$. When in state 0, the source is producing zero arrivals, and when in state 1, the source is producing (independently of the previous state) arrivals according to the distribution $\mathbb{P}[A_t = 0|J_t = 1] = 0.327144, \mathbb{P}[A_t = 1|J_t = 1] = 0$, and $\mathbb{P}[A_t = i|J_t = 1] = w/i^5, w = 18.220859, 2 \leq i \leq 350$; $\rho_1 = \mathbb{E}[A_t|J_t = 1] = 3/2$. (Note that these are bounded arrivals.) Thus, according to the previous theorem, the queue length distribution is proportional to $1/i^4$, and the constant of proportionality is easily calculated to be $c = w\pi_1/(4(1 - \rho_1\pi_1)) = w/7 = 2.603$. The comparison between the true probabilities and the approximation c/i^4 is shown in Figure 9.

Stationary subexponentially correlated arrivals. The models that we have seen in this section exhibit weak exponential autocorrelation structure and dominant subexponential marginal distributions. For modeling subexponentially correlated arrivals in [23], we introduced the following class of processes. (These processes are a particular case of semi Markov processes [10].)

Consider a point process $T = \{T_0 \leq 0, T_n, n \geq 1\}$ such that $T_n - T_{n-1}, n \geq 1$ are i.i.d. with subexponential distribution function F. Further, let $J_n, n \geq 0$ be an irreducible aperiodic Markov Chain with finite state space $\{1, \ldots, K\}$, transition matrix $\{P_{ij}\}$, and stationary probability distribution $\pi_i, 1 \leq i \leq K$. In order to make this point process stationary (see [10], section 9.3), we choose the residual time at zero until the first jump to be distributed as an integrated tail of F, i.e., $F_1(t) = \mathbb{P}[T_1 \leq t] = m_F{}^{-1} \int_{0,t} \overline{F}(u)du$, $m_F = \mathbb{E}(T_n - T_{n-1})$.

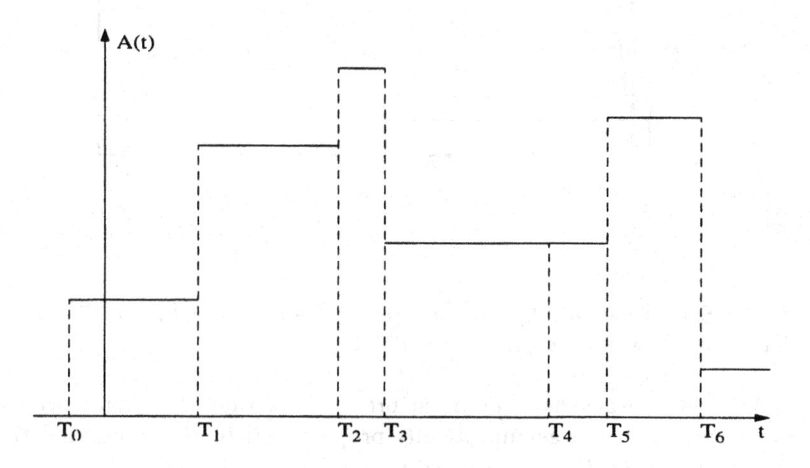

FIGURE 10. A possible realization of a Markov chain embedded into a renewal process.

Now we construct the following process:

$$A_t = J_n \quad \text{for} \quad T_n \leq t < T_{n+1}, \tag{11.14}$$

called a Markov Chain Embedded in a Stationary Subexponential Renewal Process (MCESSR). A typical sample path of this process is given in Figure 10. It is well known that under fairly general conditions, a Markov chain converges exponentially fast to its steady state distribution. These MCESSR processes have the characteristic that they, unlike finite state Markov chains, approach their steady state distributions with a subexponential rate. We illustrate this in the following example.

Example 11.3.9 Let F be a discrete distribution function with support $[1, 1000]$, $\mathbb{P}[T_2 - T_1 = 1] = 0.186532$, and $\mathbb{P}[T_2 - T_1 = i] = w/i^5, w =$

22.028625, $2 \leq i \leq 1000$; choose a two state Markov chain with transition probabilities $p_{01} = 1/3$ and $p_{10} = 3/4$. Then, the functions $(d_{i,1}(t) \stackrel{def}{=} (\mathrm{IP}_i[A_t = 1] - \pi_1)(\overline{F}_1(t)(\delta_{i1} - \pi_1))^{-1}$, $i = 0, 1$, converge to one as $t \to \infty$, with subexponential rate. This can be clearly seen in Figure 11.

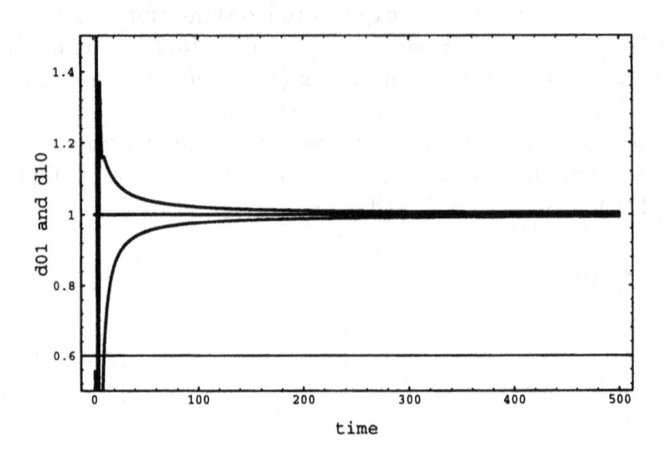

FIGURE 11. Functions $d_{i,1}(t) \stackrel{def}{=} (\mathrm{IP}_i[A_t = 1] - \pi_1)(\overline{F}_1(t)(\delta_{i1} - \pi_1))^{-1}$, $i = 0, 1$. The graph shows that $d_{i,1}(t) \to 1$ as $t \to 1$.

Another characteristic of these processes is that their autocorrelation functions, $R(t)$, are asymptotically proportional to the integrated tail of the sojourn time $T_n - T_{n-1}$ distribution, i.e. if $F, F_1 \in \mathcal{S}$, then

$$R(t) \sim \overline{F}_1(t),$$

as $t \to \infty$; this was formally proved in [23]. Combining these results, with Theorem 11.3.7, it was proven in the same paper that when the fluid flow queue is fed by these processes, its queue distribution is asymptotically proportional to its autocorrelation function, i.e.,

$$\mathrm{IP}[Q > t] \sim r\, R(t),$$

as $t \to \infty$. To the best of our knowledge, this was the first rigorous result relating the queue length distribution and the arrival process autocorrelation function.

11.4 Concluding Remarks

We have demonstrated that real-time traffic processes such as video traffic exhibit multiple time scale characteristics as well as subexponential first and second order statistics. A network multiplexer that is loaded by these processes may manifest a distinct asymptotic behavior. We summarize recent results on evaluating the asymptotic behavior of a network multiplexer in the presence of subexponential and multiple time scale arrivals. It is left to identify in practice when some of the asymptotic techniques presented here can be applied to the design of efficient admission control policies in ATM based broadband networks.

Acknowledgments: The authors wish to thank the anonymous reviewer for his/her detailed list of editing suggestions.

11.5 References

[1] J. Abate, G.L. Choudhury, and W. Whitt. Waiting-time tail probabilities in queues with long-tail service-time distributions. *Queueing systems.*, 1994.

[2] S. Asmussen, L. F. Henriksen, and C. Klüppelberg. Large claims approximations for risk processes in a markovian environment. *Stochastic Processes and their Applications*, 54:29–43, 1994.

[3] K. B. Athreya and P. E. Ney. *Branching Processes.* Springer-Verlag, 1972.

[4] F. Baccelli and P. Bremaud. *Elements of Queueing Theory: Palm-Martingale Calculus and Stochastic Recurence.* Springer Verlag, 1994.

[5] N. H. Bingham, C. M. Goldie, and J. L. Teugels. *Regular Variation.* Cambridge University Press, Cambridge, 1987.

[6] A. A. Borovkov. *Stochastic Processes in Queueing Theory.* Springer-Verlag, 1976.

[7] C. S. Chang. Stability, queue length and delay of deterministic and stochastic queueing networks. *IEEE Transactions on Automatic Control*, 39:913–931, 1994.

[8] V. P. Chistakov. A theorem on sums of independent positive random variables and its application to branching random processes. *Theor. Probab. Appl.*, 9:640–648, 1964.

[9] G. L. Choudury, D. M. Lucantoni, and W. Whitt. Squeezing the most of ATM. To appear in *IEEE Trans. on Communications*, 1995.

[10] E. Cinlar. *Introduction to Stochastic Processes.* Prentice-Hall, 1975.

[11] D. B.H. Cline. Convolution tails, product tails and domains of attraction. *Probab. Th. Rel. Fields*, 72(1):529–557, 1986.

[12] D. B.H. Cline. Convolution of distributions with exponential and subexponential tails. *J. Austral. Math. Soc. (Series A)*, 43:347–365, 1987.

[13] J. W. Cohen. Some results on regular variation for distributions in queueing and fluctuation theory. *J. Appl. Probability*, 1973.

[14] C. Courcobetis and R. Weber. Buffer overflow asymptotics for a switch handling many traffic sources. Manuscript, December 1994.

[15] A. Elwalid, D. Heyman, T. V. Lakshman, D. Mitra, and A. Weiss. Fundamental bounds and approximations for atm multiplexers with applications to video teleconferencing. *IEEE Journal on Selected Areas in Communications*, 13(6):1004–1016, August 1995.

[16] A. I. Elwalid and D. Mitra. Effective bandwidth of general markovian traffic sources and admission control of high speed networks. *IEEE/ACM Trans. on Networking*, 1(3):329–343, June 1993.

[17] P. V. Glynn and W. Whitt. Logarithmic asymptotics for steady-state tail probabilities in a single-server queue. *Studies in Appl. Prob.*, 1994.

[18] R. Guerin, H. Ahmadi, and M. Nagshineh. Equivalent capacity and its application to bandwidth allocation in high-speed networks. *IEEE J. Select. Areas Commun.*, 9:968–981, 1991.

[19] P. R. Jelenković and A. A. Lazar. On the dependence of the queue tail distribution on multiple time scales of ATM multiplexers. In *Conference on Information Sciences and Systems*, pages 435–440, Baltimore, MD, March 1995. (www: http: //www.ctr.columbia.edu-/comet/publications).

[20] P. R. Jelenković and A. A. Lazar. Evaluating the queue length distribution of an ATM multiplexer with multiple time scale arrivals. In *Proceedings of INFOCOM'96*, San Francisco, California, March 1996, to appear.

[21] P. R. Jelenković and A. A. Lazar. Asymptotic properties of a discrete time queue with multiple time scale arrivals. *Submited to Queueing Systems*, 1995.

[22] P. R. Jelenković and A. A. Lazar. Subexponential asymptotics of a network multiplexer. In *Proceedings of the 33rd Annual Allerton Conference on Communication, Control, and Computing*, Urbana-Champaign, Illinois, October 1995.

[23] P. R. Jelenković and A. A. Lazar. Subexponential asymptotics of a markov-modulated G/G/1 queue. *Submitted to Journal of Appl. Prob.*, November 1995.

[24] P. R. Jelenković, A. A. Lazar, and N. Semret. Multiple time scales and subexponentiality in MPEG video streams. In *International IFIP-IEEE Conference on Broadband Communications*, April 1996, to appear.

[25] P. R. Jelenkovic and B. Melamed. Algorithmic modeling of TES processes. *IEEE Transactions on Automatic Control*, 40(7):1305–1312, July 1995.

[26] J. Karamata. Sur un mode de croissance réguilière des fonctions. *Mathematica (Cluj)*, 1930.

[27] F. P. Kelly. Effective bandwidths at multi-class queues. *Queueing Systems*, 9:5–16, 1991.

[28] C. Kluppelberg. Subexponential distributions and integrated tails. *J. Appl. Prob.*, 25:132–141, 1988.

[29] C. Kluppelberg. Subexponential distributions and characterizations of related classes. *Probability Theory and Related Fields*, 82:259, 1989.

[30] A. A. Lazar, G. Pacifici, and D. E. Pendarakis. Modeling video sources for real-time scheduling. *Multimedia Systems*, 1(6):253–266, 1994.

[31] R. M. Loynes. The stability of a queue with non-independent interarrival and service times. *Proc. Cambridge Philos Soc.*, 58:497–520, 1968.

[32] A.G. Pakes. On the tails of waiting-time distribution. *J. Appl. Probab.*, 12:555–564, 1975.

[33] E. J. G. Pitman. Subexponential distribution functions. *J. Austral. Math. Soc. Ser. A*, 29:337–347, 1980.

[34] D. Stoyan. *Comparison Methods for Queues and Other Stochastic Models*. John Wiley & Sons, 1983.

[35] D. Tse, R. Gallager, and J. Tsitsiklis. Statistical multiplexing of multiple time-scale markov stream. *IEEE, Selected Areas in Communications*, August 1995.

[36] N. Veraverbeke. Asymptotic behavior of wiener-hopf factors of a random walk. *Stochastic Proc. and Appl.*, 5:27–37, 1977.

[37] A. Weiss and A. Shwartz. *Large Deviations for Performance Analysis: Queues, Communications, and Computing*. New York: Chapman & Hall, 1995.

[38] E. Willekens and J. L. Teugels. Asymptotic expansion for waiting time probabilities in an M/G/1 queue with long-tailed service time. *Queueing Systems*, 10:295–312, 1992.

12

Networks of Queues with Long-Range Dependent Traffic Streams

Venkat Anantharam

ABSTRACT Traditional applications of queuing theory in the analysis and design of communication networks often use the fact that quasi-reversible queues can be interconnected with Bernoulli routing to form network models whose stationary distribution is of product form. To construct such models it is necessary to assume that the external arrival process is a Poisson process (in continuous time) or an i.i.d. sequence of Poisson random variables (in discrete time). Recently, however, several empirical studies have provided strong evidence that the traffic to be carried by the next generation of communication networks exhibits long-range dependence, implying that it cannot be satisfactorily modeled as a Poisson process.

Our purpose in this paper is to exploit the features that make networks of quasi-reversible queues product form so as to construct a self-contained class of queuing network models whose external arrival processes can be long-range dependent. Throughout the paper, we restrict ourselves to discrete time. The long-range dependent arrival process models we consider can be described as follows : Sessions arrive according to a sequence of i.i.d. Poisson random variables. Each session is active for an independent duration that is positive integer valued with a regularly variying distribution having finite mean and infinite second moment. While it is active, a session brings in work at rate 1. The networks we consider are comprised of monotone quasi-reversible S-queues with Bernoulli routing. The queues handle their arrivals at the session level, i.e., once a session enters service at a queue it continues to receive service at rate 1 until all the work it brings in is completed.

Our main observation is that all the internal traffic processes of such networks are long-range dependent. This result provides a family of interesting new examples of long-range dependent processes, in addition to its potential use in studying performance issues for communication networks.

12.1 Introduction

In the past few years the telecommunication network arena has witnessed the evolution of networks that can offer a wide variety of services, such as audio, video, data, etc. using a common protocol suite. Such networks are called *integrated services networks*. Since these networks are largely based on transmission over a fibre-optic medium and thus have a bandwidth that is orders of magnitude higher than that of earlier generations of networks, they are also generally called *broad-band networks*. Several recent empirical studies of traffic to be carried by such broad-band networks have established the importance of studying queuing behaviour with traffic models having long-range dependence. For instance, Leland et al. [8] have demonstrated the self-similar nature of Ethernet traffic by a statistical analysis of Ethernet traffic measurements at BellCore; Beran et al. [3] have demonstrated long-range dependence in samples of variable bit rate video traffic generated by a number of different codecs; and Paxson and Floyd [12] have concluded the presence of long-range dependence in TELNET and other wide area network traffic.

Motivated by this empirical evidence a number of studies have recently appeared that investigate single-server queues with long-range dependent arrival models. These include the works of Norros [10] (discussed further by Duffield and O'Connell [5]) who uses a fractional Brownian arrival model; Resnick and Samorodnitsky [13] who use an arrival model derived from a stable integral; Likhanov et al. [9] who use an asymptotically self-similar discrete time arrival model (which also appears in the survey of Cox [4, page 68]); Anantharam [2] who studies extensions of the model of [9]; and Parulekar and Makowski [11] who also study the model of [9].

In the analysis and design of telecommunication networks, the concept of a *quasi-reversible queue*, see Kelly [7], Walrand [14], Walrand [15], and Section 12.3 of this paper, has been a key in enabling the effective use of queuing models. Networks constructed from quasi-reversible queues with Bernoulli routing and an external arrival process which is a Poisson process (in continuous time) or a sequence of i.i.d. Poisson random variables (in discrete time) offer considerable modeling flexibility and admit *product form* stationary distributions, which makes the computation of stationary performance quantities easy.

Our purpose in this paper is to construct queuing network models with tractable analytical features that allow the modeling of long-range dependent traffic processes. To this end, we bring together the discrete time quasi-reversible S-queue network models of Walrand [15] and the extension of the model of Likhanov et al. [9] studied in [2]. Our networks have external arrival processes which are long-range dependent processes described as follows : Sessions arrive according to a sequence of i.i.d. Poisson random variables. Each session lasts for an independent duration having the distribution of a random variable τ which is positive integer valued with finite mean and infinite variance and having regularly varying tail, i. e. $P(\tau > t) \sim t^{-\alpha} L(t)$ where $1 < \alpha < 2$ and $L(\cdot)$ is a slowly varying function (see appendix). While it is active, a session brings in work at rate 1. Our such that

$$q_k \sim k^{-\alpha} L(k) . \tag{12.1}$$

(See the appendix for the basic definitions on regular variation.) Each session active during a slot generates traffic at rate 1 during that slot. Note that, since there are many different slowly varying functions, there is considerable flexibility in the choice of arrival model.

Let $\xi_l(m)$ denote the number of sessions starting at time l that are active for exactly m slots. We have $\xi_l = \sum_{m=1}^{\infty} \xi_l(m)$, with $(\xi_l(m), l \in \mathbf{Z}, m \geq 1)$ being independent Poisson random variables, having mean λp_m for each m.

Let X_l denote the total amount of work brought in by the external arrival process during slot l. Then

$$X_l = \sum_{m=-\infty}^{0} \sum_{j=1}^{\infty} \xi_{l+m}(-m+j) .$$

Note that $(X_l, l \in \mathbf{Z})$ is a stationary stochastic process. A straightforward calculation gives

$$r(k) \triangleq \text{Cov}(X_l, X_{l+k}) = \lambda \sum_{j=1}^{\infty} q_{k+j} \sim \frac{k^{-\alpha+1}}{\alpha - 1} L(k)$$

Since $\sum_k r(k)$ is divergent when $1 < \alpha < 2$, the process $(X_l, l \in \mathbf{Z})$ is long-range dependent.

12.3 Discrete Time Quasi-Reversible Queuing Network Model

In this section we describe a class of discrete time queuing network models first introduced by Walrand [15]. We first define a kind of discrete time queue called an *S-queue*. An S-queue admits batch arrivals and has batch service. Given an arbitrary arrival sequence $\{\xi_n, n \geq 0\}$ of $\mathbf{N}$ valued random variables, the queue length process of an S-queue is given by

$$x_{n+1} = x_n + \xi_n - d_{n+1} , \tag{12.2}$$

where

$$P(d_{n+1} = j \mid x_m, d_m, 0 \leq m \leq n; \xi_k, k \geq 0, x_n + \xi_n = i) = S(i, j)$$

for $0 \leq j \leq i$ and $n \geq 0$. Here x_0 is arbitrary and $d_0 = 0$. Notice from the definition that the operation of an S-queue can be visualized as follows : there is a sequence of independent and identically distributed random variables $(d_n(i), i \geq 1)$, $-\infty < n < \infty$, with $P(d_n(i) = j) = S(i, j)$,

$0 \le j \le i$, which is independent of the arrival process. This sequence can be thought of as a virtual departure process. At time $n+1$, if the the queue size just prior to release of the departures is i, we release $d_{n+1}(i)$ customers.

An S-queue is called *quasi-reversible* if, when $\{\xi_n, n \ge 0\}$ is a sequence of i.i.d. Poisson random variables such that the state admits an equilibrium distribution, then the sequence of actual departures $\{d_n, n \ge 0\}$ is also a sequence of i.i.d. Poisson random variables in equilibrium and for all n $\{d_l, l \le n\}$ and x_n are independent. Walrand [15] gave a necessary and sufficient condition for an S-queue to be quasi-reversible. An S-queue is quasi-reversible if and only if there is a sequence of numbers $\alpha(j), j \ge 0$, where $\alpha(0) = 1$, $\alpha(j) > 0$ for $j > 0$ and a sequence of normalizing constants $c(i), i \ge 0$ such that $S(i,j)$ has the following form :

$$S(0,0) = c(0) = 1 \ , \tag{2.1a}$$

$$S(i,0) = c(i) \ , \ i > 0 \ , \tag{2.1b}$$

$$S(i,j) = \frac{c(i)}{j!}\alpha(i)\alpha(i-1)\ldots\alpha(i-j+1) \ , \quad 0 < j \le i \ , \tag{2.1c}$$

$$\sum_{j=0}^{i} S(i,j) = 1 \ . \tag{2.1d}$$

Further, the queue admits an equilibrium distribution π for a Poisson arrival sequence of mean λ if and only if the normalizing constant κ exists such that

$$\pi(i) = \kappa \frac{\lambda^i}{\alpha(0)\ldots\alpha(i)} \ , \quad i \ge 0$$

is a probability distribution.

A discrete time quasi-reversible network of queues can be constructed from J quasi-reversible S-queues (called nodes of the network) fed by an external arrival sequence of i.i.d. Poisson random variables, with Bernoulli routing. In more detail, such a network of queues operates as follows : An external arrival is routed to node p with probability r_{0p}. Such an arrival waits in queue at node p until it receives service; on receiving service it is routed to node q with probability r_{pq} and routed out of the network with probability r_{p0}. If routed to node q it joins the queue there. How many customers are served at each queue at each time is decided according to each queue's individual S-queue parameters, and is independent from queue to queue. The routing decisions are independent from customer to customer, and are also independent of the arrival process and the service decisions at the queues.

Walrand [15] has shown that, as long as the usual rate conditions are met (see below), such a network fed by an i.i.d sequence of Poisson random variables is stable and has a product form stationary distribution, where each queue in the network has a marginal stationary distribution as if it were being fed by an i.i.d. Poisson sequence of the appropriate rate. Here, by the *usual rate conditions* we mean the following : Let γ denote the rate of the external arrival process, and let λ_p, $1 \leq p \leq J$, be the unique solutions of the *flow balance equations* :

$$\gamma r_{0p} + \sum_q \lambda_q r_{qp} = \lambda_p \ , \quad 1 \leq p \leq J \ .$$

We require that for each $1 \leq p \leq J$, the S-queue p be stable when fed by an i.i.d. Poisson sequence of random variables of rate λ_p, i.e. that the normalizing constant κ_p can be defined such that

$$\pi_p(i) = \kappa_p \frac{\lambda_p^i}{\alpha_p(0)\ldots\alpha_p(i)} \ , \quad i \geq 0 \tag{12.3}$$

is a probability distribution, where $\alpha_p(\cdot)$ are the parameters of queue p. By *product form stationary distribution* we mean that the stationary distribution of the number of backlogged customers at the nodes of the network is given by

$$\pi(i_1, \ldots i_J) = \prod_{p=1}^{J} \pi_p(i_p) \ ,$$

with $\pi_p(\cdot)$ as in (12.3).

To prove our main results we need to impose an additional condition on the parameters of the S-queues used to construct our network. An S-queue is called *monotone* if the sequence $\alpha(i), i \geq 1$ is nondecreasing [1, Section IV]. In Lemma 2 of [1] it is proved that for a monotone S-queue one has

$$\sum_{j=k}^{i+1} S(i+1, j) \geq \sum_{j=k}^{i} S(i, j) \text{ for all } 0 \leq k \leq i \text{ and } i \geq 0 \ . \tag{12.4}$$

In addition, in Lemma 1 of [1] it is proved that for any S-queue (monotone or not) one has

$$\sum_{j=k}^{i+1} S(i+1, i+1-j) \geq \sum_{j=k}^{i} S(i, i-j) \text{ for all } 0 \leq k \leq i \text{ and } i \geq 0 \ . \tag{12.5}$$

Together, equations (12.4) and (12.5) imply that it is possible to construct the virtual departure variables $((d_n(i), i \geq 1), n \in \mathbf{Z})$ of a monotone S-queue in such a way that one simultaneously has

$$d_n(i+1) \geq d_n(i) \text{ for all } i \geq 0 \text{ and } n \in Z \ , \tag{12.6}$$

and

$$i + 1 - d_n(i+1) \geq i - d_n(i) \text{ for all } i \geq 0 \text{ and } n \in Z \ , \tag{12.7}$$

see Lemma 2 of [1]. This means that it is possible to couple the evolution of a network of monotone quasi-reversible S-queues from two different initial conditions, thereby using pathwise techniques to compare the evolution from these different initial conditions. What is meant by this last sentence will be made more clear in Section 12.5 where such a coupling is used.

12.4 Network Model with Long-Range Dependent Traffic

In this section we observe that it is possible to combine the long-range dependent arrival models of the kind considered in Section 12.2 with the quasi-reversible queuing networks of S-queues described in Section 12.3 to create a self-contained class of network models that handle long-range dependent arrival processes. In the model we develop, we assume that service decisions are made at the session level without reference to the duration of the session. Further, once a session enters service at a node, we assume that it continues to receive service at rate 1 until all the work that it brings in is completed.

To make this precise, let us first consider what this means for a single quasi-reversible S-queue fed by such an arrival process, cf. Eqn. (12.2). Immediately after the service decision at time n, let there be x_n backlogged sessions at the queue. Here a backlogged session means one which has arrived at a time prior to time n but has not yet entered service. A session that has already entered service by time n is not considered backlogged even though it might still be currently being served, and indeed the work that it is to bring in might not yet all have arrived. Now at time n, ξ_n new sessions enter the queue as described by the external arrival process model. Then the decision to let d_{n+1} new sessions enter service at time $n+1$ is made according to the conditional probabilities of the S-queue. Finally this leaves behind $x_{n+1} = x_n + \xi_n - d_{n+1}$ sessions still backlogged after the service decision at time $n+1$.

Once a session enters service it continues to receive service at rate 1 until all the work that it brings in is exhausted. To clarify this further, note for instance that a session of duration τ that arrived at time $m < n$ and is still backlogged immediately after the service decision at time n will have brought in a total amount of work of $\tau \wedge (n-m)$ up to time n, which will be sitting in queue throughout slot n. If $\tau > n-m$ this backlogged session will be bringing in further work at rate 1 during slot n, while if $\tau \leq n-m$ no further work will be brought in by this session during slot n. If this session now enters service at time $n+1$, and if $\tau \geq n+1-m$, a total of $n+1-m$ units of work will remain in queue until time $m+\tau$ after which the work in queue contributed by this session will decrease at rate one. If, on the other hand $\tau \leq n+1-m$, the work in queue contributed by this session will immediately begin to decrease at rate 1 starting from time $n+1$.

The decision of which sessions to serve is not allowed to refer to the duration of the session. In the case of a single quasi-reversible S-queue fed by a long-range dependent arrival process of the kind considered in Section 12.2 it is then seen that in equilibrium the departure process is once again a process of the type considered in that section. This is nothing more than the observation that the stationary departure process of a quasi-reversible S-queue is a sequence of i.i.d. Poisson random variables, which is a direct consequence of the definition of quasi-reversibility.

To completely specify the model we also assume that sessions enter service in FCFS order of their arrival at the node, with ties broken by uniform randomization. The particular priority used to serve sessions will be seen to be irrelevant for the general results that we derive, although it would be important in a more detailed analysis of the stationary distribution of the performance quantities of interest.

In our general network model each S-queue fuctions at the session level as above. The Bernoulli routing between the quasi-reversible queues is also assumed to take place at the session level, i.e. all the work in a session travels along the same route. Heuristically, one can picture a session of duration τ as consisting of a "train" of τ units of work that extends over τ slots, with the work spread out uniformly over this duration. Immediately after the service decisions at all the queues at time n the head of the train of work corresponding to the session will either be at some queue p or will have left the network; in the former case we say that the session is backlogged at node p. It may happen that only a portion of the τ units of work associated with the session may be present in the network. This could happen for two reasons : If the head of the train of work left the network at time $m \leq n$ and $\tau > n-m$, then the initial $n-m$ units of work associated with the session have left the network forever; similarly, if the session first arrived at into the network at time $m \leq n$ and $\tau > n-m$, then the last $\tau-(n-m)$ units of work associated with the session have not even entered the network yet. Of the work associated with the session that is in the network, portions may be sitting at various queues. Rather than introduce burdensome notation, we illustrate the heuristic picture by means of the example of Figure 1.

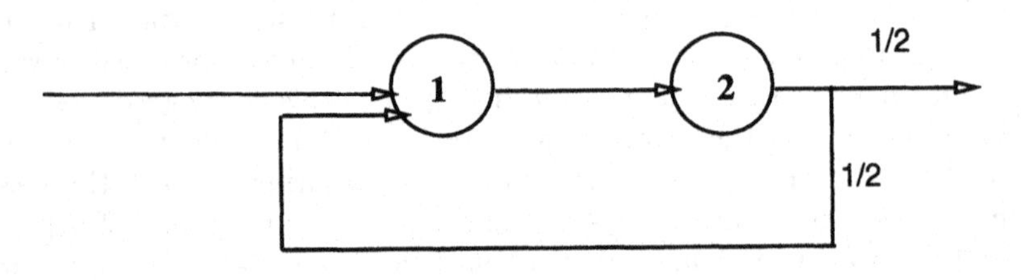

FIGURE 1. An example network.

This is a network of two S-queues, with routing parameters given by

$$r_{01} = 1 \ , \quad r_{02} = 0 \ , \quad r_{10} = r_{11} = 0 \ , \quad r_{12} = 1 \ , \quad r_{20} = r_{21} = \frac{1}{2} \ , \quad r_{22} = 0 \ .$$

The precise S-queue parameters of the two queues are not relevant to the discussion. Assume a session of duration 8 slots arrived in the network at time -6, was served by node 1 at time -5, served next by node 2 at time -4, was routed back to node 1, served next by node 1 at time -2, served next by node 2 at time -1 and routed out of the network. In this case we illustrate in Figure 2 where the work associated with this session is sitting immediately after the service decisions at time 0.

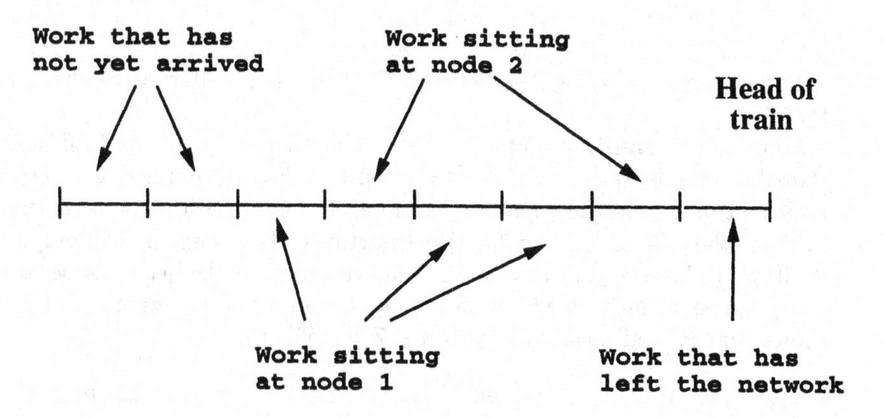

FIGURE 2. An example illustrating where the work associated with a session might be located.

We remark that our network model allows a session to visit several different nodes, with the possibility of feedback. Further, of course, a session may visit several different nodes. Such a session will have the same duration on each visit to each node. It is basically this feature that makes the analysis of the general network model of this kind more interesting than the analysis of a single S-queue that was carried out at the beginning of this section.

To clarify the network model further, and to be able to state and prove the main results, we introduce some notation. Let $(a_n, n \in \mathbf{Z})$ denote the external arrival process of sessions into the network. This is a sequence of i.i.d. Poisson random variables of rate γ. As in Section 12.2 each session is active for an independent random duration having the distribution of τ, a positive integer valued random variable that is regularly varying with finite mean and infinite variance. Let $(a_n^p, n \in \mathbf{Z})$ denote the process of external arrivals of sessions at node p; this is a sequence of i.i.d. Poisson random variables of rate γr_{0p}. Let $(\xi_n^p, n \in \mathbf{Z})$ denote the overall arrival process of sessions to node p; in stationarity this is a stationary sequence of nonnegative integer valued random variables of rate λ_p. Let $(d_n^p, n \in \mathbf{Z})$ denote the process of departures of sessions from node p and $(d_n^{pq}, n \in \mathbf{Z})$ denote the process of departing sessions from node p that are routed to node q, $0 \le q \le J$ (with $q = 0$ corresponding to the departing sessions that leave the network). Then we have

$$\xi_n^p = a_n^p + \sum_{q=1}^{J} d_n^{qp} .$$

Further, with x_n^p denoting the number of backlogged sessions at node p immediately after the service decision at time n, we have

$$x_{n+1}^p = x_n^p + \xi_n^p - d_{n+1}^p . \tag{12.8}$$

In the following, we will sometimes write d_n^{0p} as an alternative notation for a_n^p.

Some important comments are in order at this point. First of all, note that the arrivals to a node at time n consist of the external arrivals to this node at time n and the departures from all the nodes to this node at time n. Since the arrivals occur after the departures, these sessions will only be available to be served at time $n+1$. Thus at any time the head of a session can progress at most one node. Secondly, the proof of Theorem 3.1 of [15] shows that in equilibrium for each $n \in \mathbf{Z}$ it holds that

$$x_n^1, \ldots, x_n^J, d_n^{10}, \ldots, d_n^{1J}, \ldots, d_n^{J0}, \ldots, d_n^{JJ}, (a_m^1, m \ge n), \ldots, (a_m^J, m \ge n)$$

is an independent family of random variables with all of them except perhaps the x_n^p being Poisson random variables of the appropriate means. (Note that this *does not* imply that $(d_n^{pq}, n \in \mathbf{Z})$ is an i.i.d. Poisson sequence.) Finally, the existence and uniqueness of a stationary regime for the network model when the usual rate conditions are met can be argued as follows : Because all service and routing decisions are carried out at the session level, the results of Section 12.3 ensure that there is a unique stationary regime for the process of the number of backlogged sessions at the nodes. Now, in this stationary regime, let $\mathcal{D}_l$ denote the *set* of sessions that leave the network at time l (or more precisely the set of sessions whose head leaves the network at time l), and let $\mathcal{X}_l$ denote the *set* of backlogged sessions in the network at time l. Further, for any session i let $M(i)$ denote the *number* of nodes the session visits (multiple visits to the same node are counted with their multiplicity). Consider the total amount of work in the network at time n, i.e. the amount of work the network would have to handle if all fresh arrivals from time n onwards were to be deleted. This can be shown to be a.s. finite as follows : It is upper bounded by

$$\sum_{i \in \mathcal{X}_n} M(i)\tau(i) + \sum_{l \le n} \sum_{i \in \mathcal{D}_l} M(i)(\tau(i) - (n-l))^+ . \tag{12.9}$$

Since $\mathcal{X}_n$ is a.s. finite, the first term of equation (12.9) is a.s. finite. Quasi-reversibility ensures that for each $j \geq 1$ $(|\mathcal{D}_l \cap \{i : M(i) = j\}|, l \leq n)$ is a sequence of independent Poisson random variables of mean $\gamma\beta_j$, where β_j is the probability that an individual session visits j nodes during its sojourn in the network (this probability depends only on the routing probabilities). Further, since the service decisions and routing do not refer to the session durations, it follows that the number of terms of type $jm \geq 1$ that need to be summed in the second summation on the right hand side of equation (12.9) is a Poisson random variable of mean $\gamma\beta_j \sum_{j=0}^{\infty} p_{m+j} = \gamma\beta_j q_m$, and that these Poisson random variables are independent. Thus the second summation the right hand side of equation (12.9) can be rewritten as

$$\sum_{j=1}^{\infty}\sum_{m=1}^{\infty} jmZ_{jm}$$

for appropriately defined independent Poisson random variables $(Z_{jm}, j, m \geq 1)$ of means $\gamma\beta_j q_m$ respectively. This sum is a.s. finite, as can be seen by noting that

$$\begin{aligned}
& P(Z_{jm} = 0 \forall j \geq J \text{ and } \forall m \geq M) \\
= \; & \Big(\prod_{m\geq M}\prod_{j\geq 1} e^{-\gamma\beta_j q_m}\Big)\Big(\prod_{1\leq m<M}\prod_{j\geq J} e^{-\gamma\beta_j q_m}\Big) \\
\geq \; & \Big(\prod_{m\geq M} e^{-\gamma q_m}\Big)\Big(\prod_{j\geq J} e^{-\gamma\beta_j}\Big) \\
= \; & e^{-\gamma(\sum_{m\geq M} q_m + \sum_{j\geq J}\beta_j)} \\
\rightarrow \; & 1 \text{ as } M \rightarrow \infty \text{ and } J \rightarrow \infty .
\end{aligned}$$

The main results of this paper, derived in Section 12.5, refer to this stationary regime.

12.5 Main Results

For $0 \leq p, q \leq J$, let X_n^{pq} denote the total amount of work routed from node p to node q during slot n. (By convention, X_n^{0q} denotes the total amount of work arriving at node q during slot n due to external arrivals and X_n^{p0} denotes the total amount of work leaving the network from node p during slot n.) Our main result is the following theorem.

Theorem 12.5.1 Suppose all the nodes in the quasi-reversible network of S-queues defined in Section 12.4 are monotone. Fix any $0 \leq p, q \leq J$. In stationarity the process $(X_n^{pq}, n \in \mathbf{Z})$ is long-range dependent.

Theorem 12.5.1 is a consequence of the following two lemmas. The first of these lemmas needs some motivation, so we give it here before formally stating the lemma. Our idea to prove theorem 12.5.1 is to first decompose the process d_n^{pq} of customers moving from node p to node q as a sum of processes. The individual processes in this decomposition are parametrized by increasing sequences of positive integers starting at zero. Given such a sequence $\mathbf{r} = \{0 = r_0 < r_1 < \ldots < r_m\}$ for some $m \geq 0$, the process, say $d_n^{pq,\mathbf{r}}$, that is one of the constituents of this decomposition, consists of those customers who move from node p to node q a total of $m+1$ times of the first is time n, the next time $n + r_1$, and so on, the last being time $n + r_m$. This decomposition of d_n^{pq} also gives a convenient way of decomposing the process X_n^{pq}, enabling the proof of theorem 12.5.1.

Lemma 12.5.2 For $m \geq 0$, let

$$\mathcal{L}_m = \{0 = r_0 < r_1 < \ldots < r_m\} ,$$

where $r_0, \ldots, r_m \in \mathbf{N}$ be the set of all possible increasing sequences of positive integers of length $m+1$, starting at 0. Let $\mathcal{L} = \cup_{m=0}^{\infty} \mathcal{L}_m$. Suppose that for each $\mathbf{r} \in \mathcal{L}$ there is given a wide sense stationary $\mathbf{N}$-valued sequence $(\zeta_l^{\mathbf{r}}, l \in \mathbf{Z})$ such that these sequences are also jointly wide sense stationary.

Let $\mathcal{Z}_l^{\mathbf{r}}$ denote a *set* of cardinality $\zeta_l^{\mathbf{r}}$. We think of each $i \in \mathcal{Z}_l^{\mathbf{r}}$ $l \in \mathbf{Z}$, $\mathbf{r} \in \mathcal{L}$ as a session, distinct identical copies of which become active for the first time at $l + r_0, \ldots, l + r_m$. By this we mean the following : Let $\tau(i)$ be the duration of such a session i, which has regularly varying distribution as in Section 12.2; the session duration variables are assumed to be independent and independent of the processes $(\zeta_l^{\mathbf{r}}, l \in \mathbf{Z})$, $\mathbf{r} \in \mathcal{L}$. Then one copy of this session is active during the slots $l + r_0, \ldots, l + r_0 + \tau(i) - 1$, another during the slots $l + r_1, \ldots, l + r_1 + \tau(i) - 1$, and so on for a total of $m+1$ copies, the last of which is active during the slots $l + r_m, \ldots, l + r_m + \tau(i) - 1$.

Let Y_n denote the total number of copies of sessions that become active at time n. Thus

$$Y_n = \sum_{l \in \mathbf{Z}} \sum_{m \geq 0} \sum_{\mathbf{r} \in \mathcal{L}_m} \zeta_l^{\mathbf{r}} \sum_{a=0}^{m} 1(l + r_a = n) . \tag{12.10}$$

During each slot during which any copy of any session is active, it brings in one unit of work. Let X_n denote the total amount of work brought in during slot n. Thus

$$X_n = \sum_{l \in \mathbf{Z}} \sum_{m \geq 0} \sum_{\mathbf{r} \in \mathcal{L}_m} \sum_{i \in \mathcal{Z}_l^{\mathbf{r}}} \sum_{a=0}^{m} 1(l + r_a \leq n \leq l + r_a + \tau(i) - 1) \tag{12.11}$$

Assume that $(Y_n, n \in \mathbf{Z})$ and $(X_n, n \in \mathbf{Z})$ are well defined (a.s. finite) processes. Then, if $(Y_n, n \in \mathbf{Z})$ is short-range dependent, it holds that $(X_n, n \in \mathbf{Z})$ is long-range dependent. □

Lemma 12.5.3 Consider a discrete time quasi-reversible network of J monotone S-queues with Bernoulli routing, in stationarity. For each $0 \leq p, q \leq J$, the process $(d_n^{pq}, n \in \mathbf{Z})$ is short-range dependent. □

We first prove the main theorem using these lemmas, and then prove the lemmas.

Proof of Theorem 12.5.1 :

Since each $(a_n^p, n \in \mathbf{Z})$ and each $(d_n^{p0}, n \in \mathbf{Z})$, $1 \leq p \leq J$ is an i.i.d. sequence of Poisson random variables, it is immediate that each $(X_n^{0p}, n \in \mathbf{Z})$ and each $(X_n^{p0}, n \in \mathbf{Z})$, $1 \leq p \leq J$ is a process of the kind considered in Section 12.2, and is therefore long-range dependent. We therefore focus on the case $1 \leq p, q \leq J$.

It suffices to identify the appropriate sequences $(\zeta_l^{\mathbf{r}}, l \in \mathbf{Z})$, $\mathbf{r} \in \mathcal{L}$ so as to apply Lemma 5.1. Given fixed $1 \leq p, q \leq J$, for $m \geq 0$ and $\mathbf{r} \in \mathcal{L}_m$, define $\zeta_l^{\mathbf{r}}$ to be the number of sessions that are routed from node p to node q at precisely the times $l, l+r_1, \ldots, l+r_m$. When the network is in stationarity, the sequences $(\zeta_l^{\mathbf{r}}, l \in \mathbf{Z})$, $\mathbf{r} \in \mathcal{L}$ are jointly stationary. Further, for each $\mathbf{r} \in \mathcal{L}$ and $l \in \mathbf{Z}$ we have $\zeta_l^{\mathbf{r}} \leq d_l^{pq}$, and d_l^{pq} is a Poisson random variable of mean $\lambda_p r_{pq}$. From this it follows that each $\zeta_l^{\mathbf{r}}$ has finite second moment. Since the processes $(\zeta_l^{\mathbf{r}}, l \in \mathbf{Z})$, $\mathbf{r} \in \mathcal{L}$ are jointly stationary and have finite second moment, they are jointly wide sense stationary, so that Lemma 5.1 can be applied. Note that we also have, following equations (12.10) and (12.11), that

$$Y_n = d_n^{pq} \text{ and } X_n = X_n^{pq} .$$

By Lemma 5.2 we see that $(Y_n, n \in \mathbf{Z})$ is short-range dependent. By Lemma 5.1, the claim of the theorem follows.

□

Proof of Lemma 12.5.2 :

Let τ be a generic $\mathbf{N}$-valued random variable having the distribution of the session duration random variables, and let $l \in \mathbf{Z}$. We write $C(l, \tau)$ for the indicator of the event that slot 0 is one of the slots $l, \ldots, l+\tau-1$. In other words,

$$\begin{aligned} C(l, \tau) &= 1(l \leq 0 \leq l+\tau-1) \\ &= 1(l \leq 0) - 1(l+\tau \leq 0) \end{aligned} \tag{12.12}$$

Note that

$$E[C(l, \tau)] = \begin{cases} 0 & \text{if } l > 0 \\ P(l+\tau \geq 1) = q_{1-l} & \text{if } l \leq 0 \end{cases} \tag{12.13}$$

For convenience, set $q_k = 0$ if $k \leq 0$. With this notation, we may then write equation (12.13) as $E[C(l,\tau)] = q_{1-l}$ for all $l \in \mathbf{Z}$.

Let $\mathcal{F}$ denote the sigma-field generated by the variables $\zeta_l^{\mathbf{r}}$, $l \in \mathbf{Z}$, $\mathbf{r} \in \mathcal{L}$. Let $r(k) = \mathrm{Cov}(X_l, X_{l+k})$. Then we may write

$$\begin{aligned} r(k) &= E[X_0 X_k] - E[X_0]E[X_k] \\ &= E[E[X_0 X_k \mid \mathcal{F}] - E[X_0 \mid \mathcal{F}]E[X_k \mid \mathcal{F}]] \\ &\qquad + E[E[X_0 \mid \mathcal{F}]E[X_k \mid \mathcal{F}] - E[X_0]E[X_k]] \\ &= r^{(1)}(k) + r^{(2)}(k) \ . \end{aligned} \tag{12.14}$$

In the following we will handle each of the two terms on the right hand side of equation (12.14) separately.

With the notation introduced in equation (12.12), we may rewrite equation (12.11) as

$$X_n = \sum_{l \in \mathbf{Z}} \sum_{m \geq 0} \sum_{\mathbf{r} \in \mathcal{L}_m} \sum_{i \in \mathcal{Z}_l^{\mathbf{r}}} \sum_{a=0}^{m} C(l - n + r_a, \tau(i)) \tag{12.15}$$

Using equation (12.13), we see from equation (12.15) that

$$E[X_n \mid \mathcal{F}] = \sum_{l \in \mathbf{Z}} \sum_{m \geq 0} \sum_{\mathbf{r} \in \mathcal{L}_m} \zeta_l^{\mathbf{r}} \sum_{a=0}^{m} q_{n+1-l-r_a} \ . \tag{12.16}$$

We turn to the evaluation of the first term on the right hand side of equation (12.14). Note that

$$\begin{aligned} & E[X_0 X_k \mid \mathcal{F}] - E[X_0 \mid \mathcal{F}]E[X_k \mid \mathcal{F}] \\ &= E[(X_0 - E[X_0 \mid \mathcal{F}])(X_k - E[X_k \mid \mathcal{F}]) \mid \mathcal{F}] \ , \end{aligned}$$

is the conditional covariance of X_0 and X_k given $\mathcal{F}$. We find that

$$\begin{aligned} & \mathrm{Cov}\,(C(l + r_a, \tau), C(l - k + r_b, \tau)) \\ &= q_{1-l-r_a} \wedge q_{k+1-l-r_b} - q_{1-l-r_a} q_{k+1-l-r_b} \ , \end{aligned} \tag{12.17}$$

which is nonnegative. From equation (12.15) and the conditional independence of the various $\tau(i)$ given $\mathcal{F}$, we see that

$$\begin{aligned} & E[X_0 X_k \mid \mathcal{F}] - E[X_0 \mid \mathcal{F}]E[X_k \mid \mathcal{F}] \\ &= \sum_{l \in \mathbf{Z}} \sum_{m \geq 0} \sum_{\mathbf{r} \in \mathcal{L}_m} \sum_{i \in \mathcal{Z}_l^{\mathbf{r}}} \sum_{a=0}^{m} \sum_{b=0}^{m} \mathrm{Cov}(C(l + r_a, \tau(i)), C(l - k + r_b, \tau(i))) \ , \end{aligned}$$

because all the cross-covariances vanish.

Let $\lambda^{\mathbf{r}}$ denote $E[\zeta_l^{\mathbf{r}}]$. From the preceding equation and equations (12.13) and (12.17) we have

$$r^{(1)}(k) = \sum_{l\in\mathbf{Z}} \sum_{m\geq 0} \sum_{\mathbf{r}\in\mathcal{L}_m} \lambda^{\mathbf{r}} \sum_{a=0}^{m} \sum_{b=0}^{m} (q_{1-l-r_a} \wedge q_{k+1-l-r_b} - q_{1-l-r_a} q_{k+1-l-r_b}) \tag{12.18}$$

Pick some $\mathbf{r} \in \mathcal{L}$ for which $\lambda^{\mathbf{r}} = \lambda > 0$ (there must be at least one such, otherwise the situation considered in the lemma is vacuous). Noting that every term on the right hand side of equation(12.18) is nonnegative, we may write

$$r^{(1)}(k) \geq \lambda \sum_{l\in\mathbf{Z}} (q_{1-l} \wedge q_{k+1-l} - q_{1-l} q_{k+1-l}) \tag{12.19}$$

Noting that $q_{1-l} = 0$ if $l > 0$ and that $q_{k+1-l} \leq q_{1-l}$ if $k \geq 0$, we have

$$\begin{aligned} \sum_{k\geq 0} r^{(1)}(k) &\geq \lambda \sum_{l\leq 0} (1 - q_{1-l}) \sum_{k\geq 0} q_{k+1-l} \\ &= \infty , \end{aligned}$$

where we have used equation (12.1) for the tail probability asymptotics of the session duration random variables.

We turn next to the evaluation of the second term on the right hand side of equation (12.14). Note that

$$r^{(2)}(k) = \mathrm{Cov}\,(E[X_0 \mid \mathcal{F}], E[X_k \mid \mathcal{F}]) \ . \tag{12.20}$$

Using the relation

$$q_j = \sum_{t\in\mathbf{Z}} q_t 1(t = j)$$

and the definition of Y_n in equation (12.10) we may rewrite equation (12.16) to get

$$\begin{aligned} E[X_n \mid \mathcal{F}] &= \sum_{l\in\mathbf{Z}} \sum_{m\geq 0} \sum_{\mathbf{r}\in\mathcal{L}_m} \zeta_l^{\mathbf{r}} \sum_{a=0}^{m} q_{n+1-l-r_a} \\ &= \sum_{l\in\mathbf{Z}} \sum_{m\geq 0} \sum_{\mathbf{r}\in\mathcal{L}_m} \zeta_l^{\mathbf{r}} \sum_{a=0}^{m} \sum_{t\in\mathbf{Z}} q_t 1(t = n+1-l-r_a) \\ &= \sum_{t\in\mathbf{Z}} q_t \sum_{l\in\mathbf{Z}} \sum_{m\geq 0} \sum_{\mathbf{r}\in\mathcal{L}_m} \zeta_l^{\mathbf{r}} \sum_{a=0}^{m} 1(l + r_a = n+1-t) \\ &= \sum_{t\in\mathbf{Z}} q_t Y_{n+1-t} \ . \end{aligned} \tag{12.21}$$

Now let

$$\Gamma_j = \text{Cov}(Y_0, Y_j) \ . \tag{12.22}$$

From equations (12.20), (12.21), and (12.22) we have

$$\begin{aligned} r^{(2)}(k) &= \text{Cov}\left(E[X_0 \mid \mathcal{F}], E[X_k \mid \mathcal{F}]\right) \\ &= \text{Cov}\left(\sum_{t\in\mathbf{Z}} q_t Y_{1-t}, \sum_{s\in\mathbf{Z}} q_s Y_{k+1-s}\right) \\ &= \sum_{t\in\mathbf{Z}}\sum_{s\in\mathbf{Z}} q_t q_s \Gamma_{k+t-s} \\ &= \sum_{j\in\mathbf{Z}} \Gamma_j \sum_{t\in\mathbf{Z}} q_t q_{k+t-j} \ . \end{aligned}$$

Thus

$$\begin{aligned} \sum_{k\in\mathbf{Z}} |r^{(2)}(k)| &= \sum_{k\in\mathbf{Z}} \left| \sum_{j\in\mathbf{Z}} \Gamma_j \sum_{t\in\mathbf{Z}} q_t q_{k+t-j} \right| \\ &= \sum_{k\in\mathbf{Z}}\sum_{j\in\mathbf{Z}} \mid \Gamma_j \mid \sum_{t\in\mathbf{Z}} q_t q_{k+t-j} \\ &= \sum_{j\in\mathbf{Z}} \mid \Gamma_j \mid \sum_{t\in\mathbf{Z}} q_t \sum_{k\in\mathbf{Z}} q_{k+t-j} \\ &\leq (E[\tau])^2 \sum_{j\in\mathbf{Z}} \mid \Gamma_j \mid \\ &< \infty \ , \end{aligned}$$

where we have used the assumption that $\sum_{j\in\mathbf{Z}} \mid \Gamma_j \mid < \infty$.
This completes the proof of the lemma.

□

Proof of Lemma 12.5.3 :

First note that for any $1 \leq p \leq J$, each of the processes $(d_n^{0p}, n \in \mathbf{Z})$ and $(d_n^{p0}, n \in \mathbf{Z})$ is an i.i.d. sequence of Poisson random variables, and is therefore short-range dependent. Thus we focus on the case $1 \leq p, q \leq J$.

It is in the proof of this lemma that we exploit the assumption that each of the quasi-reversible S-queues used to construct our network is monotone. Fix $1 \leq p, q \leq J$. Our goal is to demonstrate that

$$\sum_{k\geq 0} \mid \text{Cov}(d_0^{pq}, d_k^{pq}) \mid < \infty \ . \tag{12.23}$$

Note that

$$\begin{aligned}\mathrm{Cov}(d_0^{pq}, d_k^{pq}) &= E[d_0^{pq} d_k^{pq}] - E[d_0^{pq}]E[d_k^{pq}] \\ &= \sum_{a \in \mathbf{N}} aP(d_0^{pq} = a)\,(E[d_k^{pq} \mid d_0^{pq} = a] - E[d_k^{pq}]) \ ,\end{aligned}$$

so that

$$\begin{aligned}&\sum_{k \geq 0} \mid \mathrm{Cov}(d_0^{pq}, d_k^{pq}) \mid \\ &\leq \sum_{a \in \mathbf{N}} aP(d_0^{pq} = a) \sum_{k \geq 0} \mid E[d_k^{pq} \mid d_0^{pq} = a] - E[d_k^{pq}] \mid \ .\end{aligned}$$

We now proceed to estimate $\sum_{k \geq 0} \mid E[d_k^{pq} \mid d_0^{pq} = a] - E[d_k^{pq}] \mid$. By the independence of

$$x_0^1, \ldots, x_0^J, d_0^{10}, \ldots, d_0^{1J}, \ldots, d_0^{J0}, \ldots, d_0^{JJ}, (a_m^1, m \geq 0), \ldots, (a_m^J, m \geq 0)$$

in stationarity, conditioning on $d_0^{pq} = a$ is equivalent to setting up an initial condition at the queues just prior to the service decisions at time 1 that differs from the equilibrium situation *only* in that node q was fed a customers from node p instead of the Poisson mean $\lambda_p r_{pq}$ customers that would have been fed in stationarity. But now we may couple the two initial conditions (the stationary situation and the situation where we have conditioned on $d_0^{pq} = a$) by the following simple expedient : If there are more customers than a fed back from node p to node q in the stationary situation, we color the extra customers in the stationary situation red, while if there are less customers than a fed back from node p to node q in the stationary situation we color the extra customers in the conditioned situation yellow. We then run the network (which consists entirely of either colorless customers and red customers or of colorless customers and yellow customers), with the convention that colorless customers have priority over colored customers.

The virtual departure variables $d_n^p(i), i \geq 0, n \in \mathbf{Z}, 1 \leq p \leq J$ obey the conditions (12.6) and (12.7). Consider first the case where there are less customers than a fed from node p to node q in the stationary situation. Then the evolution of colorless customers will be precisely as in the stationary situation, and the evolution of the sum of colored and colorless customers will be exactly as in the situation where we condition on $d_0^{pq} = a$. To understand why this follows from the properties (12.6) and (12.7), consider a node r during slot m, and suppose there are k colorless customers and l colored customers in queue just prior to time $m+1$, and the decision needs to be made as to how many customers to release at time $m+1$, in both the stationary and the conditioned situation. One needs to refer to the virtual departure variables $d_{m+1}^r(k)$ and $d_{m+1}^r(k+l)$ in order to make these decisions. Property (12.6) of the virtual departure variables ensures that $d^r_{m+1}(k+l) \geq d^r_{m+1}(k)$, and property (12.7) ensures that $d^r_{m+1}(k+l) - d^r_{m+1}(k) \leq l$, so that, giving priority to colorless customers over colored customers, we can release $d^r_{m+1}(k)$ colorless customers and $d^r_{m+1}(k+l) - d^r_{m+1}(k)$ colored customers, and simultaneously meet the requirements of both the stationary situation and the conditioned situation.

Similarly, in the case where there are less customers than a fed from node p to node q in the stationary situation, the evolution of the sum of colored and colorless customers will be precisely as in the stationary situation, while the evolution of colorless customers will be precisely as in the situation where we condition on $d_0^{pq} = a$.

In either case, one sees that $\sum_{k \geq 0} \mid E[d_k^{pq} \mid d_0^{pq} = a] - E[d_k^{pq}] \mid$ equals the mean number of times colored customers move from node p to node q at times $k \geq 0$. This is upper bounded by

$$\kappa E[\mid Z - a \mid] \tag{12.24}$$

for a constant κ related to the routing variables and where Z is a Poisson random variable of mean $\lambda_p r_{pq}$. To see this, note that the number of movements made from node p to node q by the individual customers that start in node p are independent random variables, because routing decisions are independent from customer to customer. The mean number of such movements can be taken to be the constant κ, which is finite.

The expression in (12.24) can be further upper bounded by

$$\kappa \lambda_p r_{pq} + \kappa a$$

so that we get

$$\sum_{k \geq 0} \mid \mathrm{Cov}(d_0^{pq}, d_k^{pq}) \mid \leq \sum_{a \in \mathbf{N}} a(\kappa \lambda_p r_{pq} + \kappa a) P(d_0^{pq} = a) \ .$$

But d_0^{pq} is itself a Poisson random variable of mean $\lambda_p r_{pq}$. This gives the desired equation (12.23).

□

Acknowledgments: Research supported in part by NSF grant NCR 94-22513 and by the AT & T Foundation. I would also like to thank one of the referees for an extraordinarily detailed reading of the original manuscript that led to a substantially improved presentation.

Appendix *(Regular Variation)*

For convenience we reproduce the basic definitions on nonnegative random variables with regularly varying tails. For more information, consult Feller

[6, pp. 275–284].

Definition 12.5.4 A positive (not necessarily monotone) function $L(\cdot)$ defined on an interval $[a, \infty)$ is said to be *slowly varying* if for every $x > 0$ it holds that

$$\lim_{t \to \infty} \frac{L(tx)}{L(t)} = 1 .$$

Note that there are many slowly varying functions. For instance, any power of any iterated logarithm is slowly varying; any positive function that approaches a strictly positive limit at ∞ is slowly varying.

Definition 12.5.5 A positive function $G(\cdot)$ defined on an interval $[a, \infty)$ is said to be *regularly varying* with exponent $-\alpha$ if it is of the form

$$G(x) = x^{-\alpha} L(x)$$

where $L(\cdot)$ is a slowly varying function.

In this paper we are only interested in the case $1 < \alpha < 2$.

Definition 12.5.6 A nonnegative random variable X having cumulative distribution function $F(x) = P(X \leq x)$ is said to have a *regularly varying tail* if $1 - F(x)$ is regularly varying.

12.6 References

[1] Anantharam, V., The Input-Output Map of a Monotone Discrete time Quasi-reversible Node. *IEEE Transactions on Information Theory*, Vol. 39, No. 2, pp. 543 -552, 1993. Correction published in Vol. 39, No. 4, pg. 1466, 1993.

[2] Anantharam, V., On the Sojourn Time of Sessions at an ATM Buffer with Long-Range Dependent Input Traffic. *Proceedings of the 34th IEEE Conference on Decision and Control*, New Orleans, December 1995.

[3] Beran, J., Sherman, R., Taqqu, M. S., and Willinger, W., Long-Range Dependence in Variable-Bit-Rate Video Traffic. *IEEE Transactions on Communications*, Vol. 43, No. 2/3/4, pp. 1566 -1579, 1995.

[4] Cox, D. R., Long-Range Dependence : A Review. In *Statistics : An Appraisal*, edited by H. A. David and H. T. David, Iowa State University Press, pp. 55 -74, 1984.

[5] Duffield, N. G., and O'Connell, N., Large Deviations and Overflow Probabilities for the General Single-server Queue, with Applications. To appear in *Proceedings of the Cambridge Philosophical Society*, 1995.

[6] FELLER, W., *An Introduction to Probability Theory and its Application.* John Wiley and Sons, New York, 1971.

[7] KELLY, F., *Reversibility and Stochastic Networks.* John Wiley and Sons, New York, 1979.

[8] LELAND, W. E., TAQQU, M. S., WILLINGER, W., AND WILSON, D. V., On the Self-Similar Nature of Ethernet Traffic (Extended Version). *IEEE/ACM Transactions on Networking*, Vol. 2, No. 1, pp. 1 -15, 1994.

[9] LIKHANOV, N., TSYBAKOV, B., AND GEORGANAS, N. D., Analysis of an ATM Buffer with Self-Similar ("Fractal") Input Traffic. *Proceedings of the 14 th Annual IEEE Infocom*, pp. 985 -992, 1995.

[10] NORROS, I., A storage model with self-similar input. *Queueing Systems : Theory and Applications*, Vol. 16, pp. 387 -396, 1994.

[11] PARULEKAR, M. AND MAKOWSKI, A. M., Buffer Overflow Probabilities for a Multiplexer with Self-similar Traffic. *Proceedings of the 34th IEEE Conference on Decision and Control*, New Orleans, December 1995.

[12] PAXSON, V. AND FLOYD, S., Wide Area Traffic : The Failure of Poisson Modeling. *IEEE/ACM Transactions on Networking*, Vol. 3, No. 3, pp. 226 -244, 1995.

[13] RESNICK, S. AND SAMORODNITSKY, G., The Effect of Long Range Dependence in a Simple Queuing Model. *Preprint*, Cornell University, 14 pp., 1994.

[14] WALRAND, J., *An Introduction to Queuing Networks.* Prentice Hall, Englewood Cliffs, N.J., 1988.

[15] WALRAND, J., A Discrete-Time Queuing Network. *Journal of Applied Proability*, Vol. 20, pp. 903-909, 1983.

13 Fractional Brownian Approximations of Queueing Networks

Takis Konstantopoulos and Si-Jian Lin

ABSTRACT We consider a single-class queueing network with long-range dependent arrival and service processes and show that the normalized queue length converges to a reflected d-dimensional fractional Brownian motion. We identify the covariance of the limiting process in terms of the arrival rates and asymptotic variances of the driving processes. We discuss the case of multiple Hurst parameters and point out that only the largest survive in the limit.

13.1 Introduction

Recent measurements and statistical analyses of traffic in high-speed networks have pointed out that data are best explained via traffic models possessing long-range dependence, see, for instance, Leland et al. [11] and Willinger et al. [19]. This announcement has triggered an interest in analyzing stochastic models of queueing systems with long-range dependent data. Norros [14] has worked with a model of a single queue, with "fractional Brownian motion input". A similar model has been assumed in a paper by Duffield and O'Connell [6]. Both these papers deal with the tail behavior of the marginal distribution of a stationary reflected fractonal Brownian motion. Other long-range dependent models for a queue have been treated by Likhanov et al. [12] and Anantharam [1].

But why fractional Brownian motion? The justification is to be found in its characterization: fractional Brownian motion plays the same fundamental role among self-similar processes that a standard Brownian motion does among processes with independent increments. One can define a fractional Brownian motion with Hurst parameter H as a (zero-mean) Gaussian process with continuous sample paths and stationary increments. It turns out, easily in fact, that there is essentially one such process, $\{B_H(t), t \geq 0\}$, with

$$EB_H(t)^2 = Lt^{2H} ,$$

where H has to lie in the open interval $(0, 1)$. The constant $L > 0$ is an arbitrary scaling constant. We shall denote such a process as FBM$(H, 0, L)$, reserving the symbol FBM(H, μ, L) for the process $\mu t + B_H(t)$. A fundamental property of FBM$(H, 0, L)$ is that it has long-range dependent increments (in the sense of, e.g., Cox [5]), positively or negatively correlated if $H > 1/2$ or $< 1/2$, respectively. Practical considerations [19] show that one should concentrate on models possessing the former correlation sign. Furthermore, FBM$(H, 0, L)$ appears as the domain of attraction (viz., a limit in a functional central limit theorem) of a wide range of long-range dependent processes, just as ordinary Brownian motion appears as a limit of short-range dependent processes. Examples of this can be found in the recent works of Kurtz [10] and a paper of ours [8] (see also Section 13.4).

It turns out that one can say a few things about queues driven by fractional Brownian motion as in Norros [14] and Duffield and O'Connell [6]. The natural question is: How do these models arise as limits of discrete-event driven queueing networks, which are by and large universally accepted models of high-speed communication networks? We deal with this question in what follows. We consider a single class queueing network with long-range dependent arrival times and service times associated with the servers. We make the simplifying assumption that the routing does not introduce long-range dependencies; rather, these are introduced by the sources. The limit obtained is a fractional Brownian motion, reflected appropriately so that its components are non-negative. In a sense, we try to justify the following formula, explained best for a single queue. Consider a FCFS single server queue with stationary inter-arrival times possessing long-range dependence with Hurst parameter H. Denote by λ^{-1} and α the mean and asymptotic variance (in some sense to be understood later) of the typical inter-arrival time. Suppose also that the service times possess long-range dependence with the same H and denote by μ^{-1} and β their mean and asymptotic variance. Then, provided that the system is almost fully utilized (i.e. the spare capacity $\mu - \lambda$ is small), we can approximate the queue length *process* $Q(t)$ as

$$Q(t) \approx Q(0) + (\lambda - \mu)t + (\lambda^{2+2H}\alpha + \mu^{2+2H}\beta)^{1/2}\xi(t)$$

where $\xi \sim$ FBM$(H, 0, 1)$. In other words, $Q \sim$ FBM$(H, \lambda - \mu, \lambda^{2+2H}\alpha + \mu^{2+2H}\beta)$. It is desirable to know when this approximation holds, both for purposes of simulation and, hopefully, analysis. We give simple proofs of this for the multi-dimensional case, i.e. we deal with a network. Notice that, of course, $H = 1/2$ yields the known Brownian approximation for a single queue in heavy traffic. The approximation is known as diffusion approximation. For $H > 1/2$, we can by no means call the approximation a diffusion approximation, for a FBM is neither Markov nor even a semimartingale.

Our purpose for considering these approximations of queueing networks is twofold: First, they should provide approximations for a large class of discrete-event models with long-range dependent data. Second, one should be able to pose and answer questions related to performance and control. For instance, for one-dimensional networks, it is possible to obtain some estimates for the tail of the stationary queue length distribution. These estimates are robust, in that they predict the form of the tail distribution for the original models. The tractability of the fractional Brownian networks should be considered as a challenging research problem, heavily motivated by networking applications.

The paper is organized as follows. In Section 13.2 we present an inversion theorem, namely a central limit theorem for point processes with long-range dependence. This is used in Section 13.3, extending the ideas of Reiman [15], for the proof of the main limit theorem. Several remarks and extensions are made in Section 13.4.

13.2 Long-Range Dependent Point Processes

Consider a sequence $\{A^{(N)}\}_{N=1}^{\infty}$ of simple point processes on $\mathbf{R}_+$ and denote by $T_1^{(N)} < T_2^{(N)} < \cdots$ the sequence of their points. We set

$$A^{(N)}(t) = \sum_{n=1}^{\infty} \mathcal{I}(T_n^{(N)} \leq t)$$

for the number of points in $[0,t]$ and assume that the sequence of inter-event times $\{T_k^{(N)} - T_{k-1}^{(N)}\}_{k=1}^{\infty}$ is stationary with $E(T_k^{(N)} - T_{k-1}^{(N)}) = 1/\lambda^{(N)}$ such that $\lambda^{(N)} \to \lambda \in (0,\infty)$, as $N \to \infty$. (For simplicity we take $T_0^{(N)} = 0$.)

In the particular case where the sequence of inter-event times is i.i.d., and under some additional technical assumptions arising from the applicability of the CLT (Central Limit Theorem) for triangular arrays, we have the Donsker-type limit $(T_{[Nt]}^{(N)} - [Nt]/\lambda^{(N)})/\sqrt{N} \Rightarrow W(t)$ where W is a zero mean Brownian motion. (The symbol $Z^{(N)} \Rightarrow Z$ means weak convergence of random functions $Z^{(N)}$ in the space $D[0,\infty)$ (or $D^d[0,\infty)$–in the case of $\mathbf{R}^d$-valued processes) to the random function Z, under the uniform topology (uniform convergence of functions on $[0,u]$ for any u). We occasionally abuse notation and write $Z^{(N)}(t) \Rightarrow Z(t)$; this is to be understood in the functional sense.) Furthermore, $(A^{(N)}(t) - \lambda^{(N)}t)/\sqrt{N} \Rightarrow \lambda W(\lambda t)$ (see Billingsley [3]). The same result is true for a wide class of short-range dependent point processes. (We say that a point process is short-range (or long-range) dependent if the sequence of its inter-event times is short-range (resp., long-range) dependent, in the sense defined, for instance, in Cox [5, Sec. 3].)

Consider now the case where the inter-event times $\{T_k^{(N)} - T_{k-1}^{(N)}\}_{k=1}^{\infty}$ are long-range dependent with long-range dependence index H (see Cox [5]). Then, under additional technical assumptions, we have the FCLT (Functional Central Limit Theorem)

$$(T^{(N)}_{[Nt]} - [Nt]/\lambda^{(N)})/N^H \Rightarrow W(t), \tag{13.1}$$

where W is a fractional Brownian motion with Hurst parameter H. A particular (but rather important) way to introduce long-range dependence is by taking infinite linear combinations of i.i.d. random variables ξ_n. This was studied in [8] where it was shown that if $X_k = \sum_n a_{k-j}\xi_j$, where ξ_j are i.i.d., arbitrarily distributed with $E\xi_j = 0, \operatorname{var}(\xi_j) < \infty$, then

$$\frac{\sum_{k=1}^{[Nt]} X_k}{N^H} \Rightarrow \mathrm{FBM}(H, 0, L)$$

if and only if

$$\frac{\operatorname{var}\left(\sum_{k=1}^{N} X_k\right)}{LN^{2H}} \to 1, \text{ as } N \to \infty .$$

In what follows we assume directly that (13.1) holds and show that this translates into a FCLT for the counting processes.

Theorem 13.2.1 Suppose that $\{A^{(N)}\}_{N=1}^\infty$ is a sequence of point processes, each with stationary inter-event times $\{T_k^{(N)} - T_{k-1}^{(N)}\}_{k=1}^\infty$, and assume that

1. $\lambda^{(N)} = (E(T_k^{(N)} - T_{k-1}^{(N)}))^{-1} \to \lambda$,
2. ${\sigma^{(N)}}^2 = \operatorname{var}(T_k^{(N)} - T_{k-1}^{(N)}) \le K < \infty$,
3. $N^{-H}(T^{(N)}_{[Nt]} - [Nt]/\lambda^{(N)}) \Rightarrow W(t) \sim \mathrm{FBM}(H, 0, L)$.

Then we have

$$N^{-1}(A^{(N)}(Nt) - \lambda^{(N)}Nt) \Rightarrow 0, \tag{13.2}$$

and

$$N^{-H}(A^{(N)}(Nt) - \lambda^{(N)}Nt) \Rightarrow \lambda W(\lambda t) \sim \mathrm{FBM}(H, 0, \lambda^{2+2H}L). \tag{13.3}$$

Before proving the theorem, let us comment on the assumptions: we assume that the interarrival times sequence is long-range dependent and formulate our assumption directly by the convergence to $W(t)$. The way that this can happen is something not considered in this paper, simply because we are interested in how the network behaves. It should be noted that in the literature of diffusion approximations for queueing networks driven by, say, renewal processes, essentially the only assumption required is that a renewal process is in the domain of attraction of an ordinary Brownian motion. The effort thereafter is to precisely characterize the way the network "transforms" or "operates" on such an input. It so happens that i.i.d. random variables fall in the domain of attraction of an ordinary Brownian motion. Making non-i.i.d. variables fall in the domain of attraction of a fractional Brownian motion was considered, through a particular filtering-type of approach, in [8]; but this may not be the only way that this can happen.

Proof of theorem 13.2.1. The proof follows the arguments of Billingsley [3, Thm. 17.3], one difference being that we work in the space $(D[0,\infty)$, uniform topology), whereas [3] works in $(D[0,1]$, Skorokhod topology). From assumption 3 it follows that

$$\sup_{0\le t\le u} N^{-1}(T^{(N)}_{[Nt]} - [Nt]/\lambda^{(N)}) \xrightarrow{\mathrm{P}} 0\,,$$

for all $u > 0$, which readily implies

$$\sup_{0\le t\le u} N^{-1}(A^{(N)}(Nt) - Nt\lambda^{(N)}) \xrightarrow{\mathrm{P}} 0\,,$$

for all $u > 0$. This in its turn implies (13.2). To prove (13.3) observe that

$$\begin{aligned} &\frac{A^{(N)}(Nt) - \lambda^{(N)}Nt}{N^H} \\ &= \frac{A^{(N)}(Nt) - \lambda^{(N)}T^{(N)}_{A^{(N)}(Nt)}}{N^H} + \frac{\lambda^{(N)}T^{(N)}_{A^{(N)}(Nt)} - \lambda^{(N)}Nt}{N^H}\,. \end{aligned} \tag{13.4}$$

We use assumption 2 to show that the second term tends to zero. Write

$$\begin{aligned} &\sup_{0\le t\le u} |T^{(N)}_{A^{(N)}(Nt)} - Nt| \\ &\le \sup_{0\le t\le u} |T^{(N)}_{A^{(N)}(Nt)+1} - T^{(N)}_{A^{(N)}(Nt)}| = \max_{0\le k\le A^{(N)}(Nu)} |T^{(N)}_{k+1} - T^{(N)}_{k}|\,, \end{aligned}$$

and use this to bound the probability that the supremum over $[0,u]$ of the second term of (13.4) be larger than some $\varepsilon > 0$ is below

$$P(\max_{0\le k\le[\lambda' Nu]} |T^{(N)}_{k+1} - T^{(N)}_{k}| > N^H\varepsilon) + P(A^{(N)}(Nu) > \lambda' Nu)\,,$$

(for $\lambda' > \lambda$). The latter term converges to zero by assumption 2 and (13.2), while for the former we have

$$\begin{aligned} &P(\max_{0\le k\le[\lambda' Nu]} |T^{(N)}_{k+1} - T^{(N)}_{k}| > N^H\varepsilon) \\ &\le \sum_{k=0}^{[\lambda' Nu]} P(|T^{(N)}_{k+1} - T^{(N)}_{k}| > N^H\varepsilon) \le \frac{KN}{N^{2H}} \end{aligned}$$

for some constant K, and this converges to zero since $H > 1/2$. (This seemingly trivial Boole's inequality accounts for sharp bounds in other places where fractional Brownian motion is involved; see [8] and [14].) To deal now with the first term of (13.4), observe that it is obtained by setting

$$s = \frac{A^{(N)}(Nt)}{N}$$

in

$$\frac{[Ns] - \lambda^{(N)} T^{(N)}_{[Ns]}}{N^H}.$$

Since $N^{-H}([Ns]-\lambda^{(N)}T^{(N)}_{[Ns]}) \Rightarrow \lambda W(s)$ (assumptions 1 & 3) and $A^{(N)}(Nt)/N \Rightarrow \lambda t$ (first part of this theorem) we have, by the continuity of the composition mapping, $N^{-H}(A^{(N)}(Nt) - \lambda^{(N)} T^{(N)}_{A^{(N)}(Nt)}) \Rightarrow \lambda W(\lambda t)$. □

A version of Theorem 13.2.1 holds for a collection of point processes $A_i^{(N)}$, $i = 1, \ldots, d$, with points $T^{(N)}_{n,i}$, such that the vector $\boldsymbol{T}^{(N)}_n = (T^{(N)}_{n,1}, \ldots T^{(N)}_{n,d})$ is in the domain of attraction of a d-dimensional fractional Brownian motion. We shall need the definition of the latter object in Section 13.3 as well.

Definition 13.2.2 A d-dimensional fractional Brownian motion FBM$(H, 0, R)$, where $0 < H < 1$ and $R = [r_{k\ell}]$ is a $d \times d$ positive semidefinite matrix, is a collection of zero-mean Gaussian random variables $\{Z_k(t); k = 1, \ldots, d; t \in \mathbf{R}\}$ with covariance function

$$EZ_k(t)Z_\ell(s) = \frac{r_{k\ell}}{2}(|t|^{2H} + |s|^{2H} - |t-s|^{2H}).$$

The positive-definiteness of R and the positive-definiteness of $|t|^{2H} + |s|^{2H} - |t-s|^{2H}$ (see Samorodnitsky and Taqqu [16]) imply the existence of the finite-dimensional distributions of the process. An argument based on Kolmogorov's inequality shows that there exists a version of the process with continuous sample paths. In matrix notation we use the column vector $\boldsymbol{Z} = (Z_1, \ldots, Z_d)'$ and write $E\boldsymbol{Z}(t)\boldsymbol{Z}(t)' = R|t|^{2H}$.

Consider now the collection $A_i^{(N)}$, $i = 1, \ldots, d$ with points $T^{(N)}_{n,i}$ and let $\boldsymbol{T}^{(N)}_n = (T^{(N)}_{n,1}, \ldots, T^{(N)}_{n,d})$, and $\boldsymbol{A}^{(N)}(t) = (A_1^{(N)}(t), \ldots, A_d^{(N)}(t))$. Assume that $\{T^{(N)}_{n,i} - T_{n-1,i}\}_{n=1}^{\infty}$, $i = 1, \ldots, d$, are jointly stationary. Let $\boldsymbol{\lambda}^{(N)} = (\lambda_1^{(N)}, \ldots, \lambda_d^{(N)})$ be the collection of their rates and set $\boldsymbol{\lambda}^{(N)-1} = (\lambda^{(N)}{}_1^{-1}, \ldots, \lambda^{(N)}{}_d^{-1})$.

Theorem 13.2.3 Suppose

1. $\boldsymbol{\lambda}^{(N)} \to \boldsymbol{\lambda}$,
2. $\sup_{N,i} \mathrm{var}(T^{(N)}_{n,i} - T^{(N)}_{n-1,i}) < \infty$,
3. $N^{-H}(\boldsymbol{T}^{(N)}_{[Nt]} - [Nt]\boldsymbol{\lambda}^{-1}) \Rightarrow \boldsymbol{W}(t) \sim$ FBM$(H, 0, R)$.

Then

$$N^{-H}(\boldsymbol{A}^{(N)}(Nt) - Nt\boldsymbol{\lambda}) \Rightarrow \boldsymbol{\lambda} \otimes \boldsymbol{W}(\boldsymbol{\lambda}t), \tag{13.5}$$

where diag($\boldsymbol{\lambda}$) denotes a diagonal matrix the entries of $\boldsymbol{\lambda}$ on the diagonal, and $\boldsymbol{\lambda} \otimes \boldsymbol{W}(\boldsymbol{\lambda}t) = (\lambda_1 W(\lambda_1 t), \ldots, \lambda_d W(\lambda_d t))$.

The proof of this theorem is similar to that of Theorem 13.2.1, so it is omitted. We only point out the fact that $\boldsymbol{\lambda} \otimes \boldsymbol{W}(\boldsymbol{\lambda}t)$ may fail to be a fractional Brownian motion if the components of $\boldsymbol{W}(t)$ are not independent.

13.3 Approximations for Queueing Networks

A single-class queueing network with customers associated with the servers is rigorously defined by Baccelli and Foss [2]. For our purposes, we briefly describe its functioning verbally and then proceed via balance equations for the queue length. The network consists of a collection J of d nodes. To each node i customers arrive at times $T_{1,i} < T_{2,i} < \cdots$. These times comprise the point process A_i. We shall be working with the associated counting process $A_i(t) = \sum_{n=1}^{\infty} \mathcal{I}(T_{n,i} \leq t)$. At each node there is a single buffer where customers accumulate, and a server who depletes them by picking a customer from the buffer and servicing her for an amount of time $\sigma_{n,i}$ if she is the server's n-th customer, $n = 1, 2, \ldots$. The server never idles if there are customers in the buffer. Thus, service times are associated with the servers, and customers are treated rather democratically, as no class is associated to them (single-class network). Let $R_{n,i} = \sum_{m=1}^{n} \sigma_{m,i}$ and let S_i be the point process with points $R_{1,i} < R_{2,i} < \cdots$. The associated counting process is $S_i(t) = \sum_{n=1}^{\infty} \mathcal{I}(R_{n,i} \leq t)$. When the server of node k finishes his n-th service he sends his customer to another node, chosen according to some rule, or to the outside world. We are interested in limit theorems for the population vector $\boldsymbol{Q}(t) = (Q_1(t), \ldots, Q_d(t))$. We denote by $\varphi_k(n)$ the change in the population due to the completion of the n-th service at node k. Let $\boldsymbol{\Phi}_k(n) = \sum_{m=1}^{n} \varphi_k(m)$ be the total change in the population due to the first n services of server k, provided that they were the only events in the system. The balance equations for the system then read

$$\begin{aligned}
\boldsymbol{Q}(t) &= \boldsymbol{Q}(0) + \boldsymbol{A}(t) + \sum_{k \in J} \boldsymbol{\Phi}_k \circ S_k \circ B_k(t), \qquad (13.6)\\
B_k(t) &= \int_0^t \mathcal{I}(Q_k(s) > 0)\,ds.
\end{aligned}$$

These equations are due to Harrison (see Reiman [15]) and can best be explained by thinking of local clocks associated with the server: a continuous one ($B_k(t)$) that runs at speed one while the corresponding queue length is positive but freezes otherwise, and a discrete one ($S_k(t)$) that ticks at every departure from node k.

We assume now that the arrival and service processes associated with different nodes are independent. (This is done basically for ease of exposition–but will be dropped later.) Also assume that arrivals are independent of services per node. Furthermore, suppose long-range dependence in the sequence of inter-arrival times $\{T_{n,i} - T_{n-1,i}\}_{n=1}^{\infty}$ of node i, as well as long-range dependence in the sequence of service times $\{\sigma_{n,i}\}_{n=1}^{\infty}$. In a high-speed network we see no reason for the latter hypothesis, but it does not hurt in our theoretical analysis. Assume that the routing decisions are short-range dependent. In particular, assume that the routing is completely Markovian and that a customer completing service in node k is sent to node ℓ with probability $p_{k,\ell}$ or to the outside world with probability $p_{k,0} = 1 - \sum_{\ell \in J} p_{k,\ell}$. (One should probably notice at this point that under this assumption and under the assumption that the processes $\{A_i, S_i; i = 1, \ldots, d\}$ are independent Poisson processes then (13.6) is precisely a Jackson network, as explained in Walrand [18].) In other words, the sequence $\{\varphi_k(n)\}_{n=1}^{\infty}$ is a sequence of i.i.d. d-vectors such that

$$P(\varphi_k(1) = -e_k + e_\ell) = p_{k,\ell}\,,\ \ell \in J^0 = J \cup \{0\}\,,$$

where e_k ($k \in J$) is the unit d-vector with zeros everywhere but the k-th entry, which is equal to 1; for $k = 0$, $e_0 = \mathbf{0}$. Thus, for each $k \in J$, $\{\Phi_k(n)\}_{n=1}^{\infty}$ is a random walk in $\mathbf{R}^d$ with mean drift

$$\delta_k = E\varphi_k(1) = \sum_{j \in J^0} (-e_k + e_j)p_{k,j} = -e_k + \sum_{j \in J} e_j p_{k,j} = (P' - I)e_k\,,$$

where P' is the transpose of the routing matrix $P = [p_{k,\ell}]_{k,\ell \in J}$ and I is the $d \times d$ identity matrix. Denote by C_k the covariance matrix $C_k = E(\varphi_k(1) - E\varphi_k(1)) \cdot ((\varphi_k(1) - E\varphi_k(1))'$ and recall the standard FCLT

$$\frac{1}{\sqrt{N}}(\Phi_k([Nt]) - [Nt]\delta_k) \Rightarrow C_k^{1/2}\xi_k(t) \tag{13.7}$$

where ξ_k is a standard Brownian motion in $\mathbf{R}^d$. The Brownian motions $\xi_1, \ldots, \xi_d$ are independent.

We assume further that the network is an open irreducible network, i.e. that every arriving customer can eventually exit the network, which implies that the spectral radius of P is strictly smaller than one.

Now assume that we have a sequence of networks parameterized by N, each defined on a possibly different probability space. The N-th system is completely specified by the arrival times $T_{n,i}^{(N)}$, service times $\sigma_{n,i}^{(N)}$, and routing vectors $\varphi_{n,i}$ *taken independent of* N, so that the balance equations

$$\begin{aligned} Q^{(N)}(t) &= Q^{(N)}(0) + A^{(N)}(t) + \sum_{k \in J} \Phi_k \circ S_k^{(N)} \circ B_k^{(N)}(t)\,, \qquad (13.8) \\ B_k^{(N)}(t) &= \int_0^t \mathcal{I}(Q_k^{(N)}(s) > 0)ds\,, \end{aligned}$$

introduced in (13.6), hold for each N.

In a manner similar to assumption 3 of Theorem 13.2.1, long-range dependence for the inter-arrival times is manifested directly by its consequence. Namely, the sequence $\{T^{(N)}_{n,i} - T^{(N)}_{n-1,i}\}_{n=1}^{\infty}$ is stationary and

$$N^{-H}(T^{(N)}_{[Nt],i} - [Nt]/\lambda^{(N)}_i) \Rightarrow w_i(t) \sim \mathrm{FBM}(H, 0, \alpha_i),$$

where $\lambda^{(N)}_i = 1/E(T^{(N)}_{n,i} - T^{(N)}_{n-1,i})$. Similarly, assume that the service times $\{\sigma^{(N)}_{n,i}\}_{n=1}^{\infty}$ form a stationary long-range dependent sequence:

$$N^{-H}(R^{(N)}_{[Nt],i} - [Nt]/\mu^{(N)}_i) \Rightarrow v_i(t) \sim \mathrm{FBM}(H, 0, \beta_i),$$

where $\mu^{(N)}_i = 1/E\sigma^{(N)}_{n,i}$, and $R^{(N)}_{n,i} = \sum_{m=0}^{n} \sigma^{(N)}_{m,i}$. Notice that we assume the same long-range dependence index $H > 1/2$ for each node i, both for service and inter-arrival times. The assumption can (and will) be relaxed.

Our basic theorem is stated as follows.

Theorem 13.3.1 With the above notation and assumptions, and under the additional assumptions

1. $\lambda^{(N)}_i \to \lambda_i \in (0, \infty)$,
2. $\mu^{(N)}_i \to \mu_i \in (0, \infty)$,
3. $N^{1-H}(\boldsymbol{\lambda}^{(N)} + (P' - I)\boldsymbol{\mu}^{(N)}) \to \boldsymbol{c} \in \mathbf{R}^d$,
4. $N^{-H}\boldsymbol{Q}^N(0) \Rightarrow \boldsymbol{q}(0) \in \mathbf{R}^d_+$,

the sequence $\boldsymbol{Q}^{(N)}(t)$ of population vectors, defined as in (13.6), satisfies

$$N^{-H}\boldsymbol{Q}^{(N)}(Nt) \Rightarrow \boldsymbol{q}(t) = \boldsymbol{q}(0) + \boldsymbol{c}t + \boldsymbol{x}(t) + \sum_{k \in J}(-\boldsymbol{\delta}_k)\ell_k(t), \qquad (13.9)$$

where $\boldsymbol{x}$ is a d-dimensional fractional Brownian motion,

$$\boldsymbol{x} \sim \mathrm{FBM}(H, 0, \mathrm{diag}(\boldsymbol{\lambda})^{2+2H} \cdot \mathrm{diag}(\boldsymbol{\alpha}) + (P' - I)\mathrm{diag}(\boldsymbol{\mu})^{2+2H} \cdot \mathrm{diag}(\boldsymbol{\beta})(P - I)), \qquad (13.10)$$

and $\ell_k(t)$ is a reflector in the k-th direction.

Before proving the theorem we discuss the premises and conclusions. Assumption 3 is a "heavy traffic" assumption, implying that the "interior drift" $\boldsymbol{\lambda}^{(N)} + (P' - I)\boldsymbol{\mu}^{(N)}$ converges to the zero vector (at a certain speed). This is equivalent to the condition that the "traffic intensities" converge to zero. The assumption is *not* essential and can be dropped. However, in this case, the theorem has to be interpreted as quantifying the deviations from the mean trajectory, and we shall not discuss this further here.

Given a process $\boldsymbol{x} \in D^d[0,\infty)$ and d vectors $\boldsymbol{\delta}_1, \ldots, \boldsymbol{\delta}_d$ we define the reflector $\ell_k(t)$ in the k-th direction as an increasing real-valued function with $\ell_k(0) = 0$ satisfying

$$\boldsymbol{q}(t) := \boldsymbol{x}(t) - \sum_{k=1}^{d} \boldsymbol{\delta}_k \ell_k(t) \geq 0\,, \text{ for all } t\,,$$

$$\text{and} \int_0^\infty \mathcal{I}(q_k(t) > 0)\ell_k(dt) = 0\,, \text{ for all } k\,.$$

The matrix R with columns the vectors $-\boldsymbol{\delta}_k$ is called reflection matrix. It is known that there is a *unique* reflector vector $\boldsymbol{\ell}(t) = (\ell_1(t), \ldots, \ell_d(t))'$ satisfying the above conditions if the matrix $I - R$ is non-negative and contractive (see Harrison and Reiman [7]). In our case, $R = I - P'$, so the condition is satisfied. We define the operator $\mathcal{L}_R$ on $D^d[0,\infty)$ by $\mathcal{L}_R(\boldsymbol{x})(\cdot) = -\sum_{k=1}^{d} \boldsymbol{\delta}_k \ell_k(\cdot)$ for the weighted reflector obtained starting from the process $\boldsymbol{x}$. It is known that the operator $\mathcal{L}_R$ is continuous in the uniform topology.

Proof of Theorem 13.3.1. We first show that the fact that the interior drift $\lambda^{(N)} + (P' - I)\mu^{(N)} \to 0$ implies that

$$\frac{B_k^{(N)}(Nt)}{N} \Rightarrow t\,, \text{ for all } k \in J\,.$$

Set $\kappa_N = N$ and write (13.8) as follows:

$$\begin{aligned}
\kappa_N^{-1} \boldsymbol{Q}^{(N)}(Nt) \;=\; & \kappa_N^{-1} \boldsymbol{Q}^{(N)}(0) \\
& + \kappa_N^{-1} [\boldsymbol{A}^{(N)}(Nt) - \boldsymbol{\lambda}^{(N)} Nt] \\
& + \kappa_N^{-1} \sum_{k \in J} [\boldsymbol{\Phi}_k \circ S_k^{(N)} \circ B_k^{(N)}(Nt) - \boldsymbol{\delta}_k S_k^{(N)} \circ B_k^{(N)}(Nt)] \\
& + \kappa_N^{-1} [\lambda^{(N)} Nt + \sum_{k \in J} \mu_k^{(N)} \boldsymbol{\delta}_k Nt] \\
& + \kappa_N^{-1} \sum_{k \in J} [\boldsymbol{\delta}_k S_k^{(N)} \circ B_k^{(N)}(Nt) - \mu_k^{(N)} \boldsymbol{\delta}_k B_k^{(N)}(Nt)] \\
& - \kappa_N^{-1} \sum_{k \in J} \boldsymbol{\delta}_k \mu_k^{(N)} [Nt - B_k^{(N)}(Nt)]\,.
\end{aligned} \tag{13.11}$$

The first term, $N^{-1}\boldsymbol{Q}^{(N)}(0)$ converges to zero, by assumption 4. From Theorem 13.2.1, and more specifically from (13.2), we have

$$N^{-1}[\boldsymbol{A}^{(N)}(Nt) - \boldsymbol{\lambda}^{(N)} Nt] \Rightarrow \mathbf{0}\,.$$

and, similarly,

$$N^{-1}[S_k^{(N)} \circ B_k^{(N)}(Nt) - \mu_k^{(N)} B_k^{(N)}(Nt)] \Rightarrow 0\,,$$

which follows from $N^{-1}[S_k^{(N)}(Nt) - \mu_k^{(N)}Nt] \Rightarrow 0$ and the obvious fact that

$$B_k^{(N)}(Nt) \leq Nt\,.$$

Also, from the FLLN (Functional Law of Large Numbers) for the random walk $\boldsymbol{\Phi}_k$,

$$N^{-1}\left[\boldsymbol{\Phi}_k \circ S_k^{(N)} \circ B_k^{(N)}(Nt) - \boldsymbol{\delta}_k S_k^{(N)} \circ B_k^{(N)}(Nt)\right] \Rightarrow 0\,.$$

Finally, the non-random term of (13.11) is just the interior drift which converges to zero by assumption. We conclude that all but the last term in (13.11) converge to zero. Recognizing the last term as obtained by applying the previously defined operator $\mathcal{L}_R$ on the first 5 terms, and using the continuity of $\mathcal{L}_R$, we obtain $B_k{}^{(N)}(Nt)/N \Rightarrow t$, as desired.

Set now $\kappa_N = N^H$ in (13.11). The first term in (13.11) converges to $\boldsymbol{q}(0)$, by assumption 4. From (13.3) of Theorem 13.2.1 (or, rather, from (13.5) of Theorem 13.2.3) we have

$$N^{-H}[\boldsymbol{A}^{(N)}(Nt) - \boldsymbol{\lambda}^{(N)}Nt] \Rightarrow \boldsymbol{\lambda} \otimes \boldsymbol{w}(\boldsymbol{\lambda}t)\,.$$

The fifth term in (13.11) is obtained by setting

$$s = \frac{B_k^{(N)}(Nt)}{N}$$

in

$$N^{-H}[S_k^{(N)}(Ns) - \mu_k^{(N)}Ns]$$

which, again from Theorem 13.2.1, converges to $\mu_k v_k(\mu_k t)$. Using the continuity of the composition operation and the previously established fact $B_k^{(N)}(Nt)/N \Rightarrow t$, we conclude that

$$N^{-H}[S_k^{(N)} \circ B_k^{(N)}(Nt) - \mu_k^{(N)} \circ B_k^{(N)}(Nt)] \Rightarrow \mu_k v_k(\mu_k t)\,.$$

The third term of (13.11) equals

$$N^{-H+1/2}N^{-1/2}[\boldsymbol{\Phi}_k \circ S_k^{(N)} \circ B_k^{(N)}(Nt) - \boldsymbol{\delta}_k S_k^{(N)} \circ B_k^{(N)}(Nt)]\,.$$

Note that $N^{-H+1/2} \to 0$, since $H > 1/2$. Also,

$$N^{-1/2}[\boldsymbol{\Phi}_k \circ S_k^{(N)} \circ B_k^{(N)}(Nt) - \boldsymbol{\delta}_k S_k^{(N)} \circ B_k^{(N)}(Nt)] \Rightarrow C_k^{1/2}\boldsymbol{\xi}_k(\mu_k t).$$

This follows from the standard FCLT (13.7), the FLLN $S_k^{(N)}(Nt)/N \Rightarrow \mu_k t$, the fact $B_k^{(N)}(Nt)/N \Rightarrow t$, and the continuity of the composition map (twice). Hence the third term of (13.11) converges to zero. The non-random term of (13.11) is

$$N^{1-H}(\lambda^{(N)} + (P' - I)\mu^{(N)})t\,,$$

a multiple of the interior drift, which converges to ct by assumption. Hence the sum of all but the last term of (13.11) converges to

$$\begin{aligned} & q(0) + ct + \sum_{k \in J} \lambda_k w_k(\lambda_k t) + \sum_{k \in J} \delta_k \mu_k v_k(\mu_k t) \\ = \quad & q(0) + ct + \boldsymbol{\lambda} \otimes \boldsymbol{w}(\boldsymbol{\lambda} t) + (P' - I) \cdot \boldsymbol{\mu} \otimes \boldsymbol{v}(\boldsymbol{\mu} t) \\ =: \quad & q(0) + ct + \boldsymbol{x}(t), \end{aligned} \tag{13.12}$$

and $\boldsymbol{x}$ is a fractional Brownian motion as desired in (13.10). The last term in (13.11) is obtained by applying $\mathcal{L}_R$ on the sum of the remaining terms. The continuity of the operator $\mathcal{L}_R$ thus shows the validity of (13.9) and concludes the proof of the theorem. □

13.4 Discussion

13.4.1 Survival of the Fittest

Suppose that some of the processes have different Hurst parameters. Then only the ones with the largest H survive in the limit. We first present an example of this situation.

Example: Correlated splitting of a long-range dependent point process. Let $\{T_n^{(N)}\}_{n=1}^{\infty}$ be points of a point process with rate $\lambda^{(N)}$. Assume the FCLT

$$N^{-H}[T_{[Nt]}^{(N)} - \lambda^{(N)^{-1}} Nt] \Rightarrow w(t) \sim \mathrm{FBM}(H, 0, \alpha) \text{ as } N \to \infty .$$

For each N let $\{\varphi^{(N)}(n)\}_{n=1}^{\infty}$ be stationary random vectors in $\mathbf{R}^2$, independent of $\{T_n^{(N)}\}_{n=1}^{\infty}$, with marginal distribution

$$P(\varphi^{(N)}(1) = e_k) = p_k \, , k = 1, 2,$$

with $0 \leq p_1 + p_2 = 1$. Assume a FCLT of the form

$$N^{-H'} \sum_{m=0}^{[Nt]} [\varphi^{(N)}(m) - \delta] \Rightarrow \boldsymbol{v}(t) \sim \mathrm{FBM}(H', 0, R) \, ,$$

where $\boldsymbol{\delta} = p_1 e_1 + p_2 e_2$. The Hurst parameters H, H' of the arrival and routing process, respectively, are not, in general, equal. Set $A^{(N)}(t) = \sum_{n=1}^{\infty} \mathcal{I}(T_n^{(N)} \leq t)$, $\Phi^{(N)}(n) = \sum_{m=1}^{n} \varphi^{(N)}(m)$ and define

$$\boldsymbol{A}^{(N)}(t) = \boldsymbol{\Phi}^{(N)}(A^{(N)}(t))$$

to be the vector containing the two splitted point processes, $A_1^{(N)}(t), A_2^{(N)}(t)$. To obtain the FCLT for the 2-dimensional process $\boldsymbol{A}^{(N)}$ we decompose, following the method used in all the previous theorems of this paper:

$$\begin{aligned} & \boldsymbol{\Phi}^{(N)} \circ A^{(N)}(Nt) - \boldsymbol{\delta} \lambda^{(N)} Nt \\ = \quad & [\boldsymbol{\Phi}^{(N)} \circ A^{(N)}(Nt) - \boldsymbol{\delta} A^{(N)}(Nt)] + [\boldsymbol{\delta} A^{(N)}(Nt) - \boldsymbol{\delta} \lambda^{(N)} Nt]. \end{aligned}$$

Assume also that the variances of all random variables involved are bounded by some finite constant. There are three cases. The limit in each case of $N^{-\max(H,H')}[\Phi^{(N)} \circ A^{(N)}(Nt) - \delta\lambda^{(N)}Nt]$ is written below:

Case 1, $H = H'$: $v(\lambda t) + \delta\lambda w(\lambda t) \sim \text{FBM}(H, 0, \lambda^{2H}R + \delta\delta'\lambda^{2+2H}\alpha)$.

Case 2: $H > H'$: $\delta\lambda w(\lambda t) \sim \text{FBM}(H, 0, \delta\delta'\lambda^{2+2H}\alpha)$.

Case 3: $H' > H$: $v(\lambda t) \sim \text{FBM}(H', 0, \lambda^{2H}R)$.

The network case:
Consider a queueing network with the assumptions of Theorem 13.3.1, i.e. short-range dependent routing and long-range dependent arrival and service processes, independent from node to node, with possibly different Hurst parameters; let H be the largest among them. Following the logic of the previous example we then see that the limiting fractional Brownian motion is obtained by ignoring those processes with Hurst parameter strictly less than H. The covariance matrix is then obtained by replacing in (13.10) by zeros those λ_i's or μ_i's that correspond to processes with Hurst parameter strictly smaller than H. Thus, the corresponding "second order network" may have considerably simplified form. The question then that arises regards the usefulness (and meaning!) of the approximations in such cases. This needs further study and examples arising from real networking systems.

13.4.2 Simulations

A way to generate correlated variables so that they approximate fractional Brownian motion, for purposes of simulating a queue or a network, is to consider a fractional ARIMA model of the form

$$(1 - B)^d \varphi(B)\zeta_n = \theta(B)\xi_n \,, \tag{13.13}$$

where B denotes backward differencing ($Bx_n := x_{n-1}$), $\varphi(B), \theta(B)$ are polynomials in B and d is a real number in the interval $[0, 1/2)$. The term $(1-B)^d$ is the "fractional differencing term" (a terminology stemming from the fact that $(1 - B)^d$ is ordinary d-th order difference if d is integer) and is defined through its formal Taylor expansion. The variables ξ_n are i.i.d., zero mean, with finite variance but otherwise arbitrarily distributed. The polynomials $\varphi(B), \theta(B)$ are assumed to have zeros outside the closed unit disk, an assumption which makes the model (13.13) causal. It is shown in [8] that

$$N^{-d-1/2} \sum_{k=1}^{[Nt]} \zeta_k \Rightarrow \text{FBM}(d + 1/2, 0, L_d)$$

with

$$L_d = \left| \frac{\theta(1)}{\varphi(1)} \right|^2 \frac{\sin(\pi d)}{\pi d} \frac{\Gamma(1 - 2d)}{1 + 2d} .$$

Thus the Hurst parameter is $H = d + 1/2 \in [1/2, 1)$. The case $d = 0$ corresponds to ordinary Brownian motion with $L_0 = \left|\frac{\theta(1)}{\varphi(1)}\right|^2$. Admittedly, the simulations are rather slow, but they are very similar in spirit to the ones suggested by Samorodnitsky and Taqqu [16, Ch. 7]. Efficient methods for simulating approximately long-range dependent variables are important at this point as hardly any analytical tools for performance analysis are available.

13.4.3 Steady-State

Going back to the case $d = 1$, i.e. a single server queue, our results imply that the queue length is approximated by the one-dimensional reflected fractional Brownian motion

$$q(t) = q(0) + ct + x(t) + \ell(t),$$

where $x(t) \sim \mathrm{FBM}(H, 0, \lambda^{2+2H}\alpha + \mu^{2+2H}\beta)$, with λ, μ being the limiting arrival and service rates, respectively, $c = \lim_{N\to\infty} N^{1-H}(\lambda^{(N)} - \mu^{(N)})$, and $\alpha = \lim_{N\to\infty} \mathrm{var}(N^{-H}T_N^{(N)})$, $\beta = \lim_{N\to\infty} \mathrm{var}(N^{-H}\sum_{m=1}^{N} \sigma_m^{(N)})$. The reflector $\ell(t)$ is given by $\ell(t) = -\inf_{0\le s\le t}(q(0) + cs + x(s)) \wedge 0$. Suppose now that $c < 0$. The choice of the initial queue length as a random variable defined on the probability space supporting x (extended on the time domain $-\infty < t < \infty$) so that the resulting q is *stationary* can be done in a unique way, in the spirit of Loynes [13], since the zero-mean fractional Brownian motion process x has *stationary increments* and is *ergodic*. For the explicit construction the reader is referred to [9]. The same result is of course available for the original pre-limit discrete-event system, precisely as in Loynes [13].

In a recent paper, Norros [14] provided the following estimate for the tail of the marginal of $q(t)$ for the afore-mentioned stationary model:

$$P(q(t) > x) \approx \exp(-\gamma x^{2-2H}) \tag{13.14}$$

for some $\gamma > 0$. The approximation has been justified as exact (in a large-deviations sense) by Duffield and O'Connell [6]. A similar estimate for another system was given by Brichet et al. [4]. Note that for $H = 1/2$ the above yields $P(q(t) > x) \approx \exp(-\gamma x)$, the usual exponential distribution which is exact for the Brownian diffusion model and approximate for models falling in the domain of attraction of a Brownian motion. By analogy, the tail behavior (13.14) should be valid for a large class of long-range dependent queueing systems. But not for everything: for instance, the tail behavior reported by Likhanov et al. [12] and Anantharam [1] is very different from (13.14). We conjecture that this is the case because their model does not fall in the domain of attraction of a fractional Brownian motion but, perhaps, it falls in the domain of attraction of a stable non-Gaussian self-similar process [16].

While it is (after decades of research) straightforward to talk about the steady-state of a one-dimensional system, it is not so for a multi-dimensional one. Only recently, in a fundamental paper by Baccelli and Foss [2], has a complete proof for the stationarity problem of the discrete-event system whose queue-length vector is described by (13.6) been provided. The reader should notice that while the limiting equation (13.9) forms a dynamical system in $\mathbf{R}_+^d$, the pre-limiting balance equation (13.6) does not. Baccelli and Foss [2] use a monotonicity property (see Shanthikumar and Yao [17]) to give a proof for the stationarity of (13.6). It would be interesting to obtain tail estimates for the limiting population vector $\boldsymbol{q}(t)$ of (13.9) as well. But one should be careful in choosing an (the) appropriate stationary version of (13.9).

13.5 Conclusions

We considered approximations to the population vector of a single-class queueing network with long-range dependent driving processes, and more specifically long-range dependent sequences that fall in the domain of attraction of a fractional Brownian motion. (Other limiting regimes are possible and the heavy traffic assumption may be dropped.) For purposes of illustration and applicability we chose short-range dependent routing, based on the hypothesis that switching cannot by itself introduce long-range dependence. Indeed, detailed experiments with real networking data indicate that the phenomenon occurs at the sources. The limiting scaled population vector was shown to be a fractional Brownian motion, appropriately reflected so that it stays non-negative. Delays and sojourn times of individual sessions can be approximated in a similar manner. Simulations of such systems are important for there are hardly any tools for their analysis. We indicated a method for doing so. Various Hurst parameters may be possible for various sources, especially when they are located at geographically distant places and are due to completely different operating environments. Our limiting regime preserves only those processes with the largest degree of long-range dependence. This needs further investigation. Steady-state performance analysis is possible only approximately and only in one dimension. Is it possible to extend those to the network case? We do not know yet. The functional approximations considered in this paper can hopefully be extended to cover multi-class systems and/or systems with more complicated service disciplines that depend on the buffers or other state variables of the network. Issues of control and optimization can, just as in a Markovian system, be raised and should be addressed from rather first principles.

Acknowledgments: This work was supported in part by NSF Research Initiation Award NCR 92-11343, by NSF Faculty Career Development Award NCR 95-02582, and by Grant ARP 224 of the Texas Higher Education Coordinating Board. We would like to thank Spyros Papadakis for a careful reading of the manuscript as well as his comments. The first author acknowledges support by the Royal Statistical Society during a stimulating Workshop on Stochastic Networks, Edinburgh, 1-11 August, 1995.

13.6 References

[1] Anantharam, V. (1995). On the sojourn time of sessions at an ATM buffer with long-range dependent input traffic. *Proc. IEEE CDC.*

[2] Baccelli, F. and Foss, S. (1993). Stability of Jackson-Type Queueing Networks, I. *INRIA Rapport de Recherche* No. 1945. Also in *QUESTA*, Vol. **17**, No. I-II.

[3] Billingsley, P. (1968). *Convergence of Probability Measures.* Wiley, New York.

[4] Brichet, F., Roberts, J., Simonian, A. and Veitch, D. (1995). Heavy traffic analysis of a storage model with long-range dependent on/off sources. *Preprint.*

[5] Cox, D.R. (1983). Long-range dependence: a review. In: *Statistics: An Appraisal*, Proc. 50th Anniversary Conference, Iowa State Statistical Laboratory.

[6] Duffield, N.G. and O'Connell, N. (1993). Large deviations and overflow probabilities for the general single-server queue, with applications. To appear in *Proc. Cambridge Phil. Soc.*

[7] Harrison, J.M. and Reiman, M.I. (1981). Reflected Brownian motion on an orthant. *Ann. Prob.*, **9**, 302-308.

[8] Konstantopoulos, T. and Lin, S.J. (1995). Approximating fractional Brownian motions. *Techn. Rep. ECE, U.T. Austin.*

[9] Konstantopoulos, T., Zazanis, M. and de Veciana, G. (1995). Conservation laws and reflection mappings with an application to multiclass mean value analysis for stochastic fluid queues. To appear in: *Stoch. Proc. Applic.*

[10] Kurtz, T. (1995). Talk delivered in the Royal Statistical Society *Workshop on Stochastic Networks*, Heriot-Watt University, Edinburgh.

[11] Leland, W.E., Taqqu, M.S., Willinger, W. and Wilson, D.V. (1994). On the self-similar nature of Ethernet traffic. *IEEE/ACM Trans. Netw.*, **2**, 1-15.

[12] Likhanov, N., Tsybakov, B. and Georganas, N.D. (1995). Analysis of an ATM buffer with self-similar ("fractal") input traffic. *Proc. 14th IEEE Infocom*, 985-992.

[13] Loynes, R. (1962). The stability of a queue with non-independent inter-arrival and service times. *Proc. Cambridge Phil. Soc.*, **58**, 497-520.

[14] Norros, I. (1994). A storage model with self-similar input. *Queueing Systems*, **16**, 387-396.

[15] Reiman, M.I. (1984). Open queueing networks in heavy traffic. *Math. O.R.*, **9**, 441-458.

[16] Samorodnitsky, G. and Taqqu, M. (1994). *Stable Non-Gaussian Random Processes.* Chapman & Hall, New York.

[17] Shanthikumar, J.G. and Yao, D.D. (1989). Stochastic monotonicity in general queueing networks. *J. Appl. Prob.*, **26**, 413-417.

[18] Walrand, J. (1988). *An Introduction to Queueing Networks.* Prentice-Hall, Englewood Cliffs.

[19] Willinger, W., Taqqu, M.S., Sherman, R. and Wilson, D.V. (1995). Self-similarity through high-variability: statistical analysis of Ethernet LAN traffic at the source level. *Proc. ACM/Sigcomm '95.*

14 Moderate Deviations for Queues with Long-Range Dependent Input

Cheng-Shang Chang, David D. Yao and Tim Zajic

ABSTRACT Based on the integral relationship between Brownian motion and fractional Brownian motion, we model a process with long-range dependence, Y, as the output of a time-invariant linear filter applied to a more familiar process X — one that does not have long-range dependence. We develop a general criterion for a broad class of X to satisfy a moderate deviations principle (MDP). Based on the MDP of X, we then establish the MDP for Y. Using this framework, we develop both transient and steady-state results, in terms of the MDP, regarding the asymptotic behavior of queues fed with the long-range dependent input process Y. We study two types of filter models, with bounded and unbounded time intervals, respectively.

14.1 Introduction

Recently, it has been demonstrated that queues with long-range dependent input have subexponential tails (Norros [20], and Duffield and O'Connell [11]). This phenomenon is qualitatively different from the exponential tail distribution in queues with renewal or Markov arrival processes — input that has short-range dependence. The exponential tail behavior is usually established via large deviations techniques; refer to, e.g., Chang [5], Glynn and Whitt [12, 13], Whitt [23], and the references there. (Refer to, e.g., Cox [8] and Beran [2], for preliminaries regarding processes with long-range dependence, as well as short-range dependence. Refer to Leland *et al* [16] and Willinger [24], among others, for studies of data from telecommunications traffic that exhibit long-range dependence, typically accompanied by self-similarity.)

It will be insightful to briefly review here the approach in Duffield and O'Connell [11]. Assuming that the "netput" — input minus (potential) output — satisfies a *large deviations principle* (LDP) with a speed that is slower than n or, more precisely, a *moderate deviations principle* (MDP) (refer to Section 14.2), Duffield and O'Connell showed that the maximum of the corresponding random walk also satisfies the MDP; in particular, when the netput is a fractional Brownian motion (FBM) with Hurst parameter

$H \in (\frac{1}{2}, 1)$, the MDP holds with speed $n^{2(1-H)}$. (See Section 14.5 for more details.) Hence, with the FBM as the netput, the steady-state queueing distribution has a subexponential tail; i.e., the exponential decay rate is $n^{2(1-H)}$ rather than n.

It is known that the FBM is a *self-similar* process, which exhibits long-range dependence (e.g., Mandelbrot and van Ness [18]). Recall that a stochastic process $\{X(t)\}$ is termed self-similar if the increment process, $\{X(t+\tau) - X(t)\}$, is equal in distribution to $\{a^{-H}[X(t+a\tau) - X(t)]\}$, for any t and any $a > 0$, where $H \geq 0$ is the Hurst parameter. A well-know example of a self-similar process, FBM, denoted B_H, admits the representation ([18]):

$$B_H(t) = b_0 + \frac{1}{\Gamma(H+\frac{1}{2})} \cdot \{ \int_0^t (t-s)^{H-1/2} dB(s) + \int_{-\infty}^0 [(t-s)^{H-1/2} - (-s)^{H-1/2}] dB(s) \}, \quad (14.1)$$

where $b_0 = B_H(0)$ is the initial state, $B(t)$ denotes the standard Brownian motion, and $0 < H < 1$ is the Hurst parameter. Also following [18], there is another FBM model, due to Lévy ([17]):

$$B_H(t) = \frac{1}{\Gamma(H+\frac{1}{2})} \cdot \int_0^t (t-s)^{H-1/2} dB(s). \quad (14.2)$$

Note that when $H = 1/2$, the FBM is reduced to the standard Brownian motion. On the other hand, when $H \in (\frac{1}{2}, 1)$, the FBM is known to have long-range dependence (refer to [8] and [18]). Also note from the above representations, (14.2) in particular, that the FBM can be viewed as the output of a time-invariant linear filter with the impulse response $h(t) = t^{H-1/2}$ subject to an input that is Brownian motion. (Similar types of linear filter are widely used in modeling stochastic systems; see, e.g., Wong and Hajek [25].)

In view of the above, our idea here is to use the type of filters in (14.1) and (14.2) as basic models to generate processes with long-range dependence. Specifically, with $H \in (\frac{1}{2}, 1)$, we will replace B above by a stochastic process with short-range dependence, denoted X, and denote the resulting output of the filter as Y (which replaces B_H above). The output process Y will then possess long-range dependence. (This can be checked by making use of the autocovariance function; refer to Section 14.3.) Moreover, a moderate excursion of X incurs a large excursion of Y through the amplifying effect of the integral. This way, we will have a systematic way to generate a broad class of processes with long-range dependence.

We will show that when X satisfies a MDP, so does Y; and that a wide class of processes qualify as candidates for X (i.e., satisfying a MDP). Examples include sample path processes corresponding to Markov chains,

m-dependent processes, and moving average processes. Indeed, we give a general criterion by which the MDP of X, when X belongs to a certain class of processes, may be verified.

Next, we study queues, via conditional limit theorems, that are fed with the long-range dependent input process Y. We characterize the transient queueing behavior; in particular, how large queues build up. In contrast to the large queue behavior in the case of queues with short-range dependent input (refer to the conditional limit theorems in Anantharam [1], Dembo and Zajic [9] and Chang [4]), the most likely path to a large queue build-up here is *nonlinear.* Finally, we discuss steady-state results, in terms of the maximum of the associated random walk, and compare them with the results of [11].

We now briefly comment on the individual merits of the two models based on (14.1) and (14.2). The principal advantage of a filter based on (14.2), referred to below as the first model, is the bounded interval $[0, t]$, which leads to a direct application of moderate deviations techniques. The disadvantage is that it does not preserve the stationary increment property, i.e., when X has stationary increments, Y need not have stationary increments. On the other hand, a filter based on (14.1), the second model, does preserve stationary increments, but the unbounded interval $(-\infty, t]$ requires necessary extensions of the MDP theory.

The rest of the article is organized as follows. In Section 14.2 we develop a general criterion for X to satisfy a MDP, and then derive the MDP of Y based on the MDP of X in Section 14.3. Queueing applications are presented in Section 14.4 and Section 14.5 for transient and steady-state results, respectively. In each of these sections, we start with and focus on the first model, and then present the analogous results for the second model.

This article is intended as a highlight of our recent work in [6, 7]; as such, the emphasis is on conveying the key ideas and presenting the main results. Results of a secondary or intermediate nature and their proofs are omitted. The reader is advised to consult [6, 7] for full details.

14.2 Moderate Deviations: A General Criterion

First recall the definition of the large deviations principle (LDP) on a metric space $\mathcal{S}$ equipped with the topology generated by the metric and the corresponding Borel σ-field. (Dembo and Zeitouni [10] should be consulted for more details.)

Definition 14.2.1 A sequence of probability measures $\mu^{(n)}$ on a metric space $\mathcal{S}$ satisfies the LDP with speed $\delta_n \uparrow \infty$ and a good rate function $I(\cdot) : \mathcal{S} \to [0, \infty]$ if

(i) the level sets $\{x \in \mathcal{S} : I(x) \leq \alpha\}$, $\alpha \in \mathbb{R}$, are compact,

(ii) for any closed set F,

$$\overline{\lim_{n\to\infty}} \frac{1}{\delta_n} \log \mu^{(n)}(F) \leq -\inf_{x\in F} I(x),$$

(iii) for any open set G,

$$\underline{\lim_{n\to\infty}} \frac{1}{\delta_n} \log \mu^{(n)}(G) \geq -\inf_{x\in G} I(x).$$

A sequence of random variables is said to satisfy the LDP if the corresponding sequence of distributions does. Furthermore, when the space $\mathcal{S}$ is realized as a space of functions, such as the space of continuous functions on $[0,1]$, we refer to the LDP as a *sample path* LDP, or LDP(sp).

We say that a sequence $\mu^{(n)}$ satisfies a moderate deviations principle (MDP), if $\mu^{(n)}$ satisfies the LDP with speed γ_n^2, where γ_n satisfies

$$\lim_{n\to\infty} \frac{\gamma_n}{\sqrt{n}} = 0 \quad \text{and} \quad \lim_{n\to\infty} \gamma_n = \infty. \tag{14.3}$$

Analogous to LDP(sp), we shall use MDP(sp) when the space $\mathcal{S}$ is realized as a space of functions. Allowing that the definition of $\mu^{(n)}$ may depend on γ_n, in case $\mu^{(n)}$ satisfies the MDP (resp. MDP(sp)) for every sequence satisfying (14.3), we say $\mu^{(n)}$ satisfies the *complete* MDP (resp. MDP(sp)).

Throughout, we shall refer to the following function spaces. Denote by $\mathcal{C}[a,b]$, $\mathcal{D}[a,b]$ and $\mathcal{AC}[a,b]$, the spaces of real-valued functions defined on the interval $[a,b]$ which are continuous, right continuous with left limits and absolutely continuous, respectively. Each of these spaces is equipped with the supremum norm

$$||\varphi||_\infty = \sup_{a\leq t\leq b} |\varphi(t)|$$

and the topology and Borel σ-field generated by this norm. In addition, we shall consider the spaces $\mathcal{C}$, $\mathcal{D}$ and $\mathcal{AC}$ of real-valued functions defined on $(-\infty,1]$ which are continuous, right continuous with left limits, and absolutely continuous, respectively. These latter spaces are equipped with the metric of uniform convergence on compact sets

$$d(\varphi,\psi)_\infty = \sum_{n=1}^{\infty} \frac{2^{-n}\sup_{t\in[-n,1]} |\varphi(t)-\psi(t)|}{1+\sup_{t\in[-n,1]} |\varphi(t)-\psi(t)|}\,.$$

and the topology and Borel σ-field generated by this metric.

A starting point for the main results in the following sections is that a certain process — specifically, a scaled version of the input process to a filter — satisfies a MDP(sp). We now describe a class of processes for which a useful criterion may be applied to verify whether a MDP(sp) is satisfied.

We do so for the setting of both filters mentioned in the Introduction, beginning with the first one in mind.

Consider a sequence of random variables in $\mathbb{R}$, $\{a(t), t = 0, 1, 2, \ldots\}$. Let $A(t_1, t_2) = \sum_{t=t_1+1}^{t_2} a(t)$ and define a sequence of stochastic processes as follows:

$$A^{(n)}(t) = \frac{1}{n} A(0, [nt]), \qquad t \in [0, 1]. \tag{14.4}$$

Denote by $\nu_A^{(n)}$ the distribution of the process $\{\frac{\sqrt{n}}{\gamma_n} A^{(n)}(t), 0 \le t \le 1\}$.

When $\{a(t), t = 0, 1, 2, \ldots\}$ is a sequence of *bounded* i.i.d. random variables with $\mathsf{E}a(0) = 0$ and $\mathsf{E}a^2(0) = \sigma^2 > 0$, it is known from Mogulskii [19], Theorem 1, that the sequence $\nu_A^{(n)}$ satisfies the complete MDP(sp) in $\mathcal{D}[0, 1]$ with good rate function

$$I(\varphi) = \begin{cases} \int_0^1 \frac{1}{2\sigma^2} (\varphi'(t))^2 dt & \text{if } \varphi \in \mathcal{AC}[0, 1],\ \varphi(0) = 0 \\ \infty & \text{otherwise .} \end{cases} \tag{14.5}$$

The following theorem provides a useful criterion for verifying that the *sample path process*, (14.4), of a given sequence $a(t)$ satisfies the complete MDP(sp) when $a(t)$ is not necessarily i.i.d.

Theorem 14.2.2 Let $\{a(t), t = 0, 1, 2, \ldots\}$ be a stationary sequence of bounded, mean zero, real-valued random variables. For a fixed integer m, and $0 = t_0 < t_1 < \cdots < t_m \le 1$, let

$$\begin{aligned} Z_n &= \left(\frac{\sqrt{n}}{\gamma_n} A^{(n)}(t_1), \frac{\sqrt{n}}{\gamma_n}(A^{(n)}(t_2) - A^{(n)}(t_1)), \right. \\ &\qquad \left. \ldots, \frac{\sqrt{n}}{\gamma_n}(A^{(n)}(t_m) - A^{(n)}(t_{m-1}))\right). \end{aligned}$$

Suppose that for each positive integer m and partition $\{t_i\}_{i=1}^m$, the sequence $\{Z_n\}$ satisfies the complete MDP in $\mathbb{R}^m$ with good rate function

$$I_m(z) = \sum_{i=1}^m \frac{1}{2\sigma^2} \frac{z_i^2}{(t_i - t_{i-1})}, \tag{14.6}$$

for some $\sigma^2 > 0$, where $z = (z_1, \ldots, z_m)$. In addition, assume that

$$\overline{\lim_{n\to\infty}} \frac{1}{\gamma_n^2} \log \mathsf{E}[\exp(\frac{\gamma_n \sum_{i=1}^n a(i)}{\sqrt{n}}) + \exp(\frac{-\gamma_n \sum_{i=1}^n a(i)}{\sqrt{n}})] < \infty \tag{14.7}$$

for every sequence γ_n satisfying (14.3). Then the sequence $\nu_A^{(n)}$ satisfies the complete MDP(sp) in $\mathcal{D}[0, 1]$ with the good rate function in (14.5).

The above theorem can be proved by mimicking the proof of Theorem 5.1.2 of Dembo and Zeitouni [10].

In the setting of the second model, we proceed as follows. Consider a sequence of random variables in $\mathbb{R}$, $\{a(t), t = \ldots, -2, -1, 0, 1, 2, \ldots\}$. Let $A(t_1, t_2) = \sum_{t=t_1+1}^{t_2} a(t)$ and define a sequence of stochastic processes as follows:

$$A^{(n)}(t) = \begin{cases} \frac{1}{n}A(0, [nt]), & t \in [0,1] \\ -\frac{1}{n}A(\lfloor nt \rfloor, 0) & t \in (-\infty, 0) \end{cases}.$$

Denote by $\nu_A^{(n)}$ the distribution of $\{\frac{\sqrt{n}}{\gamma_n}A^{(n)}(t), -\infty < t \le 1\}$.

The following is then the counterpart of Theorem 14.2.2:

Theorem 14.2.3 Let $\{a(t), t = \ldots, -2, -1, 0, 1, 2, \ldots\}$ be a stationary sequence of bounded, mean zero, real-valued random variables. For a fixed integer m and $t_0 < t_1 < \cdots < t_m \le 1$, let

$$\begin{aligned} Z_n &= \Big(\frac{\sqrt{n}}{\gamma_n}[A^{(n)}(t_1) - A^{(n)}(t_0)], \frac{\sqrt{n}}{\gamma_n}[A^{(n)}(t_2) - A^{(n)}(t_1)], \\ &\quad \ldots, \frac{\sqrt{n}}{\gamma_n}[A^{(n)}(t_m) - A^{(n)}(t_{m-1})]\Big). \end{aligned}$$

Suppose that for each positive integer m and partition $\{t_i\}_{i=1}^m$, $\{Z_n\}$ satisfies the complete MDP in $\mathbb{R}^m$ with rate function given by (14.6) for some $\sigma^2 > 0$ and that (14.7) holds. Then the sequence $\nu_A^{(n)}$ satisfies the complete MDP(sp) in $\mathcal{D}$ with good rate function

$$I(\varphi) = \begin{cases} \int_{-\infty}^{1} \frac{1}{2\sigma^2}(\varphi'(t))^2 dt & \text{if } \varphi \in \mathcal{AC},\ \varphi(0) = 0 \\ \infty & \text{otherwise}. \end{cases}$$

The proof of Theorem 14.2.3 is analogous to that of Theorem 14.2.2.

Below we present three classes of random processes that satisfy the hypotheses of both Theorem 14.2.2 and Theorem 14.2.3. We remark that the complete MDP(sp) in the space $\mathcal{D}[0,1]$ also holds for the sample path process of the process considered in the first example when the Markov chain there has arbitrary initial distribution (and hence is not stationary).

Example 1. Let $\{b(t), t = 0, \pm 1, \pm 2, \ldots\}$ be a stationary Markov chain taking values in a Polish space and satisfying the conditions of Theorem 2.1 of Wu [26]. Let, for $t = 0, \pm 1, \pm 2, \ldots$

$$a(t) = f(b(t)) - \mathsf{E}_\mu[f(b(t))]$$

for any bounded real-valued function f, where μ is the stationary distribution of the chain. Here $\sigma^2 = \mathsf{E}_\mu[a^2(0)] + 2\sum_{k=1}^{\infty} \mathsf{E}_\mu[a(k)a(0)]$, which we require to be positive.

Example 2. Let $\{\xi(t), t = 0, \pm 1, \pm 2, \ldots\}$ be a sequence of bounded, mean zero, i.i.d. random variables and $\{b(t), t = 0, \pm 1, \pm 2, \ldots\}$ a sequence of real numbers which is absolutely summable. Let the sequence $\{a(t), t = 0, \pm 1, \pm 2, \ldots\}$ be the infinite moving average process defined by

$$a(t) = \sum_{i=-\infty}^{\infty} b(i+t)\xi(i).$$

Here $\sigma^2 = (\sum_{i=-\infty}^{\infty} b_i)^2 \mathsf{E}[\xi^2(0)]$, which we require to be positive.

Example 3. Suppose $\{a(t), t = 0, \pm 1, \pm 2, \ldots\}$ is a stationary sequence of mean zero, bounded random variables which are *m-dependent.* By this we mean that the σ-fields $\sigma\{a(j); j \le k\}$ and $\sigma\{a(j); j \ge k+m\}$ are independent for any integer k. Examples of such sequences include sequences $\{f(b(t)) - \mathsf{E}[f(b(t))]\}$, where f is a bounded real-valued function and $\{b(t)\}$ is a stationary Gaussian sequence that is either φ-mixing or ψ-mixing (see, for example, Chapters 4 and 5 of [15]). Here $\sigma^2 = \mathsf{E}[a^2(0)] + 2\sum_{k=1}^{m-1} \mathsf{E}[a(k)a(0)]$, which we require to be positive.

14.3 MDP for the Filtered Processes

We first study the effect of feeding a process with short-range dependence into the following linear filter, our first model, which is based on (14.2):

$$Y(t) = \int_0^t (t-s)^{H-\frac{1}{2}} dX(s), \tag{14.8}$$

for some given $H \in (\frac{1}{2}, 1)$. Specifically, given that X satisfies a MDP, we want to show that Y, a process with long-range dependence, also satisfies a MDP.

As we mentioned in the Introduction, the advantage of the filter in (14.8) is the boundedness of the interval $[0, t]$, which allows the direct application of the MDP results of the last section. The disadvantage, however, is that the filter does not preserve the stationary increment property (of X). Neither does the filter preserve self-similarity: when X is a self-similar process, Y is only self-similar with respect to $t = 0$.

Since Y does not have stationary increments, the autocovariance function (of its increments)

$$\mathsf{Cov}[Y(\tau+1) - Y(\tau), Y(t+\tau+1) - Y(t+\tau)]$$

depends on τ (as well as on t). However, we can still use the standard definition for long-range dependence (e.g., Cox [8]), namely, that the autocovariance function is not summable over t (for any τ). (For instance, when X is the standard Brownian motion, one can verify that the above autocovariance is of order $t^{-(\frac{3}{2}-H)}$.)

When considering the first model we restrict ourselves to processes X which are either Brownian motion, which satisfies the conditions of Theorem 14.3.1 below, or possess sample paths of bounded variation on finite intervals, with $X(0) = 0$. When X is a process with sample paths of bounded variation on finite intervals, the integral in (14.8) can be viewed as a sample path Lebesgue-Stieltjes integral (see e.g., Wong and Hajek [25]). Integration by parts on (14.8) yields

$$Y(t) = \int_0^t (H - \frac{1}{2})(t-s)^{H-\frac{3}{2}} X(s) ds, \tag{14.9}$$

since $X(0) = 0$ and $H > \frac{1}{2}$. On the other hand, if $X = B$, the standard Brownian motion, applying Ito's formula (see, e.g., [25]) to (14.8) yields the same expression as above.

With this in mind, we may consider the transformation of X into Y given by either (14.8) or (14.9). Letting $Y^{(n)}(t) = Y(nt)/n$ and $X^{(n)}(t) = X(nt)/n$, a simple change of variable ($s = n\tau$) yields

$$Y^{(n)}(t) = \int_0^t (H - \frac{1}{2})(t-\tau)^{H-\frac{3}{2}} (n^{H-\frac{1}{2}} X^{(n)}(\tau)) d\tau. \tag{14.10}$$

Using the relation (14.10), we derive below a MDP for Y and related conditional limit theorems, assuming that X satisfies the MDP.

Theorem 14.3.1 Suppose, for some $\frac{1}{2} < H < 1$, the probability law of $\{n^{H-\frac{1}{2}} X^{(n)}(t), t \in [0,1]\}$, denoted $\mu_X^{(n)}$, satisfies the LDP(sp) on $\mathcal{D}[0,1]$ with speed $n^{2(1-H)}$ and a good rate function $I_X(\cdot) : \mathcal{D}[0,1] \mapsto I\!R$,

$$I_X(\varphi) = \begin{cases} \int_0^1 \frac{1}{2\sigma^2} (\varphi'(t))^2 dt & \text{if } \varphi \in \mathcal{AC}[0,1], \varphi(0) = 0 \\ \infty & \text{otherwise} \end{cases}.$$

Then, $Y(n)/n$ satisfies the LDP with speed $n^{2(1-H)}$ and a good rate function $\Lambda_Y^*(\gamma) : I\!R \mapsto I\!R$,

$$\Lambda_Y^*(\gamma) = \frac{H}{\sigma^2} \gamma^2.$$

Theorem 14.3.2 Suppose the conditions in Theorem 14.3.1 are in force.

(i) Let $\tilde{\mu}_X^{(n)}$ denote the law of $\{n^{H-\frac{1}{2}} X^{(n)}(\cdot) | Y^{(n)}(1) \geq \gamma\}$ for some $\gamma \geq 0$. Then, $\tilde{\mu}_X^{(n)}$ converges in distribution to $\delta_{\varphi_\gamma(\cdot)}$, where

$$\varphi_\gamma(t) = \gamma \frac{2H}{H + \frac{1}{2}} (1 - (1-t)^{H+\frac{1}{2}}), \quad t \in [0,1].$$

(ii) Let $\tilde{\mu}_Y^{(n)}$ denote the law of $\{Y^{(n)}(\cdot)|Y^{(n)}(1) \geq \gamma\}$. Then $\tilde{\mu}_Y^{(n)}$ converges in distribution to $\delta_{\psi_\gamma(\cdot)}$, where, for $t \in [0,1]$,

$$\psi_\gamma(t) = 2H\gamma \int_0^t (t-s)^{H-\frac{1}{2}}(1-s)^{H-\frac{1}{2}} ds.$$

In proving Theorem 14.3.1, we make use of Lemma 14.3.3 below, which is readily proved via the contraction principle (Dembo and Zeitouni [10], Theorem 4.2.1). The proof of Theorem 14.3.1 then yields the most likely path for the input process in Theorem 14.3.2(i). The most likely path for Y is then simply the output of the filter with the most likely input path.

Lemma 14.3.3 Suppose the conditions in Theorem 14.3.1 hold. Then, the distribution of $\{Y^{(n)}(t), t \in [0,1]\}$, denoted $\mu_Y^{(n)}$, satisfies the LDP(sp) with speed $n^{2(1-H)}$ and a good rate function $I_Y(\psi) : C[0,1] \mapsto I\!R$,

$$I_Y(\psi) = \inf\{I_X(\varphi) : \psi(t) = \int_0^t (t-s)^{H-\frac{1}{2}} \varphi'(s) ds,\ t \in [0,1]\}.$$

Proof. [Theorem 14.3.1]

Note that $Y^{(n)}(1) = Y(n)/n$ and the continuity of the mapping $\psi \mapsto \psi(1)$ from $C[0,1]$ to $I\!R$. It then follows from the contraction principle and Lemma 14.3.3 that the distribution of $Y^{(n)}(1)$ satisfies the LDP with speed $n^{2(1-H)}$ and a good rate function

$$\begin{aligned} \Lambda_Y^*(\gamma) &= \inf\{I_Y(\psi) : \psi(1) = \gamma\} \\ &= \inf\{I_X(\varphi) : \int_0^1 (1-s)^{H-\frac{1}{2}} \varphi'(s) ds = \gamma\}. \end{aligned}$$

From the definition of $I_X(\varphi)$ (in Theorem 14.3.1), the infimum above is tantamount to solving the following minimization problem:

$$\begin{aligned} \min_\varphi \quad & \int_0^1 \frac{1}{2\sigma^2} (\varphi'(s))^2 ds \\ \text{s.t.} \quad & \int_0^1 (1-s)^{H-\frac{1}{2}} \varphi'(s) ds = \gamma, \quad \varphi(0) = 0. \end{aligned}$$

Using that $\int_0^1 (1-s)^{2(H-\frac{1}{2})} ds = 1/(2H)$, it may be shown that

$$\Lambda_Y^*(\gamma) = \frac{H}{\sigma^2} \gamma^2.$$

Moreover, the infimum is uniquely achieved by

$$\varphi_\gamma'(s) = \gamma(2H)(1-s)^{H-\frac{1}{2}},$$

the derivative of the function appearing in the statement of Theorem 14.3.2. ■

Proof. [Theorem 14.3.2]
(i) Denote

$$B_\varepsilon = \{\varphi \in \mathcal{D}[0,1] : ||\varphi - \varphi_\gamma||_\infty \geq \varepsilon\},$$

for some $\varepsilon > 0$ (with φ_γ following the definition in Theorem 14.3.2), and

$$\Gamma_D = \{\varphi \in \mathcal{D}[0,1] : \int_0^1 (H - 1/2)(1-s)^{H-\frac{3}{2}}\varphi(s)ds \geq \gamma\}.$$

It is enough to show that for any given $\varepsilon > 0$ we have

$$\lim_{n\to\infty} \tilde{\mu}_X^{(n)}(B_\varepsilon) = \lim_{n\to\infty} \frac{\mu_X^{(n)}(B_\varepsilon \cap \Gamma_D)}{\mu_X^{(n)}(\Gamma_D)} = 0. \tag{14.11}$$

Since $\mu_X^{(n)}(\Gamma_D) = \mathsf{P}[Y(n)/n \geq \gamma]$, it follows from Theorem 14.3.1 and $\gamma \geq 0$ that

$$\lim_{n\to\infty} \frac{1}{n^{2(1-H)}} \log \mu_X^{(n)}(\Gamma_D) = -\Lambda_Y^*(\gamma). \tag{14.12}$$

On the other hand, from the LDP(sp) upper bound for $\mu_X^{(n)}$, we have

$$\lim_{n\to\infty} \frac{1}{n^{2(1-H)}} \log \mu_X^{(n)}(B_\varepsilon \cap \Gamma_D) \leq - \inf_{\varphi \in B_\varepsilon \cap \Gamma_D} I_X(\varphi)\ . \tag{14.13}$$

Combining (14.12) and (14.13), and making use of Lemma 2.8 of Shwartz and Weiss [21], we have, for some $\delta > 0$,

$$\overline{\lim_{n\to\infty}} \frac{1}{n^{2(1-H)}} \log \left(\frac{\mu_X^{(n)}(B_\varepsilon \cap \Gamma_D)}{\mu_X^{(n)}(\Gamma_D)} \right) \leq -\delta^2,$$

which clearly implies (14.11).
(ii) Using the continuity of the mapping appearing in (14.9), the desired weak convergence follows from (i). ■

We now consider the second model, the filter based on (14.1):

$$\begin{aligned} Y(t) &= \int_0^t (t-s)^{H-1/2} dX(s) \\ &\quad + \int_{-\infty}^0 [(t-s)^{H-1/2} - (-s)^{H-1/2}] dX(s)\ , \end{aligned} \tag{14.14}$$

for some given $H \in (\frac{1}{2}, 1)$.

Before going further, we make precise the processes we shall consider as possible inputs, X, to the filter (14.14). In order to allow our results to apply to FBM we allow X to be a standard Brownian motion, which satisfies the conditions of Theorem 14.3.4 below. Aside from this, we restrict ourselves to sample path processes of the type considered in Section 14.2.

Similar to the first model, given the process $X(t)$, letting $X^{(n)}(t) = X(nt)/n$ and $Y^{(n)}(t) = Y(nt)/n$, where Y relates to X through (14.14), we have

$$\begin{aligned} Y^{(n)}(t) &= \int_0^t (t-s)^{H-1/2} d(n^{H-1/2} X^{(n)}(s)) \\ &\quad + \int_{-\infty}^0 [(t-s)^{H-1/2} - (-s)^{H-1/2}] d(n^{H-1/2} X^{(n)}(s)) . \end{aligned}$$

Given this, we see that, in case X is taken to be a sample path process, meaning needs to be given to the integrals appearing in the definition of each $Y^{(n)}$. We do so as follows. Let $Y_a^{(n)}$ denote the process $Y^{(n)}$ modified so that the second integral begins at a instead of $-\infty$ and note the this process takes value in $C[0,1]$ and, moreover, belongs to $L^2[C[0,1]]$, i.e., $\mathsf{E}[(\sup_{t\in[0,1]} |Y_a^{(n)}(t)|)^2] < \infty$. We require of the process X that, for each $n = 1, 2, \ldots$ fixed, the sequence $\{Y_{-j/n}^{(n)}\}_{j=1}^{\infty}$ be a Cauchy sequence in $L^2[C[0,1]]$. We then take $Y^{(n)}$ to be the limit of this sequence.

We shall also require the assumption that, for any $\eta > 0$

$$\begin{aligned} \lim_{m\to\infty} \overline{\lim_{n\to\infty}} \frac{1}{n^{2(1-H)}} \log \mathsf{P}\{ \sup_{t\in[0,1]} | \int_{-\infty}^{-m} [(t-s)^{H-1/2} - (-s)^{H-1/2}] \\ d(n^{H-1/2} X^{(n)}(s))| \geq \eta \} = -\infty . \end{aligned} \tag{14.15}$$

We remark that this assumption holds when X is Brownian motion. In addition, it can be shown that the processes $Y^{(n)}$ exist in the sense specified in the previous paragraph and that (14.15) holds when X is a sample path process of the type considered in the examples of Section 14.2, with the following additional conditions being satisfied for the first two examples. For Example 1, it suffices to ask that the Markov chain be *reversible* and that there exists a bounded solution, h say, of the equation $Ph - h = f$, i.e. Poisson's equation, where P is the transition kernel of the chain. For Example 2, we ask that

$$\sum_{k=1}^{\infty} (\sum_{j=-\infty}^{-k} |b(j)| + \sum_{j=k}^{\infty} |b(j)|) < \infty .$$

The following two theorems are the analogues of Theorem 14.3.1 and Theorem 14.3.2, respectively:

Theorem 14.3.4 Suppose, for some $\frac{1}{2} < H < 1$, the probability law of $\{n^{H-\frac{1}{2}}X^{(n)}(t), t \in (-\infty, 1]\}$, denoted $\mu_X^{(n)}$, satisfies the LDP(sp) on $\mathcal{D}$ with speed $n^{2(1-H)}$ and a good rate function $I_X(\cdot) : \mathcal{D} \mapsto I\!R$,

$$I_X(\varphi) = \begin{cases} \int_{-\infty}^{1} \frac{1}{2\sigma^2}(\varphi'(t))^2 dt & \text{if } \varphi \in AC, \varphi(0) = 0 \\ \infty & \text{otherwise} \end{cases}.$$

Also, assume that (14.15) holds for any $\eta > 0$. Then, $Y(n)/n$ satisfies the LDP with speed $n^{2(1-H)}$ and a good rate function $\Lambda^*{}_Y(\gamma) : I\!R \mapsto I\!R$,

$$\Lambda^*{}_Y(\gamma) = \frac{\tilde{H}}{\sigma^2}\gamma^2,$$

where

$$\tilde{H} = \frac{1}{2}\Big[\int_0^1 (1-s)^{2(H-\frac{1}{2})}ds + \int_{-\infty}^{0} [(1-s)^{H-\frac{1}{2}} - (-s)^{H-\frac{1}{2}}]^2 ds\Big]^{-1}. \tag{14.16}$$

Theorem 14.3.5 Suppose the conditions in Theorem 14.3.4 are in force.

(i) Let $\tilde{\mu}_X^{(n)}$ denote the law of $\{n^{H-\frac{1}{2}}X^{(n)}(\cdot)|Y^{(n)}(1) \geq \gamma\}$ for some $\gamma \geq 0$. Then, $\tilde{\mu}_X^{(n)}$ converges in distribution to $\delta_{\varphi_\gamma(\cdot)}$, where

$$\varphi_\gamma(t) = \begin{cases} \gamma\frac{2\tilde{H}}{H+\frac{1}{2}}(1-(1-t)^{H+\frac{1}{2}}) & t \in [0,1] \\ \gamma\frac{2\tilde{H}}{H+\frac{1}{2}}(1-(1-t)^{H+\frac{1}{2}} + (-t)^{H+1/2}) & t \in (-\infty, 0] \end{cases}.$$

(ii) Let $\tilde{\mu}_Y^{(n)}$ denote the law of $\{Y^{(n)}(\cdot)|Y^{(n)}(1) \geq \gamma\}$. Then $\tilde{\mu}_Y^{(n)}$ converges in distribution to $\delta_{\psi_\gamma(\cdot)}$, where, for $t \in [0,1]$,

$$\psi_\gamma(t) = \int_0^t (t-s)^{H-1/2} d\varphi_\gamma(s) + \int_{-\infty}^{0} [(t-s)^{H-1/2} - (-s)^{H-1/2}] d\varphi_\gamma(s).$$

To facilitate the proof of Theorem 14.3.4, we require a lemma analogous to Lemma 14.3.3. In proving such a lemma we make use of condition (14.15). More precisely, for each fixed integer m, we apply the contraction principle to yield the LDP(sp) in $C[0,1]$ for the sequence $Y_m^{(n)}$, where

$$Y_m^{(n)}(t) = \int_0^t (t-s)^{H-1/2} d(n^{H-1/2}X^{(n)}(s))$$
$$+ \int_{-m}^{0} [(t-s)^{H-1/2} - (-s)^{H-1/2}] d(n^{H-1/2}X^{(n)}(s)), \quad t \in [0,1],$$

with a good rate function given by

$$I_Y^m(\psi) = \inf\{I_X(\varphi) : \psi(t) = \int_0^t (t-s)^{H-1/2}\varphi'(s)ds$$
$$+ \int_{-m}^0 [(t-s)^{H-1/2} - (-s)^{H-1/2}]\varphi'(s)ds,\ t \in [0,1]\}.$$

Furthermore, by (14.15), the $Y_m^{(n)}$ are exponentially good approximations of $Y^{(n)}$ (cf. [10, Definition 4.2.14]), and therefore [10, Theorem 4.2.23] allows to conclude the following lemma.

Lemma 14.3.6 Suppose the conditions in Theorem 14.3.4 hold. Then, the distribution of $\{Y^{(n)}(t), t \in [0,1]\}$, denoted $\mu_Y^{(n)}$, satisfies the LDP(sp) with speed $n^{2(1-H)}$ and good rate function $I_Y(\psi) : \mathcal{C}[0,1] \mapsto \mathbb{R}$,

$$I_Y(\psi) = \inf\{I_X(\varphi) : \psi(t) = \int_0^t (t-s)^{H-1/2}\varphi'(s)ds$$
$$+ \int_{-\infty}^0 [(t-s)^{H-1/2} - (-s)^{H-1/2}]\varphi'(s)ds,\ t \in [0,1]\}.$$

Proof. [Theorem 14.3.4]
Given Lemma 14.3.6, the proof may be carried out as was that of Theorem 14.3.1. ■

Proof. [Theorem 14.3.5]
(i) As in the proof of Theorem 14.3.2 (i), it suffices to show

$$\inf_{\varphi \in B_\varepsilon \cap \Gamma} \int_{-\infty}^1 \frac{1}{2\sigma^2}(\varphi'(s))^2 ds > \Lambda^*{}_Y(\gamma).$$

for all $\varepsilon > 0$, where

$$B_\varepsilon = \{\varphi \in \mathcal{D} : d(\varphi, \varphi_\gamma)_\infty \geq \varepsilon\},$$

and

$$\Gamma = \{\varphi \in \mathcal{AC} : \int_0^1 (1-s)^{H-\frac{1}{2}}\varphi'(s)ds$$
$$+ \int_{-\infty}^0 [(1-s)^{H-\frac{1}{2}} - (-s)^{H-\frac{1}{2}}]\varphi'(s)ds \geq \gamma\}.$$

That this is the case follows as in the proof of Theorem 14.3.2.
(ii) Note that the map $\tilde{F}^m : \mathcal{D} \to \mathcal{C}[0,1]$, defined by

$$\begin{aligned}\tilde{F}^m(\varphi)(t) &= \int_0^t (H-1/2)(t-s)^{H-3/2}\varphi(s)ds \\ &+\int_{-m}^0 (H-1/2)[(t-s)^{H-3/2}-(-s)^{H-3/2}]\varphi(s)ds \\ &-((t+m)^{(H-1/2)}-m^{(H-1/2)})\varphi(-m)\,, \qquad (14.17)\end{aligned}$$

is continuous. Therefore the law of $\{\tilde{F}^m(n^{H-1/2}X^{(n)}(\cdot))|Y^{(n)}(1)\geq\gamma\}$ converges in distribution to $\tilde{F}^m(\varphi_\gamma(\cdot))$. When the process $\{X(t)\}$ has sample paths of bounded variation on finite intervals, we conclude that the law of $\{F^m(n^{H-1/2}X^{(n)}(\cdot))|Y^{(n)}(1)\geq\gamma\}$ converges in distribution to $F^m(\varphi_\gamma(\cdot))$, where

$$\begin{aligned}F^m(\varphi)(t) &= \int_0^t (t-s)^{H-1/2}d\varphi(s) \\ &+\int_{-m}^0 [(t-s)^{H-1/2}-(-s)^{H-1/2}]d\varphi(s)\,. \qquad (14.18)\end{aligned}$$

To see this note that (14.17) and (14.18) are equivalent, via integration by parts, when φ is of bounded variation. Using (14.15) then allows us to conclude the desired result for such $\{X(t)\}$. In case $\{X(t)\}$ is Brownian motion a similar argument may be used. ■

14.4 Queueing Applications: Transient Results

Consider a queue with an input process $\{Y(t)\}$, which, in turn, is the output of a linear filter subject to the input process $\{X(t)\}$, following (14.8) or (14.14). Hence in general, the input to the queue (the Y process) has long-range dependence.

We start with the first model and assume that the conditions in Theorem 14.3.1 are in force. Consider $V(t)=\sup_{0\leq s\leq t}(Y(t)-Y(s)-c(t-s))$. Note that $V(t)$ is the *state* process of a queue with input Y and a deterministic service mechanism that depletes work at rate c. In queueing literature, $V(t)$ is often referred to as the virtual waiting time process, or the workload/inventory process. See, for instance, Chang [3], Duffield and O'Connell [11], Harrison [14], and Sigman and Yao [22], among others.

Let $V^{(n)}(t)=V(nt)/n$; in particular, $V^{(n)}(1)=V(n)/n$.

Theorem 14.4.1 The process $\{V(n)/n, n\geq 0\}$ satisfies the LDP with speed $n^{2(1-H)}$ and good rate function $\Lambda_V^*(\gamma):\mathbb{R}\mapsto\mathbb{R}$,

$$\Lambda_V^*(\gamma)=\frac{H}{\sigma^2}\inf_{0\leq s\leq 1}\frac{(c(1-s)+\gamma)^2}{v(s)+(1-s)^{2H}}, \qquad (14.19)$$

where

$$v(s) = 2H \int_0^s [(1-\tau)^{H-\frac{1}{2}} - (s-\tau)^{H-\frac{1}{2}}]^2 d\tau.$$

Theorem 14.4.2 (i) Let $\hat{\mu}_X^{(n)}$ denote the law of

$$\{n^{H-\frac{1}{2}} X^{(n)}(\cdot) \mid V^{(n)}(1) \geq \gamma\}$$

for some $\gamma \geq 0$. Suppose s^* achieves the minimum in (14.19). Then, $\hat{\mu}_X^{(n)}$ converges in distribution to $\delta_{\hat{\varphi}_\gamma(\cdot)}$, where

$$\hat{\varphi}_\gamma(t) =$$

$$\begin{cases} 2H\frac{c(1-s^*)+\gamma}{v(s^*)+(1-s^*)^{2H}} \int_0^t [(1-\tau)^{H-\frac{1}{2}} - (s^*-\tau)^{H-\frac{1}{2}}] d\tau & \text{if } 0 \leq t \leq s^* \\ \hat{\varphi}_\gamma(s^*) + 2H\frac{c(1-s^*)+\gamma}{v(s^*)+(1-s^*)^{2H}} \int_{s^*}^t (1-\tau)^{H-\frac{1}{2}} d\tau & \text{if } s^* < t \leq 1 \end{cases}.$$

(ii) Let $\hat{\mu}_V^{(n)}$ denote the law of

$$\{V^{(n)}(\cdot) \mid V^{(n)}(1) \geq \gamma\}.$$

Then $\hat{\mu}_V^{(n)}$ converges in distribution to $\delta_{\hat{\psi}_\gamma(\cdot)}$, where, for $t \in [0,1]$,

$$\hat{\psi}_\gamma(t) = \sup_{0 \leq s \leq t} \left[\int_0^t (t-\tau)^{H-1/2} d\hat{\varphi}_\gamma(\tau) - \int_0^s (s-\tau)^{H-1/2} d\hat{\varphi}_\gamma(\tau) - c(t-s)\right].$$

Note that the conditional limit in Theorem 14.4.2 (ii), which illustrates how the queue builds up when the input has long-range dependence, is qualitatively different from the more standard case of input with short-range dependence (e.g., [1, 4, 9]), where the most likely path is linear. Here, the most likely path is *nonlinear.*

Proof. [Theorem 14.4.1]
Using the continuity of the reflection mapping,

$$\psi(t) = \sup_{0 \leq s \leq t} \{\varphi(t) - \varphi(s) - c(t-s)\}, \quad t \in [0,1],$$

it follows from the contraction principle that the distribution of $V^{(n)}(1)$ satisfies the LDP with speed $n^{2(1-H)}$ and a good rate function

$$\Lambda_V^*(\gamma) = \inf\{I_X(\varphi) : \varphi \in \hat{F}\},$$

where

$$\begin{aligned} \hat{F} &= \{\varphi \in AC : \sup_{0 \leq s \leq 1} \left[\int_0^1 (1-\tau)^{H-\frac{1}{2}} \varphi'(\tau) d\tau \right. \\ &\quad \left. - \int_0^s (s-\tau)^{H-\frac{1}{2}} \varphi'(\tau) d\tau - c(1-s)\right] = \gamma\}. \end{aligned}$$

It remains to further identify the rate function. Let

$$\hat{F} = \cup_{\{0 \le s \le 1\}} \hat{F}_s ,$$

where

$$\begin{aligned} \hat{F}_s &= \{\varphi \in AC : \varphi(0) = 0, \int_0^s [(1-\tau)^{H-\frac{1}{2}} - (s-\tau)^{H-\frac{1}{2}}]\varphi'(\tau)d\tau \\ &\quad + \int_s^1 (1-\tau)^{H-\frac{1}{2}} \varphi'(\tau)d\tau = c(1-s) + \gamma\} . \end{aligned}$$

It may be shown that

$$\inf_{\varphi \in \hat{F}_s} \int_0^1 \frac{1}{2\sigma^2} (\varphi'(\tau))^2 d\tau = \frac{H}{\sigma^2} \frac{(c(1-s)+\gamma)^2}{v(s) + (1-s)^{2H}},$$

where

$$v(s) = 2H \int_0^s [(1-\tau)^{H-\frac{1}{2}} - (s-\tau)^{H-\frac{1}{2}}]^2 d\tau.$$

Furthermore, the minimum is uniquely achieved by

$$\hat{\varphi}'_\gamma(\tau) = \begin{cases} 2H \frac{c(1-s)+\gamma}{v(s)+(1-s)^{2H}} [(1-\tau)^{H-\frac{1}{2}} - (s-\tau)^{H-\frac{1}{2}}] & \text{if } 0 \le \tau < s \\ 2H \frac{c(1-s)+\gamma}{v(s)+(1-s)^{2H}} (1-\tau)^{H-\frac{1}{2}} & \text{if } s \le \tau \le 1 \end{cases} .$$

Given the above, the representation (14.19) for the rate function is clear. ■

To prove Theorem 14.4.2, we need the following lemma (proved in [6]).

Lemma 14.4.3 The solution to the minimization problem in (14.19), s^*, is unique.

Proof. [Theorem 14.4.2]
As in the proof of Theorem 14.3.2 and Theorem 14.3.5, it suffices to show that

$$\inf_{\varphi \in B_\varepsilon \cap \hat{F}} I_X(\varphi) > \Lambda_V^*(\gamma) ,$$

for all $\varepsilon > 0$, where

$$B_\varepsilon = \{\varphi \in \mathcal{D}[0,1] : ||\varphi - \hat{\varphi}_\gamma||_\infty \ge \varepsilon\} ,$$

and $\hat{F}$ is defined as in the proof of Theorem 14.4.1. The proof may be carried out as were those of these earlier theorems.

(ii) The result follows from the fact that the mappings under consideration are continuous. ■

For the second model, i.e., when Y follows (14.14), the following theorem replaces the transient results in Theorem 14.4.2, and the proof is analogous. The counterpart of Theorem 14.4.1 is presented in the next section. The statement of the following theorem makes use of the quantity s^* given by

$$s^* = \begin{cases} -\frac{H\gamma}{c(1-H)} & \text{if } \gamma < c(1-H)/H \\ -1 & \text{otherwise} \end{cases}.$$

It is assumed that the process X satisfies the conditions of Theorem 14.3.4.

Theorem 14.4.4 (i) Let $\hat{\mu}_X^{(n)}$ denote the law of

$$\{n^{H-\frac{1}{2}}X^{(n)}(\cdot)\,|\,V^{(n)}(1) \geq \gamma\}$$

for some $\gamma \geq 0$. Then, $\hat{\mu}_X^{(n)}$ converges in distribution to $\delta_{\hat{\varphi}_\gamma}$, where $\hat{\varphi}_\gamma(t) = \varphi_\gamma(t-1) - \varphi_\gamma(-1)$ and

$$\varphi_\gamma(t) =$$

$$\begin{cases} -\frac{2\tilde{H}(\gamma - cs^*)}{|s^*|^{2H}} \int_t^0 (-\tau)^{H-1/2} d\tau & \text{if } s^* \leq t \leq 0 \\ \varphi_\gamma(s^*) + \frac{2\tilde{H}(\gamma - cs^*)}{|s^*|^{2H}} \int_t^{s^*} [(s-\tau)^{H-1/2} - (-\tau)^{H-1/2}] d\tau & \text{if } -\infty < t \leq s^* \end{cases}.$$

(ii) Let $\hat{\mu}_V^{(n)}$ denote the law of $\{V^{(n)}(\cdot)|V^{(n)}(1) \geq \gamma\}$. Then $\hat{\mu}_V^{(n)}$ converges in distribution to $\delta_{\hat{\psi}_\gamma(\cdot)}$, where, for $t \in [0,1]$,

$$\hat{\psi}_\gamma(t) = \sup_{0 \leq s \leq t} \Big[\int_0^t (t-\tau)^{H-1/2} d\hat{\varphi}_\gamma(\tau) + \int_{-\infty}^0 [(t-\tau)^{H-1/2} - (-\tau)^{H-1/2}] d\hat{\varphi}_\gamma(\tau)$$

$$- \int_0^s (s-\tau)^{H-1/2} d\hat{\varphi}_\gamma(\tau) - \int_{-\infty}^0 [(s-\tau)^{H-1/2} - (-\tau)^{H-1/2}] d\hat{\varphi}_\gamma(\tau) - c(t-s)\Big].$$

14.5 Queueing Applications: Steady-State Results

In queueing theory, the steady-state limit, $V(\infty)$, is typically obtained through the limit of another process, $\tilde{M}(t) = \sup_{0 \leq s \leq t}(-Y(-s) - cs)$, the maximum of a random walk. The key linkage between the two is the stationary increment property of the process Y, which guarantees that for each t, $V(t)$ and $\tilde{M}(t)$ are equal in distribution, and hence so are $V(\infty)$ and $\tilde{M}(\infty)$. This fact is crucial in characterizing the steady-state limit, since whereas $\tilde{M}(t)$ is an increasing process (in t), $V(t)$ is not; and short of the linkage to $\tilde{M}(\infty)$, there is no easy way to even ensure the existence of $V(\infty)$.

The filter in our first model unfortunately does not yield a process, Y, with stationary increments even when X has stationary increments. Hence, in this context, we cannot claim that $\tilde{M}(\infty)$ gives a characterization of the steady-state limit of the queueing process. We do, however, in Theorem 14.5.1 below, consider a process, $M(t)$, in the setting of the first model whose definition is analogous to that of $\tilde{M}(t)$. Like the process $\tilde{M}(t)$, considered in Theorem 14.5.4 below, associated to the process $M(t)$ is a limiting random variable, $M(\infty)$. As seen from the statements of Theorem 14.5.1 and Theorem 14.5.4, the random variables $M(\infty)$ and $\tilde{M}(\infty)$ both exhibit subexponential tail behavior. Moreover, the proofs of the two theorems are carried out in a similar fashion.

Throughout this section we assume that the process X satisfies the conditions of Theorem 14.3.1 (resp. Theorem 14.3.4) when considering the first (resp. second) model.

Theorem 14.5.1 With $M(t) = \sup_{0\leq s\leq t}(Y(s) - cs)$ for some $c > 0$, $M(t)$ converges almost surely to a finite random variable $M(\infty)$, which has a subexponential tail distribution. Specifically,

$$\lim_{x\to\infty} \frac{1}{x^{2(1-H)}} \log \mathsf{P}[M(\infty) > x] = -\theta^*, \tag{14.20}$$

where

$$\theta^* = \frac{H}{\sigma^2} \frac{c^{2H}}{H^{2H}(1-H)^{2(1-H)}}. \tag{14.21}$$

To prove Theorem 14.5.1, we need two lemmas (the proofs are quite similar to the proof of Theorem 14.3.1).

Lemma 14.5.2 For all $\gamma \geq 0$

$$\lim_{n\to\infty} \frac{1}{n^{2(1-H)}} \log \mathsf{P}[M(n)/n > \gamma] = -g(\gamma),$$

where

$$g(\gamma) = \begin{cases} \frac{H}{\sigma^2}(c+\gamma)^2 & \text{if } \gamma \geq \frac{c(1-H)}{H} \\ \frac{H}{\sigma^2}\frac{c^{2H}}{H^{2H}(1-H)^{2(1-H)}}\gamma^{2(1-H)} & \text{if } \gamma < \frac{c(1-H)}{H} \end{cases}.$$

Lemma 14.5.3 Let $Z(n) = \sup_{n-1<s\leq n}(Y(s) - cs)$, $n = 1, 2, \ldots$. For all $\gamma \geq 0$,

$$\lim_{n\to\infty} \frac{1}{n^{2(1-H)}} \log \mathsf{P}[Z(n)/n > \gamma] = -\frac{H}{\sigma^2}(c+\gamma)^2.$$

Proof. [Theorem 14.5.1]
Since $M(t)$ is an increasing sequence, it converges almost surely to a random variable $M(\infty)$.

We prove the limit (14.20) by arguing that the liminf and the limsup are dominated from below and from above by the same bound.

To establish the lower bound, we follow the approach in Chang [3]. Note that for all $\gamma > 0$

$$\begin{aligned} & \varliminf_{x\to\infty} \frac{1}{x^{2(1-H)}} \log \mathsf{P}[M(\infty) > x] \\ \geq \; & \varliminf_{n\to\infty} \frac{1}{(\gamma n)^{2(1-H)}} \log \mathsf{P}[Y(n) - cn > \gamma n] \\ \geq \; & -\frac{H}{\sigma^2}\frac{(\gamma+c)^2}{\gamma^{2(1-H)}}, \end{aligned}$$

where we have used Theorem 14.3.1. Thus,

$$\begin{aligned} & \varliminf_{x\to\infty} \frac{1}{x^{2(1-H)}} \log \mathsf{P}[M(\infty) > x] \\ \geq \; & -\inf_{\gamma>0} \frac{H}{\sigma^2}\frac{(\gamma+c)^2}{\gamma^{2(1-H)}} \\ = \; & -\theta^*. \end{aligned}$$

To establish the upper bound, note that for all $\gamma > 0$

$$\begin{aligned} & \varlimsup_{x\to\infty} \frac{1}{x^{2(1-H)}} \log \mathsf{P}[M(\infty) \geq x] \\ = \; & \varlimsup_{n\to\infty} \frac{1}{(\gamma n)^{2(1-H)}} \log \mathsf{P}[M(\infty) \geq \gamma n]. \end{aligned}$$

Write

$$\begin{aligned} & \mathsf{P}[M(\infty) \geq \gamma n] \\ = \; & \mathsf{P}[M(n) \geq \gamma n] + \mathsf{P}[M(n) < \gamma n, \; \sup_{s>n}(Y(s) - cs) \geq \gamma n] \\ \leq \; & \mathsf{P}[M(n) \geq \gamma n] + \mathsf{P}[\sup_{m\geq n+1} Z(m) \geq 0], \end{aligned} \tag{14.22}$$

where $Z(n)$ follows the definition in Lemma 14.5.3.

The key now is to argue (cf. [11]) that γ can be made small enough so that "large excursions" occur at least as likely in the first term on the right side of (14.22) as in the second term. Regarding the first term, if $\gamma < \frac{c(1-H)}{H}$, then from Lemma 14.5.2,

$$\varlimsup_{n\to\infty} \frac{1}{(\gamma n)^{2(1-H)}} \log \mathsf{P}[M(n) \geq \gamma n] = -\theta^*. \tag{14.23}$$

It remains to show that γ can also be chosen small enough such that the second term grows no faster than the first. In view of the upper bound for $\{Z(n)/n, n \geq 0\}$ in Lemma 14.5.3, for every $\varepsilon > 0$ there exists an n_0 such that for all $n \geq n_0$ we have

$$\mathsf{P}(Z(n)/n \geq 0) \leq \exp\Big(- n^{2(1-H)}(\frac{H}{\sigma^2}c^2 - \varepsilon)\Big).$$

Now choose ε such that $0 < \varepsilon < \frac{H}{2\sigma^2}c^2$, and then choose γ small enough such that

$$\gamma^{2(1-H)}\theta^* + \varepsilon < \frac{H}{\sigma^2}c^2 - \varepsilon \quad \text{and} \quad \gamma < \frac{c(1-H)}{H}.$$

This way, we have, for $n \geq n_0$,

$$\begin{aligned} &\mathsf{P}[\sup_{m \geq n+1} Z(m) \geq 0] \\ \leq &\sum_{m=n+1}^{\infty} \mathsf{P}[Z(m)/m \geq 0] \\ \leq &\sum_{m=n+1}^{\infty} \exp\Big(- n^{2(1-H)}(\gamma^{2(1-H)}\theta^* + \varepsilon)\Big). \end{aligned} \tag{14.24}$$

Simple algebra yields

$$\overline{\lim_{n\to\infty}} \frac{1}{(\gamma n)^{2(1-H)}} \log \sum_{m=n+1}^{\infty} \exp\Big(- n^{2(1-H)}(\gamma^{2(1-H)}\theta^* + \varepsilon)\Big) \leq -\theta^*. \tag{14.25}$$

(Note that for $\alpha > 0, z > 0, n > (\frac{2}{\alpha z})^{1/\alpha}$

$$ne^{-n^\alpha z} = \int_n^\infty (\alpha z s^\alpha - 1)e^{-s^\alpha z}ds \geq \int_n^\infty e^{-s^\alpha z}ds.$$

Choose $\alpha = 2(1-H)$, $z = \gamma^{2(1-H)}\theta^* + \varepsilon$, with $\varepsilon > 0$ arbitrarily small.) Therefore, combining (14.22), (14.23), (14.24) and (14.25), we have established the desired upper bound,

$$\overline{\lim_{x\to\infty}} \frac{1}{x^{2(1-H)}} \log \mathsf{P}(M(\infty) \geq x) \leq -\theta^*.$$

■

Now, consider the second model. The following theorem and lemma replace Theorem 14.5.1 and Lemma 14.5.2

Theorem 14.5.4 With $\tilde{M}(t) = \sup_{0 \le s \le t}(-Y(-s) - cs)$ for some $c > 0$, $\tilde{M}(t)$ converges almost surely to a finite random variable $\tilde{M}(\infty)$, which has a subexponential tail distribution. Specifically,

$$\lim_{x \to \infty} \frac{1}{x^{2(1-H)}} \log \mathsf{P}[\tilde{M}(\infty) > x] = -\frac{\tilde{H}}{H}\theta^*,$$

where θ^* follows (14.21), and $\tilde{H}$ follows (14.16).

Lemma 14.5.5 For all $\gamma \ge 0$, under the conditions of Theorem 14.3.4,

$$\lim_{n \to \infty} \frac{1}{n^{2(1-H)}} \log \mathsf{P}[\tilde{M}(n)/n > \gamma] = -\tilde{g}(\gamma),$$

where

$$\tilde{g}(\gamma) = \begin{cases} \frac{\tilde{H}}{\sigma^2}(c+\gamma)^2 & \text{if } \gamma \ge \frac{c(1-H)}{H} \\ \frac{\tilde{H}}{\sigma^2}\frac{c^{2H}}{H^{2H}(1-H)^{2(1-H)}}\gamma^{2(1-H)} & \text{if } \gamma < \frac{c(1-H)}{H} \end{cases}.$$

Recall that

$$V(t) = \sup_{0 \le s \le t} (Y(t) - Y(s) - c(t-s)), \tag{14.26}$$

and $V^{(n)}(t) = V(nt)/n$. Recall also that when Y has stationary increments, as is the case when X is the standard Brownian motion, the random variable $V(t)$ has the same distribution as $\tilde{M}(t)$. We claim that it still holds that $V(n) =_d \tilde{M}(n)$ for all integer n, when X is a sample path process of the type considered in Section 14.2. To see this claim, note that, for $u \in [0,1]$ arbitrary

$$\begin{aligned} & n^{3/2-H}(Y^{(n)}(1) - Y^{(n)}(u)) \\ = \ & \sum_{j=-\infty}^{n} (1-j/n)^{H-1/2}a(j) - \sum_{j=-\infty}^{[nu]} (u-j/n)^{H-1/2}a(j), \end{aligned}$$

which, by stationarity,

$$\begin{aligned} =_d \ & \sum_{j=-\infty}^{n} (1-j/n)^{H-1/2}a(j-n) - \sum_{j=-\infty}^{[nu]} (u-j/n)^{H-1/2}a(j-n) \\ = \ & \sum_{j=-\infty}^{0} (-j/n)^{H-1/2}a(j) - \sum_{j=-\infty}^{[nu]-n} ((u-1)-j/n)^{H-1/2}a(j) \\ = \ & n^{3/2-H}(Y^{(n)}(0) - Y^{(n)}(u-1)) \ . \end{aligned}$$

Given this, in part (ii) of the following theorem the values of t should be taken as integers when X is a sample path process. The proof of the theorem is a direct consequence of Theorem 14.5.4 and Lemma 14.5.5.

Theorem 14.5.6 Suppose that the conditions in Theorem 14.5.4 hold.

(i) The process $\{V(n)/n, n \geq 0\}$ satisfies the LDP with speed $n^{2(1-H)}$ and a good rate function $\Lambda_V^*(\gamma) : \mathbb{R} \mapsto \mathbb{R}$, where $\Lambda_V^*(\gamma) = \tilde{g}(\gamma)$.

(ii) The random variable $V(t)$ converges in distribution to a finite random variable $V(\infty)$, which has a subexponential tail distribution. Specifically,

$$\lim_{x\to\infty} \frac{1}{x^{2(1-H)}} \log \mathsf{P}[V(\infty) > x] = -\frac{\tilde{H}}{H}\theta^*.$$

Remark. Note that when Y is the FBM in (14.1), the result in Theorem 14.5.4 is consistent with the result in Duffield and O'Connell [11] for FBM. To see this, note that when $X(t) \equiv \sqrt{2\tilde{H}}B(t)$, where $B(t)$ is standard Brownian motion, $\mathsf{E}[Y^2(t)] = t^{2H}$, which is consistent with the relation (57) of [11]. In addition, X satisfies the conditions in Theorem 14.3.1 with $\sigma^2 = 2\tilde{H}$. It then suffices to note that the minimization problem appearing in (60) of [11] has value θ^*.

Acknowledgments: The authors acknowledge support for their research from the following sources — Cheng-Shang Chang: the National Science Council, Taiwan, R.O.C., under the contract NSC 85-2121-M007-035; David D. Yao: NSF Grant MSS-92-16490 and a matching grant from EPRI; Tim Zajic: NSF Grant DMS-9508709.

14.6 References

[1] V. Anantharam, "How large delays build up in a GI/G/1 queue," *Queueing Systems*, Vol. 5, pp. 345-368, 1989.

[2] J. Beran. *Statistics for Long-Memory Processes.* New York: Chapman and Hall, 1994.

[3] C.S. Chang, "Stability, queue length and delay of deterministic and stochastic queueing networks," *IEEE Transactions on Automatic Control*, Vol. 39, pp. 913-931, 1994.

[4] C.S. Chang, "Sample path large deviations and intree networks," *Queueing Systems*, Vol. 20, pp. 7-, 1995.

[5] C.S. Chang and J.A. Thomas, "Effective bandwidth in high speed digital networks," *IEEE JSAC*, Vol. 13, pp. 1091-1100, 1995.

[6] C.S. Chang, D.D. Yao and T. Zajic, "Large deviations, moderate deviations, and queues with long-range dependent input," preprint, 1995.

[7] C.S. Chang, D.D. Yao and T. Zajic, "Moderate deviations on unbounded intervals, with applications to queues with long-range dependent input," preprint, 1996.

[8] D.R. Cox, "Long-range dependence: a review," *Statistics: An Appraisal,* H.A. David and H.T. David (eds.), Iowa State University Press, Ames, Iowa, 55-74, 1984.

[9] A. Dembo and T. Zajic, "Large deviations from empirical mean and measure to partial sums processes," *Stoch. Proc. and Appl.*, Vol. 57, No. 2, pp. 191-224, 1995.

[10] A. Dembo and O. Zeitouni. *Large Deviations Techniques and Applications.* Boston: Jones and Barlett Publishers, 1992.

[11] N.G. Duffield and N. O'Connell, "Large deviations and overflow probabilities for the general single-server queue, with applications," *preprint,* 1993.

[12] P.W. Glynn and W. Whitt, "Logarithmic asymptotics for steady-state tail probabilities in a single-server queue," *J. Appl. Prob*, Vol. 31A, pp. 131-156, 1994.

[13] P.W. Glynn and W. Whitt, "Large deviations behavior of counting processes and their inverses," *Queueing Systems*, Vol. 17, pp. 107-128, 1994.

[14] J.M. Harrison, *Brownian Motion and Stochastic Flow Systems.* Wiley, New York, 1985.

[15] I. Ibragimov and Y. Rozanov, *Gaussian Random Processes.* Springer-Verlag, New York, 1978.

[16] W.E. Leland, M.S. Taqqu, W. Willinger and D.V. Wilson, "On the self-similar Nature of Ethernet Traffic," *IEEE/ACM Trans. on Networking*, Vol. 2, pp. 1-15, 1994.

[17] P. Lévy, "Random functions: general theory with special reference to Laplacian random functions," *Univ. California Publ. Statist.*, Vol. 1, pp. 331-390, 1953.

[18] B.B. Mandelbrot and J. van Ness, "Fractional Brownian motions, fractional noises and applications," *SIAM Review*, Vol. 10, No. 4, pp. 422-437, 1968.

[19] A.A. Mogulskii, "Large deviations for trajectories of multidimensional random walk," *Th. Prob. Appl.*, Vol. 21, pp. 300-315, 1976.

[20] I. Norros, "A storage model with self-similar input," *Queueing Systems*, Vol. 16, pp. 387-396, 1994.

[21] A. Shwartz and A. Weiss, *Large deviations for performance analysis.* Chapman & Hall, London, 1995.

[22] K. Sigman and D.D. Yao, "Finite moments for inventory processes," *Annals of Appl. Prob.*, Vol. 4, pp. 765-778, 1994.

[23] W. Whitt, "Tail probability with statistical multiplexing and effective bandwidths in multi-class queues," *Telecommunication Systems*, Vol. 2, pp. 71-107, 1993.

[24] W. Willinger, "Traffic modeling for high-speed networks: theory versus practice," *Stochastic Networks*, F. Kelly and R. Williams (eds.), IMA Volume in Mathematics and Its Applications. Springer-Verlag, New York, 1994.

[25] E. Wong and B. Hajek. *Stochastic processes in engineering systems.* New York: Springer-Verlag, 1984.

[26] L. Wu, "Moderate deviations of dependent random variables related to CLT," *Ann. of Prob.*, Vol. 23, pp. 420-445, 1995.

Lecture Notes in Statistics

For information about Volumes 1 to 34 please contact Springer-Verlag

Vol. 35: T. Calinski and W. Klonecki (Editors), Linear Statistical Inference. Proceedings, 1984. vi, 318 pages, 1985.

Vol. 36: B. Matern, Spatial Variation. Second Edition. 151 pages, 1986.

Vol. 37: R. Dykstra, T. Robertson and F.T. Wright (Editors), Advances in Order Restricted Statistical Inference. Proceedings, 1985. viii, 295 pages, 1986.

Vol. 38: R.W. Pearson and R.F. Boruch (Editors), Survey Research Designs: Towards a Better Understanding of Their Costs and Benefits. v, 129 pages, 1986.

Vol. 39: J.D. Malley, Optimal Unbiased Estimation of Variance Components. ix, 146 pages, 1986.

Vol. 40: H.R. Lerche, Boundary Crossing of Brownian Motion. v, 142 pages, 1986.

Vol. 41: F. Baccelli, P. Brémaud, Palm Probabilities and Stationary Queues. vii, 106 pages, 1987.

Vol. 42: S. Kullback, J.C. Keegel, J.H. Kullback, Topics in Statistical Information Theory. ix, 158 pages, 1987.

Vol. 43: B.C. Arnold, Majorization and the Lorenz Order: A Brief Introduction. vi, 122 pages, 1987.

Vol. 44: D.L. McLeish, Christopher G. Small, The Theory and Applications of Statistical Inference Functions. vi, 124 pages, 1987.

Vol. 45: J.K. Ghosh (Editor), Statistical Information and Likelihood. 384 pages, 1988.

Vol. 46: H.-G. Müller, Nonparametric Regression Analysis of Longitudinal Data. vi, 199 pages, 1988.

Vol. 47: A.J. Getson, F.C. Hsuan, {2}-Inverses and Their Statistical Application. viii, 110 pages, 1988.

Vol. 48: G.L. Bretthorst, Bayesian Spectrum Analysis and Parameter Estimation. xii, 209 pages, 1988.

Vol. 49: S.L. Lauritzen, Extremal Families and Systems of Sufficient Statistics. xv, 268 pages, 1988.

Vol. 50: O.E. Barndorff-Nielsen, Parametric Statistical Models and Likelihood. vii, 276 pages, 1988.

Vol. 51: J. Hüsler, R.-D. Reiss (Editors). Extreme Value Theory, Proceedings, 1987. x, 279 pages, 1989.

Vol. 52: P.K. Goel, T. Ramalingam, The Matching Methodology: Some Statistical Properties. viii, 152 pages, 1989.

Vol. 53: B.C. Arnold, N. Balakrishnan, Relations, Bounds and Approximations for Order Statistics. ix, 173 pages, 1989.

Vol. 54: K.R. Shah, B.K. Sinha, Theory of Optimal Designs. viii, 171 pages, 1989.

Vol. 55: L. McDonald, B. Manly, J. Lockwood, J. Logan (Editors), Estimation and Analysis of Insect Populations. Proceedings, 1988. xiv, 492 pages, 1989.

Vol. 56: J.K. Lindsey, The Analysis of Categorical Data Using GLIM. v, 168 pages, 1989.

Vol. 57: A. Decarli, B.J. Francis, R. Gilchrist, G.U.H. Seeber (Editors), Statistical Modelling. Proceedings, 1989. ix, 343 pages, 1989.

Vol. 58: O.E. Barndorff-Nielsen, P. Bl¾sild, P.S. Eriksen, Decomposition and Invariance of Measures, and Statistical Transformation Models. v, 147 pages, 1989.

Vol. 59: S. Gupta, R. Mukerjee, A Calculus for Factorial Arrangements. vi, 126 pages, 1989.

Vol. 60: L. Gyorfi, W. Härdle, P. Sarda, Ph. Vieu, Nonparametric Curve Estimation from Time Series. viii, 153 pages, 1989.

Vol. 61: J. Breckling, The Analysis of Directional Time Series: Applications to Wind Speed and Direction. viii, 238 pages, 1989.

Vol. 62: J.C. Akkerboom, Testing Problems with Linear or Angular Inequality Constraints. xii, 291 pages, 1990.

Vol. 63: J. Pfanzagl, Estimation in Semiparametric Models: Some Recent Developments. iii, 112 pages, 1990.

Vol. 64: S. Gabler, Minimax Solutions in Sampling from Finite Populations. v, 132 pages, 1990.

Vol. 65: A. Janssen, D.M. Mason, Non-Standard Rank Tests. vi, 252 pages, 1990.

Vol 66: T. Wright, Exact Confidence Bounds when Sampling from Small Finite Universes. xvi, 431 pages, 1991.

Vol. 67: M.A. Tanner, Tools for Statistical Inference: Observed Data and Data Augmentation Methods. vi, 110 pages, 1991.

Vol. 68: M. Taniguchi, Higher Order Asymptotic Theory for Time Series Analysis. viii, 160 pages, 1991.

Vol. 69: N.J.D. Nagelkerke, Maximum Likelihood Estimation of Functional Relationships. V, 110 pages, 1992.

Vol. 70: K. Iida, Studies on the Optimal Search Plan. viii, 130 pages, 1992.

Vol. 71: E.M.R.A. Engel, A Road to Randomness in Physical Systems. ix, 155 pages, 1992.

Vol. 72: J.K. Lindsey, The Analysis of Stochastic Processes using GLIM. vi, 294 pages, 1992.

Vol. 73: B.C. Arnold, E. Castillo, J.-M. Sarabia, Conditionally Specified Distributions. xiii, 151 pages, 1992.

Vol. 74: P. Barone, A. Frigessi, M. Piccioni, Stochastic Models, Statistical Methods, and Algorithms in Image Analysis. vi, 258 pages, 1992.

Vol. 75: P.K. Goel, N.S. Iyengar (Eds.), Bayesian Analysis in Statistics and Econometrics. xi, 410 pages, 1992.

Vol. 76: L. Bondesson, Generalized Gamma Convolutions and Related Classes of Distributions and Densities. viii, 173 pages, 1992.

Vol. 77: E. Mammen, When Does Bootstrap Work? Asymptotic Results and Simulations. vi, 196 pages, 1992.

Vol. 78: L. Fahrmeir, B. Francis, R. Gilchrist, G. Tutz (Eds.), Advances in GLIM and Statistical Modelling: Proceedings of the GLIM92 Conference and the 7th International Workshop on Statistical Modelling, Munich, 13-17 July 1992. ix, 225 pages, 1992.

Vol. 79: N. Schmitz, Optimal Sequentially Planned Decision Procedures. xii, 209 pages, 1992.

Vol. 80: M. Fligner, J. Verducci (Eds.), Probability Models and Statistical Analyses for Ranking Data. xxii, 306 pages, 1992.

Vol. 81: P. Spirtes, C. Glymour, R. Scheines, Causation, Prediction, and Search. xxiii, 526 pages, 1993.

Vol. 82: A. Korostelev and A. Tsybakov, Minimax Theory of Image Reconstruction. xii, 268 pages, 1993.

Vol. 83: C. Gatsonis, J. Hodges, R. Kass, N. Singpurwalla (Editors), Case Studies in Bayesian Statistics. xii, 437 pages, 1993.

Vol. 84: S. Yamada, Pivotal Measures in Statistical Experiments and Sufficiency. vii, 129 pages, 1994.

Vol. 85: P. Doukhan, Mixing: Properties and Examples. xi, 142 pages, 1994.

Vol. 86: W. Vach, Logistic Regression with Missing Values in the Covariates. xi, 139 pages, 1994.

Vol. 87: J. Müller, Lectures on Random Voronoi Tessellations.vii, 134 pages, 1994.

Vol. 88: J. E. Kolassa, Series Approximation Methods in Statistics.viii, 150 pages, 1994.

Vol. 89: P. Cheeseman, R.W. Oldford (Editors), Selecting Models From Data: AI and Statistics IV. xii, 487 pages, 1994.

Vol. 90: A. Csenki, Dependability for Systems with a Partitioned State Space: Markov and Semi-Markov Theory and Computational Implementation. x, 241 pages, 1994.

Vol. 91: J.D. Malley, Statistical Applications of Jordan Algebras. viii, 101 pages, 1994.

Vol. 92: M. Eerola, Probabilistic Causality in Longitudinal Studies. vii, 133 pages, 1994.

Vol. 93: Bernard Van Cutsem (Editor), Classification and Dissimilarity Analysis. xiv, 238 pages, 1994.

Vol. 94: Jane F. Gentleman and G.A. Whitmore (Editors), Case Studies in Data Analysis. viii, 262 pages, 1994.

Vol. 95: Shelemyahu Zacks, Stochastic Visibility in Random Fields. x, 175 pages, 1994.

Vol. 96: Ibrahim Rahimov, Random Sums and Branching Stochastic Processes. viii, 195 pages, 1995.

Vol. 97: R. Szekli, Stochastic Ordering and Dependence in Applied Probability. viii, 194 pages, 1995.

Vol. 98: Philippe Barbe and Patrice Bertail, The Weighted Bootstrap. viii, 230 pages, 1995.

Vol. 99: C.C. Heyde (Editor), Branching Processes: Proceedings of the First World Congress. viii, 185 pages, 19

Vol. 100: Wlodzimierz Bryc, The Normal Distribution: Characterizations with Applications. viii, 139 pages, 1995.

Vol. 101: H.H. Andersen, M.Højbjerre, D. Sørensen, P.S.Eriksen, Linear and Graphical Models: for the Multivariate Complex Normal Distribution. x, 184 pages, 1995.

Vol. 102: A.M. Mathai, Serge B. Provost, Takesi Hayakawa Bilinear Forms and Zonal Polynomials. x, 378 pages, 1995.

Vol. 103: Anestis Antoniadis and Georges Oppenheim (Editors), Wavelets and Statistics. vi, 411 pages, 1995.

Vol. 104: Gilg U.H. Seeber, Brian J. Francis, Reinhold Hatzinger, Gabriele Steckel-Berger (Editors), Statistical Modelling: 10th International Workshop, Innsbruck, July 10 14th 1995. x, 327 pages, 1995.

Vol. 105: Constantine Gatsonis, James S. Hodges, Robert E. Kass, Nozer D. Singpurwalla(Editors), Case Studies in Bayesian Statistics, Volume II. x, 354 pages, 1995.

Vol. 106: Harald Niederreiter, Peter Jau-Shyong Shiue (Editors), Monte Carlo and Quasi-Monte Carlo Methods in Scientific Computing. xiv, 372 pages, 1995.

Vol. 107: Masafumi Akahira, Kei Takeuchi, Non-Regular Statistical Estimation. vii, 183 pages, 1995.

Vol. 108: Wesley L. Schaible (Editor), Indirect Estimators ir U.S. Federal Programs. iix, 195 pages, 1995.

Vol. 109: Helmut Rieder (Editor), Robust Statistics, Data Analysis, and Computer Intensive Methods. xiv, 427 pages, 1996.

Vol. 110: D. Bosq, Nonparametric Statistics for Stochastic Processes. xii, 169 pages, 1996.

Vol. 111: Leon Willenborg, Ton de Waal, Statistical Disclosure Control in Practice. xiv, 152 pages, 1996.

Vol. 112: Doug Fischer, Hans-J. Lenz (Editors), Learning fr Data. xii, 450 pages, 1996.

Vol. 113: Rainer Schwabe, Optimum Designs for Multi-Fac Models. viii, 124 pages, 1996.

Vol. 114: C.C. Heyde, Yu. V. Prohorov, R. Pyke, and S. T. Rachev (Editors), Proceedings of the Athens Conference on Appplied Probability and Time Series Volume I: Applied Probability. viii, 424 pages, 1996.

Vol. 115: P.M. Robinson, M. Rosenblatt (Editors), Proceedings of the Athens Conference on Applied Probabilit and Time Series Volume II: Time Series. viii, 448 pages, 19

Vol. 116: Genshiro Kitagawa and Will Gersch, Smoothness Priors Analysis of Time Series. x, 261 pages, 1996.

Vol. 117: Paul Glasserman, Karl Sigman, David D. Yao (Editors), Stochastic Networks. xii, 298, 1996.